THE SECRET WIRELESS WAR

A catalogue record of this book is available from the British Library

Paperback Edition: August 2008

ISBN: 978-09560515-2-3

To order additional copies of this book please visit: http://www.geoffreypidgeon.com

Published by: Arundel Books
3 Arundel House, Courtlands, Sheen Road, Richmond. TW10 5AS Email: info@geoffreypidgeon.com

Designed by: Prestige Press Web: http://www.prestige-press.com

The Secret Wireless War

by

Geoffrey Pidgeon

Arundel Books

Dedication

I dedicate this book to the many fine men and women who worked in MI6 (Section VIII) and its Special Communication Units during World War II, under their founder Brigadier Sir Richard Gambier-Parry, KCMG, Head of SIS Communications (MI6 Section VIII) – from 1938 to 1946.

Brigadier Sir Richard Gambier-Parry KCMG

I should also like to express my deep appreciation here to my wife, Jane, who has been so patient and supportive in my work on this book as in every other sphere of life in our many years together.

Contents

Part III
Their story: by some of those involved in the Secret Wireless War 1939-1945

Preface

In the years following World War II, with a growing family and a business to run, I did not spend much time thinking about my wartime experiences. Then, in 1995, as part of the fiftieth anniversary celebrations to mark the end of the war in Europe, a number of newspapers suggested that readers write to the editor with their story of VE Night, 1945. I did not write to a newspaper; instead I wrote to the owner of the Cock Hotel in Stony Stratford (now part of Milton Keynes), where I had lived and worked during the war. I tried to put down in writing for him how we had so very energetically celebrated VE night in the hotel with the owners of the time.

This effort awakened my interest in the important part played during the War by three local stately homes – Bletchley Park, Hanslope Park and Whaddon Hall – which formed the angles of a triangle covering North Buckinghamshire.

Towards the end of 1996, my wife Jane and I visited Bletchley Park where my mother had worked as a Red Cross nurse in the clinic. This was one of Bletchley Park's then fortnightly 'Open Days' and I found the whole place quite fascinating. A room in one of the brick buildings was described as 'The DWS Room' which was mainly devoted to wireless equipment used during the War, including sets used by our agents abroad. It also told the story of the wireless side of 'The Ultra Secret' which emanated from the brilliant code-breakers who worked in the neighbouring huts – many of them still to be seen in the grounds.

While viewing the agents' sets on display, I noticed a mistake in one description and commented on it to the Curator of the DWS Room, David White. Naturally, he asked me how I knew about the mistake. I explained that I had helped to build the chassis for that particular model during the early production runs, both at Whaddon and in the later workshops started in 1943, at nearby Little Horwood.

Mr. White was evidently delighted. He told me I was the first person he had met who had actually worked inside the Whaddon Hall complex, and please would I write about it for him and provide a drawing of the layout.

Some weeks later, I sent him a sketch from memory of the layout of the original buildings and the new huts built in the grounds. I also gave him a description of some of the work that was done at Whaddon and the people involved in it. From that small beginning, he urged and encouraged me to expand the record into a wider coverage of the role of the SCUs and their staff – and that has slowly developed into this book.

This is not an attempt to set down my personal role during the war, nor is it to boast of my part in these great events. I am hardly in a position to do that since I was about the youngest (and most junior) of the staff, working inside the security zone of Whaddon Hall between 1942 and 1945. Being the youngest at the time, also leaves me still the youngest of a sadly dwindling band of those who worked at Whaddon during the war, or who worked for its units elsewhere.

I know I owe a great debt to the many old colleagues whom I have since contacted, and to the additional information that has come to light through them. Therefore, this book is actually a collective effort by wartime companions from the various Special Communication Units, formed by MI6 (Section VIII), under its head, Richard Gambier-Parry.

I have tried to be selective, in choosing from the mass of material that has been so generously made available to me. However, there may be an element of duplication in some of the stories as they are about the same unit, but each will tell the story from a different perspective.

One or two of the chapters are semi-technical, while others deal with our operations abroad. Martin Shaw's story is quite different from these, however, dealing with the minutiae of his life at Gees and Hanslope. It is a fascinating insight into the military section of the unit that some – myself included – hardly knew existed.

I am particularly grateful to David White who had to continuously prod and encourage me, until I finally came to realise the importance of the Section VIII story he so wanted me to write. I was convinced by David that these stories simply had to be told to record the great work done by a devoted bunch of highly motivated men, whose achievements and technical brilliance has largely gone unnoticed. The book is also very much about the people – just as much as about their work – or the operations involved.

Another reason for putting all this 'on the record' is that when I started, I discovered that there were already a number of books on the subject of Ultra, Enigma, Bletchley Park, MI6, GCHQ, SOE, Stewart Menzies, and 'Y' stations. Unfortunately, very few mentioned Whaddon Hall, and even those that did so, usually wrote as if it were a satellite of Bletchley Park, and not a key player in its own right.

The book The Ultra Secret is a case in point. It was written by Group Captain Fred. W. Winterbotham who had been head of MI6 Section IV (Air) during the War. He seemed to break ranks with senior SIS officers in writing the book back in 1974, when Bletchley Park and the word 'Ultra' were still unknown to the public. He constantly refers to 'Special Liaison Units', in fact they appear in the index over thirty times. Yet although he takes a proprietary approach to the whole SLU network, not once does he mention that the wireless units were designed, built and run by MI6 (Section VIII). Furthermore, there is no mention of Brigadier Gambier-Parry, of Whaddon, of Section VIII, or even of SCUs.

Writing about the Special Liaison Unit staff at the headquarters of the 9th US Tactical Air Force, he says *'I had put an American officer in charge of the SLU unit here, with the usual complement of RAF cypher sergeants and W/T personnel.'* He omits to say that the *'W/T personnel',* and all their equipment, were provided entirely by Gambier-Parry's SCU organisation. One of them was Bernard Gildersleve of SCU8, in our Dodge wireless vehicle, who was attached to the 9th. (See Chapter 36).

In 1978, Ronald Lewin wrote Ultra Goes to War, mentioning Whaddon Hall and its transmission station 'Windy Ridge' for the first time in print. Nigel West, in his excellent book MI6, written in 1983, briefly mentions Section VIII, Gambier-Parry and other aspects of our work.

Unfortunately, my narrative has to include expressions such as 'I believe' or 'to the best of my knowledge'. Please forgive these caveats which are partially due to the passage of time but also to the obvious 'need-to-know' regime that existed in this most secret of wartime units.

Obviously few photographs were taken at the time but, with the aid of friends, we have found a small but interesting selection. Even this small number is surprising since we were so security-conscious.

Today, after almost sixty years, it is virtually impossible to compile a detailed and accurate account of the work of Section VIII, and the situation is unlikely to change in future, even if more archival material from SIS finds its way to the Public Records Office. I am told, with some authority, that records of the wartime activities of

Section VIII were pulped many years ago. Memories are fallible and, sadly, only a dwindling band of witnesses is left.

I have been extremely lucky in having had unstinting help from a number of colleagues from the wartime SCU organisation or their relatives. They are listed at the end of this preface, but I would like to make special mention of a few who made an outstanding contribution.

Edgar Harrison (see Chapter 31 – 'Wireless operator to Winston Churchill'), was one of the earliest members of the unit and continued in DWS after the War, becoming its Principal Signals Officer. His is a most authoritative voice in recalling those days.

Pat Hawker was enrolled as a VI (Voluntary Interceptor) in 1940, and then, from 1941, employed directly by MI6 Section VIII. He has been an absolute tower of strength throughout my research, and supplied me most generously with vital information on aspects of his work in the various SCU units to which he belonged. Pat has read and commented on a number of the chapters, and advised on other spheres of the Secret Wireless War, arising from his own technical knowledge and investigations. His own fascinating story, 'Pat Hawker and his many roles in the Secret Wireless War', is told in Chapter 34.

Norman Walton, who was in the unit from 1940, continued patiently to answer my many questions until he sadly died in late 2001. I shall always cherish the memory of taking tea with Norman at the Different Drummer Hotel in Stony Stratford. There, on several occasions, he gave me the benefit of his knowledge of the early history of the unit and its personnel. We also talked of colleagues and events, from our time together in 1945 at SCU 11/12 in Delhi and Calcutta.

Ex-SCU and RSS members hold an annual meeting at Bletchley Park. The organiser is **Bob King** who was a VI, then a wartime member of RSS, and is now regarded as a leading authority on its work. He has made a valuable contribution to the book and most especially to Chapter 15.

I could not leave **David White** out of this list. Whilst he is the man who persuaded me to write this book in the first place, he did not then cast me adrift. David has been there to help every time I had a question, and all of us refer to him from time to time on the history of the SCUs. He lived near Whaddon but was too young to be in our wartime units, but he later joined Diplomatic Wireless Service which was formed out of them. He is now the Curator of the Wireless Museum at Bletchley Park. A visit there is a must for those interested in the work of the SCUs, and is an opportunity to see some of the equipment we used.

It really is invidious to select any more from those who have helped, and are listed at the foot of the Preface, but it would also be wrong not to mention **Wilf Neal** in this context. He has been a constant support in the enterprise and his story is told in Chapter 25.

John Riley was one of the first to come forward and sent me all his notes on his wartime work with the unit, which was mostly at Hanslope. (See Chapter 36).

Additionally, I have the great good fortune to be writing this book, just as several previously unknown documents have come to light, written by some of the most senior men in Section VIII. Firstly, the reader will be astonished to learn that one of the 'founding fathers' of the new Section VIII was writing a personal diary of his travels and meetings throughout the months leading up to the War, until he left the unit in 1940. That man was **John Darwin** and his story is told in Chapter 21.

It was difficult to know where to start but I had found Steve Dorman's address and Jane and I went down to see him at his home at Looe in Cornwall. He had retired as head of the technical department of HMGCC based at Hanslope, after a lifetime in the service. Steve had joined the unit in 1942, around the same time as I did, and although very much my senior, he was always friendly and helpful.

When I met him some fifty five years later, the clock was turned back and there he was, totally willing to help where he could. Sadly, Steve died a year or two later but he had already given me some of his material, and we taped over two hours of our conversation together. His story appears in Chapter 32.

Shortly before his death in 1992, Lt. Col. Bob Hornby, our most senior wireless engineer, dictated some notes about his time with the unit, to his daughter Antoinette Messenger, and these are used in his story, in Chapter 22.

Just before Christmas 2001, a detailed account of Major 'Spuggy' Newton's service in Section VIII, written by himself, was discovered by one of his sons in a box in his loft. It contains an amazing amount of material that has never been seen before. Spuggy Newton's story is in Chapter 23.

In his book Ultra goes to war, Ronald Lewin writes about Rommel's own secret traffic, sent by his No. 10 Signals Regiment at Afrika Corps HQ in North Africa to the German base in Rome, and then onward to OKW, the German High Command in Berlin. My good fortune continued when last year, I made contact with Siegfried Maruhn who was a member of that same signals unit handling Rommel's Enigma traffic, and his story is told in Chapter 37.

I have found great difficulty in deciding whether to refer to some aspects of our work as being 'Wireless' or 'Radio'. As a boy at home before the war, we had a *wireless* set although we looked up the programmes in *Radio* Times. The term 'radio' was coming into common use at the outbreak of the War but our transmitting stations at Whaddon were always known as 'Wireless stations'. My father ran the 'Wireless stores' at Whaddon and we built 'wireless' sets, not 'radios', at Whaddon and Little Horwood. Many of my wartime colleagues regarded themselves at that time as being 'Wireless operators' not 'Radio operators'.

There was the important Radio Society of Great Britain but, so far as I recall, the only regular use of the word 'radio' in our units, was with RSS the 'Radio Security Service' – which became absorbed into SCU3. Even then, we only referred to it as 'RSS'. In an attempt to be consistent, I have used the word *wireless* wherever I can, and only *radio* when referring to RSS – for example. Some may quarrel with my decision, which was very finely balanced, but those who can cast their minds back to the late thirties, will perhaps sympathise with my ultimate choice of word.

Apart from a general clearance to speak about wartime Whaddon from the Ministry of Defence, my only direct approach to the 'powers-that-be' at Hanslope for more information and their cooperation met with no response. However, I have to acknowledge the help of the Archives Section of the Foreign and Commonwealth Office in the research on Barnes.

In my own story (Chapter 38), I have talked in depth about my experiences with SCU up until 1945, and then briefly about my final months with the unit until I left in early 1947. All the other stories have ended in 1945, even though a number of contributors continued to work for DWS or other post-war parts of our unit. Only a few agents are mentioned by name, but after the passage of sixty years, their exploits should be of interest only to the reader and historians.

The wireless apparatus we used at that time was probably closer to the 'cat's whisker' wireless sets of the 1920s than to the sophisticated communications equipment now used by our secret services. Today, the transmission of intelligence around the world is based on computers and satellites and is a technology I can hardly comprehend. However, I hope that the modern generation will welcome a record of this important, and largely successful part of SIS communication history.

Our Windy Ridge and Main Line stations at Whaddon handled some of the most important wireless traffic of World War II. I hope this book will demonstrate the vital role played by Section VIII, not only in the dissemination of the Ultra intelligence, but in other important work of the Secret Wireless War. The omission or overlooking of its role requires correction that is long overdue; I trust it will be noted by historians who will be able to deal with the subject more fully than I have managed, especially if other records should become available. The files on my study shelves eventually measured 1900 mm across, and I have two filing cabinet drawers full as well – so there is a lot left for someone to work on if they wish, even among my own papers.

I appreciate that the many sets of initials used in the book might be confusing. Because of their sheer number and complexity, I have included a detailed glossary here, instead of the more usual place at the back of the book. You will find the book easier to read if you first read the glossary and acquaint yourself with some of the terminology.

Throughout the book, and especially in the personal recollections in Part 3, I have left the various ranks as they were at the time of the story in question. For example, a person might appear as a Captain in one chapter and as a Major in another. I felt it is unnecessary to use the formula, 'Captain (later Major)' etc.

Some of the stories given to me were almost complete books in themselves, such as those by Bill Miller and Martin Shaw, and the extensive work done by John Riley – amongst others. I have had to shorten them but with the permission of the authors, I hope eventually to deposit their papers at the Imperial War Museum, along with a copy of this book. They already have a bound copy of my compilation of the remaining Stable Gossip – the house magazine of Whaddon Hall – on file (See Chapter 20).

To put the work of Section VIII during World War II into context, I have written a brief background to the development of the British secret services, leading up to 1939. For an overview of British espionage and counter-espionage, there are many excellent books on other aspects of the work of the SIS, some of which are listed in the acknowledgements.

This book is written with a deep sense of my responsibility to the importance of the subject, to those fine men who are no longer with us, to my wartime colleagues, and their direct kin and others who have so generously assisted me in my research. I list below those to whom I am so tremendously in debt.

Geoffrey Pidgeon
Richmond, Surrey
June 2003

Pat Hawker / David White / Edgar Harrison / Bob King / Phil Luck / Wilf Neal / Martin Shaw / Steve and Tony Newton, sons of 'Spuggy' Newton / Ben Hornby and Antoinette Messenger, son and daughter of Bob Hornby / Joyce Lilburn / Lawson Mann / Bill Miller / Maurice Richardson / Ken Rymer / Siegfried Maruhn / Dudley Bradford / Walter Dunkley / Dennis Herbert / Ted Cooper / 'Mac' McLean / Jack Whitley / John Lloyd / Tom Chandler / John Riley / Bert Fry / Ron Unwin / Griselda Brook, (née Darwin) niece of John Darwin / Jimmy Gee / Max Houghting / Charles Tracey / Christopher Maltby, son of Ted Maltby / Irene Healey (née West) / George Hainsworth / Hugh Humphreys / Alan Stuff / Stuart Hill / Evelyn Watts / Pamela Bedford (née Tricker) / Bernard Gildersleve / Tessa Holden / Ray Small.

Acknowledgements and Bibliography

I am deeply grateful to a large number of people and organisations for their assistance in producing this book. I hope I have included them all here but ask forgiveness if anyone has been missed out.

Books with general information on Ultra and the wider SIS scene

There are a number of books that have been a great assistance to me in my research and there are over thirty on my study shelves. However, several are quite outstanding for those seeking more information on the wider aspects of the British Intelligence services. I acknowledge the help they have provided.

'A History of The British Secret Service'
by Richard Deacon, published by Frederick Muller Ltd.

'MI6' and 'GCHQ'
Both are by Nigel West and published by Weidenfeld & Nicolson. They are full of detail and are a necessary read for those wishing to know more about the general history of both organisations.

'The Secret War'
by Brian Johnson published by BBC Publications in 1978.

'Spycatcher'
by Peter Wright, first published by Viking Penguin in 1987. Whilst this is largely about his own career in MI5, the early chapters refer to his father's close connection with the Marconi organisation.

'The Secret Servant – The Life of Sir Stewart Menzies – Churchill's Spymaster'
by Anthony Cave Brown published by Sphere Books Limited.

Books directly connected with Ultra and/or the work of Section VIII

These specifically cover the work at Bletchley Park, the 'Y' service and other operations that had connections with MI6 Section VIII.

'Ultra goes to War'
Quotes from this excellent book by Ronald Lewin are reproduced with the kind permission of Curtis Brown Ltd, London, on behalf of Rosemary Lewin Copyright (c) Ronald Lewin.

'Enigma – the Battle for the Code'
by Hugh Sebag-Montefiore published by Phoenix. This explains the vital part played by the Royal Navy in providing the codebreakers with Enigma machines and German cypher books. It gives a more global view of the Ultra story than most others, and is therefore, a must for those interested in understanding a wider picture. I am grateful to the author for permission to quote from his book.

'Britain's best kept Secret – Ultra's base at Bletchley Park'
This excellent publication by Ted Enever is on sale at the Bletchley Park book shop. A very comprehensive book covering the history of Bletchley Park itself, the arrival of SIS, a description of the work done during the war, the Huts, the Bombes, and Colossus.

With his kind permission, the sections dealing with the early history of Bletchley Park and the Leons in Chapter 8, have been taken directly from his book.

'The Codebreakers – The inside story of Bletchley Park'
by F. H. Hinsley and Alan Striff. Quite technical in parts but a rewarding book to read; containing stories of the successes, and failures of the code breakers.

'England Needs You – The story of Beaumanor'
by Joan Nicholls who was in the ATS as an operator in the 'Y' (Wireless Interception) unit at Beaumanor. Her book is available from the Bletchley Park book shop. This is an important book for those wishing to learn about the Y service and its importance as part of the Ultra Trilogy. Joan Nicholls has graciously written the section on 'Wireless intercept – the 'Y' service' – in chapter 8, especially for this book.

'After the Battle'
Winston Ramsey is the Editor of the Magazine and I am deeply indebted to him for his kind permission to use one of his articles in Chapter 18 on Black Propaganda.

'The Ultra Secret' by F. W. Winterbotham
I am grateful to Weidenfeld and Nicolson for permission to quote from the book.

'British Intelligence in the Second World War' (Vol. IV)
by F.H. Hinsley, and C.A.G. Simkins. HMSO, 1990.

Organisations

English Heritage for permission to include the aerial photograph of Whaddon.

Imperial War Museum for their encouragement and support

Eton College for information on the schooling of Michael Gambier-Parry and his brother Richard Gambier-Parry.

Tough Bros. Ltd. at Teddington, Robert Tough for his help with details of the MFU

Foreign Office Archives for searches on the early station at Barnes

Ordnance Survey who have kindly allowed me to use the map of Whaddon and district for the cover of the book, and to illustrate points inside.

Bletchley Park Trust for their help. My special thanks to Christine Large, Director of the Bletchley Park Trust, for her great personal support to me when I started the book. At that time, the Section VIII and SCU contribution to the Ultra story were not appreciated, or seemingly understood, at Bletchley Park. Christine has helped me to balance the picture.

People

I am indebted to many colleagues who were directly or indirectly employed by Section VIII and they are listed with my grateful thanks in the Preface. Some are relatives of wartime colleagues. However, I must reinforce my thanks to one or two of those, and to express my gratitude to the others who have been of great assistance in the production of this book.

Pat Hawker has undoubtedly been the greatest support to me in this project. He has been unstinting with both his time and his great knowledge of the events I have tried to portray. Without Pat, this would be a leaner and less authoritative book.

Edgar Harrison for his wonderful memory and the stories he relates of years long ago in the 1930s, when he started working the Peking to Barnes wireless traffic.

Bob King has given his time and personal recollections of our wartime organisation. In particular, the story of the VIs and RSS in chapter 15 is largely his work, and I am immensely grateful for his input.

David White without whose prodding this would never have happened anyway; for his unstinted assistance in research, and in checking aspects of my work.

John Lloyd who was unfailing with his help to me.

Joyce Lilburn (née Hill) for her continued support and encouragement as well as providing material for her family's fascinating story.

John Riley was at Hanslope with RSS and was a tower of strength in the early days when I had lots of folders with chapter headings – but nothing in them.

Ray Herbert who wrote about 'Funny Neuk' for the Bourne Magazine and has helped me with my chapter on this part of the history of SIS wireless, particularly by putting me in touch with Ron Humphrey.

Ron Humphrey has provided me with pictures of Funny Neuk, the story of the Public Schools Battalion of the Middlesex Regiment, and the building of the World War I army camp that became Woldingham Garden Village.

Siegfried Maruhn for the enthralling chapter in which he relates his story as an Enigma operator with Rommel in the Afrika Corps.

Richard and Pauline Winward owners of Whaddon Hall today who have helped me in every possible way. I am so very grateful to them.

Peggy Martin for her kind help in researching place names for me in Stony Stratford.

David and Debra Rixon proprietors of **Grindelwald Productions** who made the documentary film 'The Secret Wireless War' and who have helped me with photographs – and encouragement.

Mrs. Maisie Brown Chairman of Barnes and Mortlake History Society for giving her time so generously to help me unravel the Barnes puzzle.

Mr. Steven Earl who searched the records for me at the Metropolitan Police Historical Museum at Charlton in London.

Mr. Kewal Rai in the Foreign Office Archives Department was very helpful in my search for the site of the station at Barnes.

Vivian Kindersley and her niece Griselda Brook (née Darwin), for generously allowing me to use extracts from John Darwin's diary, and especially to Griselda Brook for providing an edited version for me.

Lt. Col. Peter Crocker – Curator of The Royal Welch Fusiliers Regimental Museum, Caernarfon, for information about Richard and Michael Gambier-Parry's service with the Regiment in World War I. In the chapter on Richard Gambier-Parry, you will see I have used the Regiment's preferred way of spelling 'Welch' with a 'C' – although this was only confirmed in an Army Order No. 56 in 1920. The title comes up again in Chapter 31 (referring to action in World War II), by which time it was correct to use the word 'Welch'.

The County Archivist of **Surrey History Centre** where Jane Tombe kindly supplied me with the electoral registers of Funny Neuk. These confirmed that 'C' – Admiral Sir Hugh Sinclair – was on the register as the householder in 1938 and 1939.

The late Hugh Trevor-Roper (Lord Dacre) for graciously allowing me to quote from his broadcast in 1979 on the work of the RSS – in chapter 15.

Peter Spooner for his assistance with the chapters on the history of Whaddon Hall and the village itself.

Rosiland Mottram for permission to use her copyright pictures of Gawcott and Potsgrove.

Richmond Museum The Museum's Local Studies Librarian – Jane Baxter and Phillip Jones – have been very helpful in supplying maps of Barnes and the pictures of the Barnes police station.

Damien Horn co-owner of The Channel Islands Military Museum, St. Ouen, Jersey. Damien has a large collection of German memorabilia and I am very grateful to him for allowing me to use photographs of his Enigma machines.

Photographs and Illustrations

Considering the secrecy that surrounded the unit's work it is truly surprising how many photographs have come to light. Almost all of these were supplied by colleagues or their relatives, whose names appear in the Preface, or above. Some books add the source of each picture under the caption but I believe that detracts from the story being told. I am sure those of my colleagues who supplied pictures will accept that view, and the professional libraries accept that I mean no disrespect. However, I now list those libraries who have provided me with images.

National Portrait Gallery I must thank Matthew Bailey and Jennifer Mumujie of the National Portrait Gallery. They were untiring in their search for pictures of Sir Francis Walsingham and Stewart Menzies – 'C' Chief of SIS – for almost all the war. Both appear in Chapter 1.

Getty Images They found the picture of Baden-Powell for me which is in Chapter 1.

The Imperial War Museum, London Thank you to everyone in the Photographic Archives department for finding the painting of Mansfield-Cumming and the photograph of Rear-Admiral Hugh Sinclair.

Bletchley Park Trust kindly sent me a picture of the Mansion and of one of their Enigma machines.

There are others where I cannot now trace the source in order to express my appreciation. If I have failed to express my gratitude to any individual or to an organisation, then I can only apologise profusely.

I want to thank my editor Josephine Bacon of American Pie Limited for all her hard work in reading through and correcting the text. She has also dealt well with my firmly held views on composition, and usually managed to show me the error of my ways.

Finally, I was hugely fortunate in finding Prestige Press as my publisher and its CEO David Pearman who has been a tower of strength in seeing my book came together broadly as envisaged. I truly believe without his enthusiasm for the task, his skills, and our growing friendship, this book would still be a series of notes piling up in my study. Thank you David.

Prologue

Just before the start of World War II, Admiral Sir Hugh Sinclair, then Chief of the Secret Intelligence Service (SIS), realised that his existing wireless network would be quite inadequate to deal with the volume of traffic in the war, that was so clearly looming. He decided to enlarge the small wireless section based at the SIS headquarters at 54 Broadway Buildings in Westminster, and at its wireless station shared with the Foreign Office, out at Barnes.

The new organisation was required to provide a secret wireless communication network, independent of the Foreign Office. In 1938, he recruited Richard Gambier-Parry, the sales manager of the UK division of Philco, to set it up.

Amongst a number of other departments of the organisation, Broadway housed the Government Code and Cypher School (GC&CS), under the control of Commander Alastair Denniston.

In 1938, Admiral Sinclair purchased Bletchley Park, a rambling Victorian mansion just outside the town of Bletchley, in North Buckinghamshire. Government funds were not made available so Sinclair purchased the property himself from his own funds. It was to be the 'War Station' for Broadway where various sections of SIS could work away from the risk of bombing in the likely event of war. These sections included GC&CS, Section V, and Section VIII. In books written after the war, Bletchley Park has been given various names, including GC&CS, Station X, and GCHQ.

However, it was set up by Sinclair and known at Broadway simply as 'War Station'. For those who worked there, or were involved in its aspects of its operations, it was often referred to as 'The Park'. However, throughout the war it was almost universally referred to as – 'BP'– by us who worked for Section VIII outside, those who worked inside 'The Park', and certainly by the 'locals!'.

With the severe financial constraints that forced Sinclair to purchase Bletchley Park with his money, the idea that BP was called 'Station X' because it was the tenth property purchased by SIS seems unlikely. I therefore suggest SIS were hardly in a position to have purchased nine others! As you will see later, the selection of the new SIS wireless station called 'Funny Neuk' at Woldingham in Surrey, was possibly because Sinclair was already using it as a weekend retreat himself.

A few miles north of Bletchley was another country mansion called Hanslope Park belonging to the Hesketh family, also acquired for use by SIS. In the context of the '10th property concept' it is worth noting that this was only *leased* by SIS. At no time did that, or any other SIS property, have a number allocated to it. If SIS properties were to be numbered, then where was Station IX or Station XI?

With our later ability to intercept the German intelligence (Abwehr) wireless network, Hanslope was then used as a base for Radio Security Service (RSS). Richard Gambier Parry appointed Lt. Col. Ted Maltby as its Commanding Officer and it later became known as Special Communications Unit No. 3 (SCU3).

The growth of Section VIII was accelerated as they handled increased SIS wireless traffic at Barnes, at Funny Neuk, and at the new stations at Bletchley Park. This was at the same time as the codebreakers at Bletchley

Park began to decypher German military traffic from Enigma intercepts, leading to the need to rush the freshly gathered intelligence out to military commanders.

It was decided to acquire a new base for the Section VIII wireless stations away from Bletchley Park. This was in yet another Buckinghamshire mansion – Whaddon Hall – in the nearby village of Whaddon, only a few miles west from Bletchley Park.

Again, we should note that Whaddon Hall was also only taken by SIS on a *lease* from the owners, the Selby-Lowndes family by SIS, and not purchased. The rent was paid monthly by cheque to the family, often handed over personally by such as Norman Walton, then in our accounts department.

Thus, there was a triangle formed by three large country mansions, across North Buckinghamshire, Bletchley Park, Whaddon Hall, and Hanslope. These three, together with various 'Y' or intercept stations, played a major part in the war effort and materially contributed to its successful prosecution.

Admiral Sinclair died towards the end of 1939 and his place, as Head of MI6, was taken by Colonel Stewart Menzies who had previously been Head of MI6 Section II (The Military section). He remained as 'C' for many years, indeed, he only retired from the position as Head of MI6, in 1951. There is evidence that Menzies was very strongly supportive of Gambier-Parry in all aspects of the work of Section VIII.

Whaddon Hall and its extensive grounds were first taken over in 1939. Under the guidance of Richard Gambier-Parry, it now became the headquarters of MI6 Section VIII (i.e. SIS Communications) known to the outside world by its militarised title as Special Communications Unit No. 1. (SCU1).

Section VIII had multiple and rapidly developing roles. It provided wireless communications with our embassies and with our agents abroad, in addition to being involved in almost every other aspect of the Secret Wireless War.

One of its most important tasks was the operation of its wireless station at Windy Ridge at Whaddon, to disseminate the 'Ultra' intelligence traffic to military commands. Another, was its vital part in the 'Black Propaganda' war against the Germans.

Overlooked in the publicity given in recent years to Bletchley Park and its codebreakers, is the intelligence gathering already being carried out at the beginning of the war by SIS agents across the world. This source of raw intelligence was collated at embassies – usually by the local Passport Control Officer – and then sent on to SIS in London. This would be by telegram, by letter, or in the last year or two leading up to the war, increasingly by wireless communications installed by the 'reborn' Section VIII.

Before the war, the London end of the SIS world-wide wireless network was directed from its Station X located at Barnes in West London, which was shared with the Foreign Office. Later, when Whaddon started up, the bulk of the Barnes contingent moved there in mid-October 1939, as reported by John Darwin in Chapter 21. No doubt, a skeleton staff kept the station at Barnes running for a short time afterwards to ensure an overlap.

The SIS wireless station was installed right up in the tower of the mansion but only handled SIS traffic – it had nothing whatever to do with interception, codebreaking, Enigma, or Commander Alastair Denniston.

All other wireless traffic, including embassy traffic, was directed from the building now known as Hut 1. Later it moved to Whaddon, in new buildings in the fields in front of Whaddon Hall, in what became known as 'Main line' station.

Near the tower, on the top floor of Bletchley Park Mansion, was the cypher room for the SIS operation –

working behind locked doors near to its station in the tower. It was not connected in any way with the work being done by GC&CS, elsewhere in the mansion, or its various outbuildings.

Locally collected intelligence by SIS agents, and representatives abroad, continued to be important throughout the war, even when Bletchley Park broke the German codes created by the Enigma cypher machines. After the closure of the wireless station X in the tower at Bletchley Park, SIS traffic was handled by its own new wireless stations at Nash, Weald, and later at Forfar.

The later brilliant success of Bletchley Park in decoding the German wireless traffic collected by the 'Y' service (including that from the German army, navy, air force, and Abwehr), meant that safe channels had to be established to pass that information to the strictly limited number of recipients, whether they be military commanders, or government officials.

To guarantee the distribution would be under totally secure control, 'Special Liaison Units' (SLUs) were created by Group Captain Fred Winterbotham Head of MI6 Section IV (Air). He had earlier been moved to 'War Station' at Bletchley Park under orders from 'C'. He was given the enormous task of ensuring the secrecy and speedy distribution of this vast amount of vital information.

These SLUs were located by Winterbotham in such vital organisations as the Admiralty and Fighter Command, to receive what became known as 'Ultra Traffic'. Others were attached to field commanders and headquarters of the services at home and overseas, and were usually mobile units in lorries or cars. The design, construction, and manning of the wireless content of these units was under the control of Section VIII, whether fixed, as in St. James' Park, or in mobile units.

Whaddon Hall contained workshops making wireless sets for our agents and covert stations abroad. These workshops also fitted out the wireless vehicles for use as mobile SLUs. It housed substantial and important research departments, and from those came brilliant new wireless equipment used in many aspects of the wireless war. Our unit worked on aircraft and ships, trained wireless operators for use in a multitude of fields, provided basic wireless and Morse skills to secret agents before being sent abroad – the list of the roles undertaken is seemingly endless.

Wars are fought on many levels – on the battlefield, in the air and at sea. This is the story of the Secret Wireless War that went on – more often in quiet corners – than in open conflict. However, there can be no doubt the part Section VIII played in the successful prosecution of the war, would make any member proud to have served under Richard Gambier-Parry.

It has been claimed that the Ultra Secret code-breakers of Bletchley Park helped shorten the war by at least a year, and saved the lives of hundreds of thousands of the allied forces. These claims are almost certainly true, but the task of intercepting those messages was carried out by the 'Y' service, and of ensuring the safe dissemination of much of the vital knowledge gleaned at Bletchley Park, was in the hands of MI6 Section VIII, based at Whaddon Hall.

However, MI6 Section VIII played a major part in the secret wireless war, in many more ways than just handling Ultra, and I trust I can explain some of this to you in this book.

I hope you find it interesting and that it is a fair record of those exciting days. Certainly, I have done all I can to ensure it is accurate as possible, considering the absence of proper records, and the passage of over sixty years.

Bletchley Park is universally recognised as the natural home of the Enigma and codebreaker's story. It is also singularly qualified to be the repository for the story, records, and equipment of Section VIII's

wartime history. It is my fervent wish that the Wireless Museum, at present there in Hut 1, should be expanded to make room for the wider selection of artifacts and records that exist. Only then will there be a broader public knowledge and appreciation of its work.

Geoffrey Pidgeon

Glossary

Abwehr. (Literally 'defence'). This was Admiral Canaris's military intelligence organisation similar to our MI6.

AR88. (*ACK-R-88!*) A wireless receiver made by RCA (Radio Corporation of America). Used by Section VIII and by 'Y' intercept stations across the country. Very reliable and easy to tune. Whaddon also had some of the earlier model the AR77.

Ascension. This was a device for speaking directly to agents from an aircraft flying nearby. It was FM speech instead of Morse, so that it was easier for less skilled operatives to make contact and pass on information. It was also more difficult to locate the agent by DF as he did not have to go through the long call-sign procedures and the message could be passed more quickly. It was devised by a number of Section VIII engineers and, at various times, technical input was made by Bob Hornby, Spuggy Newton, Steve Dorman, Wilf Lilburn, Alfie Willis and by Dennis Smith. It was regarded as a great success.

ATS. Auxiliary Territorial Service. The women's service in the army.

BCRA. Bureau Central de Renseignements et d'Action. The French central office of intelligence and operations.

B-Dienst. German wartime interception system with the 'B' being the equivalent of our 'Y' services.

Boniface. A fictitious name given to decyphered Enigma traffic early on, implying that the information was being received from an agent named Boniface. The intelligence was referred to as coming from 'Source Boniface' thus ensuring that nobody suspected we were actually breaking Enigma itself. The information was later designated 'Ultra'.

Boss. You might also be intrigued to see the word 'Boss' used in the chapters that follow. I do not recall hearing anyone refer to a person above you as your 'superior officer' although with differing ranks, that was technically the case.

Let me give you an example. For part of 1943 I worked in the 'factory' at Little Horwood making wireless chassis, Morse keys and other components for the agent's sets being assembled there. There must have been some forty people involved when I left, of all ranks, and we just used Christian names or nicknames. Only Hugh Castleman (who was in charge of us), was addressed directly as 'Sir'. In SCU circles Gambier-Parry was '**The** Boss'.

B2. A wireless set used widely by SOE. Designed by John Brown at the Frythe, Welwyn Garden City and made at Coronation Works in Birmingham. The cover was Interservice Research Bureau (ISRB) and its B2 was used by SOE all over Europe. It had a range of up to 1000 miles and some were supplied to our units for use in special areas – see Tom Kennerly's story chapter 30.

'C' is the Head of SIS. Also sometimes shown as 'CSS'.

CPO is capable of two interpretations in the context of this book:

The first use of the initials **CPO** is Chief Petty Officer, a senior non-commissioned officer in the Royal Navy. Here, it will mainly refer to those who joined Section VIII from the Navy's wireless units.

Secondly, the initials **CPO** can also refer to Chief Passport Officer. This was the title of the head of the Passport Office before the war, as a cover for SIS, attached to embassies in major cities across the world.

DMI. Director of Military Intelligence.

DWS. Diplomatic Wireless Service. This was created out of Section VIII by Gambier-Parry at the end of the war and 'sold' as a complete communication system to the Foreign Office to handle both diplomatic and covert traffic.

DX-er. An amateur wireless operator who was used to working with Morse signals over long distances.

Enigma. This was a commercially available cypher machine used by the German services: army, navy and air force – as well as the Abwehr, police and Gestapo. It is important to this story and so I have provided a more detailed section called 'Enigma' which immediately follows this glossary.

GC&CS. Government Code & Cypher School. It was a descendant of Room 40 of the Admiralty which did such sterling work on codes in World War I. This pre-war code-breaking organisation, was first based in Queens Gate in London, but in the 1930s moved into 54, Broadway, the headquarters of SIS.

Gees. The almost universally applied name for the units base at Little Horwood, part of which was housed in the premises built by Mr. Gee of Gee Walker Slater, a prominent building contractor in pre-war London.

G-P. A commonly used shorthand description for Brigadier Richard Gambier-Parry, Head of MI6 Section VIII throughout the war. However, I will not be using it in the chapters that follow.

Head of Station. The title given to the chief SIS officer, usually at an embassy. Before the war he was often referred to as the CPO or Chief Passport Officer.

HRO. A high grade American wireless receiver made by the National Company which had a distinctive removable coil bank change. Used by Section VIII in great numbers, by interceptors around the country, and by our stations abroad.

HMGCC. This was His Majesty's Government Communication Centre. It was used as a cover name for Section VIII, and also used to sometimes disguise its connection. The Black Propaganda operation at Crowborough for example, used this title for the wireless side of the project. After the war, it became the name covering some of the functions of the Whaddon operation, including technical workshops.

Hudson. The Lockheed Hudson was supplied to the RAF for a variety of duties, including Coastal Command. It was used by squadrons based at Tempsford in Cambridgeshire to carry SOE & SIS agents into and out of Europe.

The SIS used them to work to its agents on the ground through the wireless equipment called 'Ascension'. It also delivered weapons and supplies dropped from the air, in the early days before the Tempsford had the larger ex-bomber command planes, like the four engine Stirling and Halifax.

ISLD. Inter Services Liaison Department. The cover name for the SIS presence in active military zones like the Mediterranean, North Africa, the Middle East and India/Burma.

ISK. Intelligence Services Knox. A section at Bletchley Park under Dilly Knox, concerned with special machine-based cyphers

ISOS. Intelligence Services (Oliver) Strachey. A section at Bletchley Park concerned with hand-encyphered signals.

Lysander. The Westland Lysander was designed as an army co-operation aircraft before the war but served in many capacities. Some went to France in support of the BEF but proved unsuitable against enemy fighters and many were lost.

It came into its own however, when used to ferry agents and supplies into occupied Europe and France in particular. Its short take-off and landing capabilities proved enormously useful and was one of the main ways of landing agents – and vitally in picking them up – since everything else needed longer runways.

Whilst most of its 'passengers' would have been from SOE, it was also used by SIS personnel. The two RAF squadrons involved in this work were 138 and 161 based at Tempsford. It was on such a flight returning from France, that Major Jack Saunders (then in charge of Main Line at Whaddon), was lost over the Channel.

Magic. Code name for the intelligence obtained from the ability of the American secret services to decypher signals from Japanese sources.

MDJ Sets. This is a term used to describe agent's wireless sets in the early days, a kind of 'code word'. It has come up several times in my research and is mentioned in John Darwin's story (chapter 21), but no one seems to know its true meaning.

I offer my own personal solution to the mystery. On the outbreak of war, there were three competing 'would-be' deputies to Gambier-Parry – Maltby, Darwin, and Jourdain. They were men of considerable self-esteem, and so used their own initials – ***M***altby, ***D***arwin, ***J***ourdain – **MDJ** to describe the new agent's sets that had actually been designed by Bob Hornby, and others!

MFU. Mobile Flotation Unit. One of the closest guarded secrets, see Chapters 19 and 38.

MGB. Motor Gun Boat. Used by Slocum's Navy. See Chapter 38.

Mitchell. The B25 Mitchell was made by the North American Company as a front line medium bomber for the US forces. A large number of these aircraft were also supplied to squadrons of the British and allied air forces for various duties. We fitted our Ascension air-to-ground wireless equipment into (I believe), two squadrons based at Hartford Bridge (now called Blackbushe) near Camberley in Surrey. One was Free French and another (on which I worked a number of times with Dennis Smith), I recall seemed to have a largely Commonwealth crew.

MFV. Motor fishing vessel. Our unit supplied the wireless sets to a number of fishing vessels and similar small craft used for ferrying agents in and out of occupied territory and for interception.

MI5. The security branch of Britain's intelligence services. Mostly concerned with home affairs.

MI6. This is the external branch of Britain's Secret Intelligence Services. The public use the term MI6 but within the service it calls itself 'The Firm' or SIS and fellow members as 'Friends'.

MkI / MkIII / MkV / MkVII / MkIX and MkX. Numbers of various agent's sets produced by Section VIII at its Whaddon or Little Horwood workshops. In the case of the MkI it was first produced at Barnes and then later manufactured at Funny Neuk.

Morse. Signalling by code in which each letter is formed by a combination of dots and dashes. The best known letters in Morse are SOS (· · · ‾ ‾ ‾ · · ·). In the context of this book, it was sent by wireless telegraphy.

MTB. Motor Torpedo Boat. A number of these were kept to ferry SIS agents and equipment into and out of occupied Europe. Some were based at Dartmouth and moored alongside a pre-war River Dart pleasure steamer, the 'Westward Ho'. Others were in Brixham Harbour and I worked on them in both places with my boss – Dennis Smith.

NCO. Non-commissioned officer. Any rank between a lance-corporal and a Regimental Sergeant-Major.

NPAF. Not paid army funds.

One-time pads. A method of encoding messages that made them virtually indecypherable by the enemy because they required both sender and receiver have an identical pad of tear-off sheets. The sender indicates the relevant sheet to the receiver and the sheets are destroyed after use. The one-time pad was probably the safest of all cypher systems from the point of view of security, although slow and cumbersome in use. Later on, SLUs used the RAF-designed Typex machine which was somewhat similar to the Enigma. It was faster to use than a one-time pad and very safe.

Passport Office. Cover name for SIS units abroad, usually attached to an embassy.

PCO. Passport Control Officer. A member of the staff at the Passport Office.

Pop. Perhaps surprisingly, this is the way we usually referred to Brigadier Richard Gambier-Parry amongst ourselves, although not to his face, and absolutely not by the most junior staff there – like me! It arose from his own constant use of the name in such things as his Christmas message to us and in personal letters to his senior staff. Also see 'Boss'.

PT Boat. This is the American equivalent of the British MTB. There were several alongside the 'Westward Ho' steamer moored in the harbour at Dartmouth – as well as our own. I think they carried out similar work to ours. i.e. ferrying agents and material into and out of France.

Rockex. A cypher machine manufactured under utmost secrecy at Hanslope Park (SCU3). Phased out in the early 1980s.

SBS. Special Boat Squadron of the Royal Marines. An elite force used on special missions. I believe it was SBS personnel who manned the MFU craft being fitted out at Teignmouth – see Chapter 38.

Section VIII was the communication department (or 8th Section), of MI6 during the war years. From time to time, I will say MI6 (Section VIII), and sometimes just Section VIII.

Note: SIS then had a habit of using Roman numerals, so even today I find it difficult, when describing one of our wireless sets, to type it as Mk 3 – it simply has to be MkIII.

Shetland Bus. This was the name given to motor fishing vessels of Slocum's Navy (see below) and they were fitted with Whaddon made transmitters and HRO receivers. The role was the delivery and collection of agents to Scandinavia and some interception.

SIGINT. Signals intelligence. High-grade Sigint later became known as Ultra.

Sigm. Signalman, the Royal Corps of Signals equivalent of a private soldier.

SIS. Secret Intelligence Service.

Slocum's Navy. SIS controlled a number of boats like MFVs and MTBs whose role was the ferrying of agents and some interception. The officer in charge for SIS was Commander Frank Slocum RN – hence the name 'Slocum's Navy'.

SOE. Several departments of the Foreign Office and the War Office were merged in July 1940 to form an organisation to counter the Nazis by subversion. This became known as Special Operations Executive or SOE for short. The very existence of this unit was one of the best kept secrets of the war. Its headquarters was at 64 Baker Street, London W1.

SOE's purpose was to train agents to be sent into France, and other occupied countries, to link with local resistance units and to cause sabotage. There was therefore, a fundamental difference between SIS agents, who were trained to quietly collect intelligence and pass it back to London, and SOE agents who were largely intent on creating havoc. There was a long-standing and tangible friction between the two organisations.

Their agents were issued with special suitcase wireless transmitters and receivers that looked like a normal piece of luggage. Initially, these were provided by Section VIII at Whaddon but later they produced their own sets.

SOG. Special Operations Group. A division of SCU1 specially responsible for handling traffic to commands in the field. Its main transmitter was Windy Ridge.

SSU – SCU – SLU.

SSU. Special Signals Unit. The first name given to the militarised Section VIII in 1940 but it was changed in 1941, to SCU – Special Communications Unit. It was thought the initials SSU might be regarded as standing for ***Secret Service Unit.*** There were also unpleasant connotations, with the German use of SS. In 1940 the letters SSU appeared painted – as the units army designation – over a blue-and-white badge on the wing of its military vehicles. These were later removed and replaced with the initials **SCU**.

SCU. Special Communication Unit. The later pseudo-military name for Section VIII's wartime communications to disguise both its origin and its tasks. It is important to remember, that although SSU (now SCU), were created to be the conduit for Ultra traffic, Section VIII already had a myriad other tasks to perform in the communication field. However, these were then mostly conducted under the umbrella name of Special Communication Units.

SLU. Special Liaison Unit. The name given by Fred. W. Winterbotham, to the units handling Ultra traffic. These were often mobile and would then be coupled with an SCU wireless unit. But, there were a number of fixed ones in the UK, as in the St. James Park underground complex, and at RAF Fighter Command.

Some fixed SLUs received their information directly from Bletchley Park by teleprinter but sometimes had a support team from SCU as a wireless backup and the 'Dugout' in St. James Park is a good example where this existed.
'Special Liaison Unit' or SLU later became the name for the unit handling Ultra which would contain a wireless facility (usually but not always staffed by the army), controlled by Section VIII, and a cypher facility supplied by Winterbotham from RAF cypher training schools. Quite frequently, the mobile units were now being described as SCU/SLU.

In the Mediterranean area of operations there would have been a greater mix of the three services but the control always remained the same: the wireless side by Gambier-Parry's Section VIII, and the cypher via Winterbotham's own connection with MI6 – Section IV (Air).

Station X. The name of Barnes wireless station before the war. Later used for a few months as the name of the SIS station at Bletchley Park, before it was transferred to Whaddon Hall, in late 1939. See also Appendix 2.

Stay-behind-army. An organisation intended to remain behind the enemy lines in the event of an invasion.

Typex. Cypher machine in use at Bletchley Park and in SLUs, along with one-time pads.

Ultra. Is the name given to the messages gathered by our various Y services from the German use of Enigma cypher machines, and the intelligence then emanating from the codebreakers at Bletchley Park. Their output was first marked 'Most Secret' then 'Top Secret' then 'Ultra Secret' and it is from the latter that the name 'Ultra' was derived. This name only appeared later in the war but I have used it through most of the book since we all know what is meant by the word. It is succinct and understood by everyone.

Venlo affair. An embarrassing situation arising from the abduction of two SIS men from the Hague, by the German security forces at Venlo on the German – Dutch border, on 9th November 1939. It lead to the dissolution of the 'Z Organisation' and other ramifications. See Chapter 14.

WAAF. Women's Auxiliary Air Force.

WD. War Department. Often the initials were combined on a notice reading 'WD Property – No admittance' – or sometimes – 'WD Property Keep Out'. In wartime these words on a sign were generally very effective deterrents to would-be trespassers.

Winterbotham. Group Captain Frederick Winterbotham, Head of MI6 (Section IV) Air, who was charged by 'C' (Stewart Menzies) to ensure the total security of Ultra traffic.

WRNS. Women's Royal Naval Service.

Y Service. This is the Wireless Intercept service which collected the raw wireless messages transmitted by the German forces, where the traffic was being encyphered by Enigma machines. The initial 'Y' comes from the two words ***W****ireless* ***I****nterception* – ***wi*** or 'Y'.

XW. This was the call sign for Whaddon used by all stations from around the world after MI6 had taken over Whaddon Hall in November 1939. I suggest 'X' refers to the Main Line station in the grounds of the Hall, with the 'W' standing for Whaddon.

Z Organisation. This was a secret service unit run in parallel with the Passport Office system from 1934 until the outbreak of war. It was funded by SIS but run by Claude Dansey who had seemingly left SIS under a cloud in 1934. See chapter 1.

Enigma

A Brief Outline

The Section VIII story covers a wide spectrum of our secret wireless communications. However, there can be little doubt that the dissemination of the intelligence arising from the breaking of the Enigma cypher machine codes, became its most important role. Although this book is not about the code breakers of Bletchley Park it certainly is about the distribution of its product – Ultra.

So much has been written about the Enigma machine that it might seem unnecessary to explain it in depth but a brief outline might be helpful to readers. Much more detail can be found in GCHQ by Nigel West and in Code Breakers by Hinsley and Stripp, amongst many others more qualified than I am.

Enigma

This was invented by a Dutchman, Hugo Koch, who took out a patent for an encyphering machine in 1919. It was substantially improved by an engineer from Berlin, Dr. Arthur Scherbius. He and friends formed a company called The Cypher Machine Corporation to market the machine, which he christened 'Enigma'.

It was made commercially available and used by firms to transmit confidential information. Banks and large international organisations found it particularly useful, but it was the potential to handle military information in total secrecy, that brought it to the notice of the German military in the late 1920s. In fact it was the German Navy who first, introduced it and that was followed by the resurgent German Army by the early 1930s. Around this time, the commercial model was withdrawn from sale and the company concentrated on supplying the German military machine. It is likely that there were around 80,000 Enigma machines in use by Germany during the war. Different versions might be used by the army, navy, air force, Abwehr, police and Gestapo.

Enigma machine Number A11011 produced in 1940. This is one of two owned by Damien Horn. They were given to him after the war, by someone who had taken them from the Germans, thinking they were typewriters!

However, it is known that the American secret service had already purchased samples of the latest models as did the French, the Dutch and the Poles. It is suggested that the Admiralty also bought two for experimental work and that may well have been the case – leading to our own Typex machines. However, the first to realise the true menace of Enigma were the Poles. They were deeply suspicious of the Germans who they were sure were aiming to eventually overturn the Versailles Treaty. They had established themselves as being brilliant code-breakers but the Polish General Staff became deeply puzzled by wireless signals they simply could not decypher coming from Germany.

They financed a unit at the University of Poznan where a number of outstanding students of mathematics were recruited to study cryptology. They then started their attempts to unravel the mystery of the increasing Enigma traffic. The leader Marian Rejewski, with two close colleagues, Henryk Zygalski and Jerzy Rozycki, were now enlisted into the Polish Secret Service. Running parallel to these studies, the French Intelligence cryptology department, under its head – Captain Gustave Bertrand – were liaising with Warsaw, since both France and Poland were monitoring the growth of the German military machine, by now under Hitler.

The Enigma machine looks like a large typewriter except it cannot print. It has a keyboard with keys, similar to a typewriter, and above that a lamp board with 26 round windows each representing a letter of the alphabet, in the same layout and sequence as the keyboard.

When a letter is depressed, for example an A on the keyboard, it produces an entirely different letter, say a P, which was lit up on the lamp board by a bulb mounted underneath. Each machine had a 4.5 volt battery but outside power could be used.

Pressing the key for the original letter A had started a most elaborate trail. In the earlier machines, there were three wheels or rotors, and an electric impulse went through each of the rotors in turn, which were divided into the 26 letters of the alphabet. The impulse went via a reflector back through the chain of rotor wheels but via a different channel. To add to the incredible number of permutations, the wheels also rotated, so that added yet more possibilities.

A close-up of Enigma rotors

This is a closer shot of the rotors and light board of an Enigma

---

To add to the number of permutations, on the later armed services version, they added a plugboard (*Steckerbrett*) which had series of wires with plugs relating to the letters of the alphabet. The possible permutations on a three-wheel machine was 26 x 25 x 26, meaning the original sequence would only repeat after 16,900 times. Add the plugboard and it is sufficient for the reader to know that the there are no less than 159, million, million, million possible daily key settings.

To add to the confusion, the machine had spare rotors, so the setting instruction for the operator could require the use of one of the spare rotors, with its different operational sequence.

So far as the actual operation of Enigma was concerned, it was difficult, but not impossible, for one man to handle, but normally it was done by two men. The operator read his signal text which was in plain language, and pressed the appropriate letter key. His number two would read off and write down, the resulting letter illuminated on the lamp board. When the message was complete and thus in cypher form, it would then be transmitted by the operator on the wireless set in Morse.

His original letter A would be picked up by the receiving wireless operator as a P, which he would then punch into his Enigma machine when it would then produce an A in the illuminated panel, as part of the original message.

However, as you will have seen, that would entirely depend upon the settings chosen for the wheels and connecting wires. The letter will only come back to the intended and correct letter, if both sender and receiver have set their machines to the same setting or 'Key.' That 'key' setting was originally done only once a month then in shorter periods of time and eventually daily.

The Poles made replica Enigma machines and handed one over to the French and one personally to Stewart

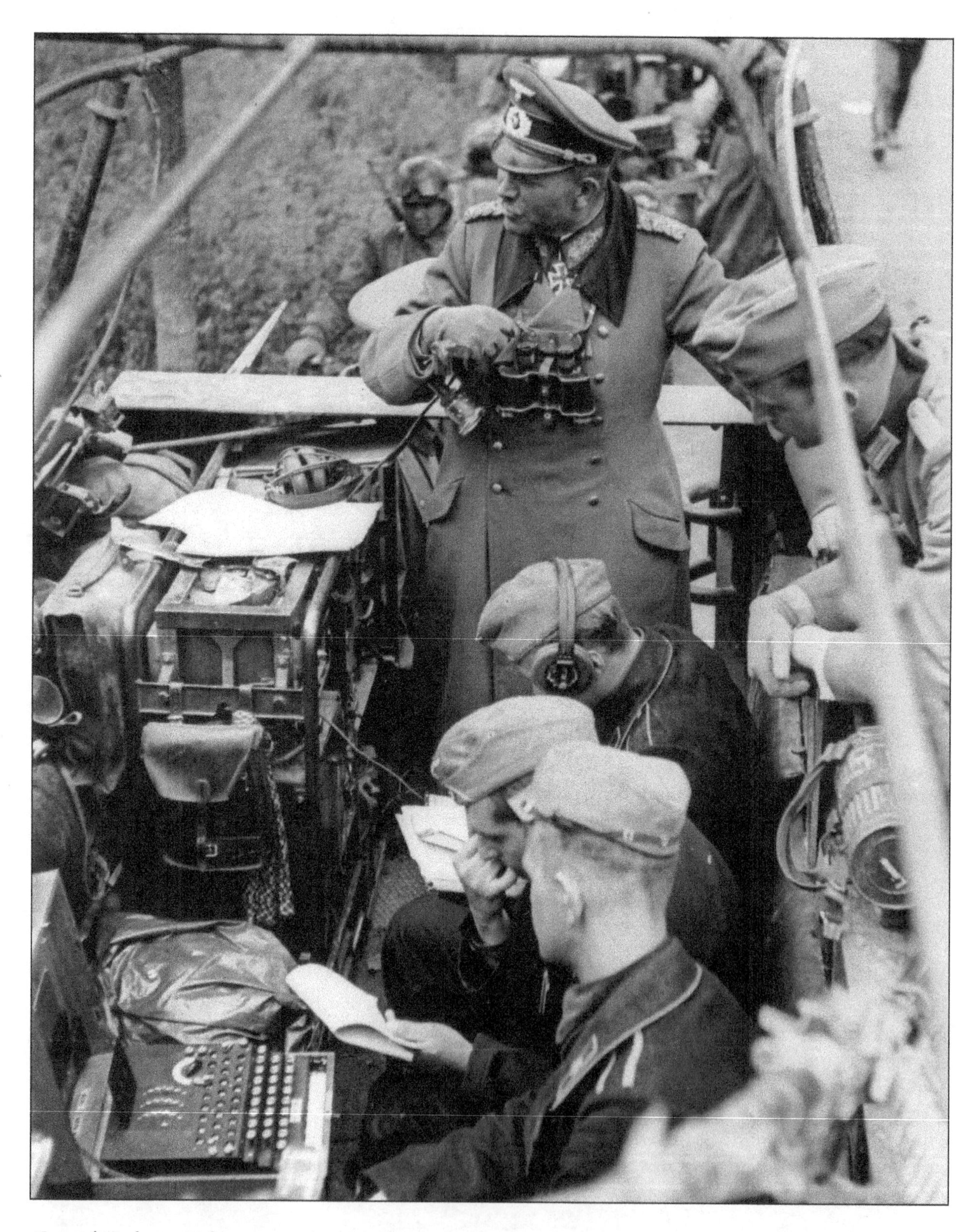

General Guderian in his command vehicle during the blitzkrieg across France in 1940. German forces covered, in just a few days, countryside that had been fought over for four years in World War I. To control such a vast mobile army – travelling at high speed – required extensive and secure wireless communications. His Enigma machine, wireless set and crew are right beside him. This picture demonstrates its importance to Guderian whose choice of command vehicle was effectively a travelling wireless van.

Menzies representing the British Secret Service. They had also designed Bombes, a mechanical device to break the codes, and gave their secrets to SIS. These were much improved by Messers Turing and Welchman, so that the final bombes made in the UK bore little resemblence to the Polish version.

Later during the war, there were four rotor machines, and further developments that ensured the code breakers at Bletchley Park were continuously having to catch up, not only to find the daily 'key' for Enigma, but to deal with the increasingly difficult situations arising.

The Enigma traffic was limited during the 'phoney war' – the time from the outbreak of war in September 1939 – until Hitler unleashed his divisions in 1940. Internal communication inside Germany was partially by telephone but once the Blitzkrieg started rolling into Holland, Belgium and France, the Enigma came into its own and its volume of traffic soared. Communication on the move then depended upon wireless, and in the case of German armed forces, that meant the use of codes created on their Enigma cypher machines.

One illustration shows General Guderian in France in his command vehicle. This was June 1940 and his Enigma team and cypher machine are close beside him.

The work was hugely stressful for the senior personnel at Bletchley Park, who alone knew the results of their failure to break the key at any time. This was especially true in the case of the German Naval codes, when they were informed of the enormous losses to merchant shipping arising from our inability to crack the coded messages going out to submarines in the Atlantic.

Perhaps the most important breakthrough came as a result of acquiring actual Enigma machines and code books as related in 'Enigma' by Hugh Sebag-Montefiore.

Slowly, the Enigma cypher machine was made to give up its secrets but it is incredible that to the very end, the entire German military machine believed its coded wireless traffic was totally secure.

Part I

The Background

Chapter 1

The Historical Background to SIS

A brief history of intelligence-gathering, the formation of the British Intelligence services before, during and after World War II and the heads of the SIS.

'Intelligence is an activity which consists, essentially, of three functions. Information has to be acquired; it has to be analysed and interpreted; and it has to be put into the hands of those who can use it.' (British Intelligence in World War II, Volume I)

It may be helpful to readers to have some background knowledge of the secret service organisation in this country, and how the main sections came to be formed. For those seeking more information, there is a large selection of books available on the subject some of which I refer to in the acknowledgements section.

The Secret Intelligence Service (SIS) is perhaps better known as MI6 (Military Intelligence, Department 6). It is concerned with external intelligence, i.e. counter-intelligence outside the UK, and it is closely associated with the Foreign and Commonwealth Office. It is known within the organisation as SIS or 'the firm'. Those who worked for it referred to one another as 'The Friends' and their HQ is known simply as 'The Office'.

The other widely known secret service is MI5 which during World War II, concerned itself with counter-espionage in the UK and the British Empire. It is officially known as the Security Service and is associated with the Home Office. More recently, it has concentrated on the IRA and other terrorist organisations in addition to other aspects of home security.

The initials 'MI' stand for Military Intelligence but this is a misnomer, as neither organisation has ever been part of the armed forces. It arose because in earlier days the military had been closely involved with it but by 1939, MI6 had direct links with the Foreign Office and MI5 to the Home Office. The army and navy had their own quite separate intelligence services, but were not empowered to run secret agents.

Secret agents for intelligence-gathering, security and espionage have existed through the ages. Alexander the Great used agents who had a form of cypher for conveying secret messages but the Chinese, Ancient Egyptians and the Romans all employed agents. They had various secret methods of passing information such as invisible inks, carrier-pigeons and secret couriers.

Sir Francis Walsingham is believed to have created the first formal secret service in England and used it to support Queen Elizabeth during her reign. He spent some time as her Ambassador in Paris – an early example of the link between the diplomatic service and intelligence-gleaning which continues to the present day.

John Thurloe was Cromwell's spymaster and is said to have had agents in every court in Europe. Daniel Defoe, the author of *Robinson Crusoe* and *Moll Flanders*, was employed in Queen Anne's secret service and was involved in many adventures. The list of distinguished citizens who worked for the British Secret Service

Sir Francis Walsingham was a diplomat and spymaster to Queen Elizabeth I. In addition to intelligence gathering he was also involved in espionage. For these services he received grants directly from the Queen. However, like Hugh Sinclair some 350 years later, he also had to draw on his own resources to manage all the many aspects of his intelligence operations.

included William Eden, First Baron Auckland, as well as Sir Richard Burton, the famous explorer. During the Napoleonic Wars, spies were active on both sides, particularly during the Peninsular Campaigns.

Down the years, most countries have sought to protect themselves by knowing the advance plans of friends and potential foes alike. However, during Queen Victoria's reign Britain allowed its own intelligence networks to decay and any serious reporting was left to the diplomatic corps.

The dangers of inadequate intelligence manifested themselves during the Crimean War when many military blunders were caused by a lack of information. So bad had things become that a military intelligence division was established at the War Office, which later transferred to Queen Anne's Gate, Westminster in the 1870s.

Interestingly, those same premises have continued in use by various intelligence agencies and in the years between the wars they were the home of the Passport Office (See Chapter 2). At the top of the building, there was a small wireless station. The Head of SIS kept a flat in the building for his own use.

The Boer War was the first to be fought in the modern style. Intelligence officers were attached to the flying columns of reconnaissance cavalry. Even so, they were poorly organised and the great distances meant that much of the information arrived too late to be of strategic advantage. One of the best intelligence officers in the Boer War was Lord Baden Powell, founder of the Scout movement, He was also a distinguished spy for his country in parts of Europe, and particularly in the Mediterranean area.

Poor intelligence during the Boer War was responsible for a number of embarrassing defeats for the British army. It was widely recognised that something had to be done to make up for the deficiencies of the British secret services so between 1902 and 1905, a series of discussions took place at high level to see how to improve performance for the future.

At that time, we seemed to have little use for professional spies except in Ireland, then the responsibility of the Special Branch (of the Metropolitan Police) and in the constantly turbulent North-west frontier between India and Afghanistan. Here, the threat was always from Russia, said to be seeking a passage to India. In other countries, the British Consul was expected to report on any unusual events and gather worthwhile intelligence, but this seemed to be a very haphazard affair.

Because of the worldwide activities of its ever-growing naval fleet at the turn of the century, Britain was apparently content to leave serious intelligence-gathering to captains of Royal Navy ships who were encouraged to report on fortifications, foreign ships and other events that were likely to prove useful in time of war. It is for this reason that the country's principle intelligence-gathering unit was the Naval Intelligence Department (NID), one of whose earlier directors had been Captain William Henry Hall. This predominance of naval officers at the top of Britain's intelligence service continued right through until the death of Admiral Sinclair in the early months of World War II.

Later, at the time of the reorganisation of the secret services, Hall's son, Commander William Reginald Hall, began to take an interest in intelligence matters. He joined a band of officers who were concerned about the growing menace of the rapidly expanding German naval fleet. As captain of a cadet training ship, the *HMS Cornwall*, Hall set off in May, 1909 on a training cruise, visiting German ports. He decided to go ashore himself at Kiel and made detailed plans of the dock facilities there. During the cruise, he discovered that the naval charts were hopelessly out-of-date, and returned to the Admiralty to impress upon them the urgent need to overhaul their intelligence service.

As a result of Hall's pressure, in 1910, two Marine officers, Brandon and Trench, were ordered to make a covert examination of the German sea coast defences, concentrating on the Frisian Islands. Unfortunately, they were caught, arrested and sentenced to four years' imprisonment. However, they were pardoned by the Kaiser before they were due to be released, to mark the visit of King George V to Berlin.

This amateur enterprise hastened a high level investigation into our intelligence services. Astonishingly, when plans were mooted in 1909 for a formal secret service covering all aspects of national intelligence work, it was discovered that not a single British agent was operating in continental Europe.

Lord Baden Powell is known as the founder of the Scout Movement but was a highly respected soldier, and a member of British intelligence before, during and after the Boer War. He is shown here at a levy at St. James – just prior to World War II – in uniform as a much decorated Major-General.

It was proposed that the new secret intelligence service be formed as a single body but a year later it was divided into two separate sections, the first was to concentrate on home security and the second on foreign intelligence.

Captain Vernon Kell, first head of the home section, had been in the army and his division was known initially as MO5, and later MI5. The first head of the foreign section of the new SIS organisation was Captain Mansfield George Smith R.N. In a reshuffle in 1915, SIS was redesignated as MI 1(c). This was later to become MI6. Perhaps the fact that the head of both organisations was from the navy, demonstrated the existing predominance of the Senior Service in intelligence-gathering.

Mansfield Smith had graduated from Dartmouth Naval College in the 1870s and saw service in Her Majesty's ships in different parts of the world. However, he suffered from sea-sickness and was clearly unlikely to reach senior rank on active service. Ashore, he was used in number of roles in intelligence-gathering and was considered to be brilliant at his work. He married twice, the second time to the daughter of a wealthy family from Scotland named Cumming. To keep the name in existence, he changed his name to Mansfield George

Smith-Cumming and when knighted by King George V he became Mansfield Cumming. He went on to sign his letters with the initial 'C'.

Mansfield-Cumming was the first Chief of MI6 having been involved in our intelligence services since before World War I. He had a peg leg as a result of a motoring accident, and sported a monocle, so he created quite a distinctive figure. He signed all documents with his initial 'C' for Cumming. Since that time, all Chiefs of the Secret Intelligence Service have signed in the same way, and indeed are known only as 'C' to press and public alike.

It is widely assumed that the head of MI6 is known by the first initial of his surname. Whilst that was certainly true for Mansfield Cumming – Sinclair continued to use the same initial 'C' throughout his long career as head of MI6 – as did Menzies after him. There is a suggestion that Ian Fleming, when writing his James Bond stories, was referring to Menzies when describing his fictional head of SIS as 'M'. However, throughout the years, from the start of Mansfield Cumming's career as head of MI6 until today, the head of the British secret services has been known to the press and public alike by the initial 'C'.

Mansfield Cumming had become well known in official circles for his intelligence work, but he became a legend as a result of an accident in October 1914. He was driving with his son, Lieutenant Alexander Smith-Cumming, to an appointment in France when the car crashed to into a tree and overturned. Mansfield Smith-Cumming was pinned down under the car by his leg. He realised that his son was badly injured but could not free himself from the wreck. He took out his pocket-knife and cut off his own smashed leg to free himself, in an attempt to help his fatally injured son. He later had an operation to remove the remains of the damaged leg and for the rest of his life used a wooden leg. His habit of tapping it from time to time was very alarming for those in conversation with him!

The new Secret Service had only a few years in which to organise staff, agents and all the other paraphernalia, before the country was at war with Germany. Until then, the service had been housed at the War Office but SIS was now moved to the Admiralty. There was a Naval Intelligence Department there already that had years of experience behind it, and little time for the new service foisted upon them.

Cummings and his department had worked hard in the years since 1909 and at the outbreak of war there were agents in place in a number of embassies in Europe. Much of this was due to the spirit and tenacity of Cumming himself. Some agents were listed as military attachés, some as diplomatic staff. It was a dramatic improvement in a short time, achieved with little financial support.

SIS clearly established a good reputation for itself in World War I, having earlier realised that the likely enemy would be Germany. British intelligence was in better condition than ever before. There were a number of successes and agents were active in all parts of Europe, contributing to the intelligence-gathering process and taking part in actual operations.

In France, as in other major commands, there were SIS men in place in the British Expeditionary Force headquarters. They were used as interrogators and to run agents who were behind the enemy lines. Intelligence-gathering by the military itself was in the hands of its own intelligence service. The best-known officer in military intelligence is probably John Charteris who became Director of Intelligence to Sir Douglas Haig who was, by then, Commander-in-Chief of the British Expeditionary Force (BEF). Over the years, Charteris' reputation has been damaged for seemingly providing information to Haig that was slanted to please his commanding officer, but not necessarily reflecting the true situation on the ground.

All that was separate to the SIS appointees who largely kept themselves apart from the pure army intelligence services except, perhaps, when it suited them. One army officer on counter-espionage service was to later become the most famous of Head of Britain's Secret Intelligence Services (SIS). His name was Captain Stewart Menzies DSO, MC, who had been seconded from the Life Guards where he served as Adjutant. Already a fine soldier, he quickly grasped the basics of intelligence warfare as demonstrated by his brilliant work later during World War II.

Whilst Cummings dominated the relatively new SIS, naval intelligence continued to operate quite separately under the Naval Intelligence Department (NID), carrying on the role it had traditionally held as a gatherer of information from around the world. Its forceful approach was supported by their chief, Captain (later Admiral) William Reginald Hall, Director of Naval Intelligence (DNI). Hall was a ruthless character who caused the NID to become regarded by many foreign secret organisations as the leading British intelligence agency, and certainly the most feared. Perhaps the Americans held Hall's NID in the highest regard. Unfortunately, Hall made his own rules, ignored protocol and, at some time or other, upset almost every British government department.

In spite of Hall's obvious genius and the successes of his units, it is a reflection on his growing power in the land that Lloyd George refused to allow him to be present at the Versailles Conference. It is said that the prime minister was becoming increasingly worried about Hall's possible political ambitions but he certainly ensured that no honours were ever bestowed upon him. In contrast, Mansfield Cumming was knighted by King George V after the war and had earlier received awards from the Czar of Russia, Belgium and Italy, amongst others.

Nevertheless, when World War II started, the NID was still a separate intelligence organisation and continued to rival the SIS throughout the war, often with serious consequences. Perhaps it did not act in such a cavalier way as it had under Hall but he had left his mark, ensuring that it was a highly effective unit.

By 1923, Sir Mansfield Cumming had died and his place as 'C' – Head of SIS – had been taken over by Admiral Hugh Sinclair, also known to friends as 'Quex'. Sinclair's work was hampered by the limited financial resources made available to him in the 1920s and 1930s, and from a lack of authority over the increasingly disparate sections of his organisation. There was also the uncertainty as to where such resources as he had should be directed, since Russia was seen by many to be the main potential 'enemy' and only a few regarded Germany with suspicion until the mid-thirties. By that time, it was not easy to install an intelligence network in the heartland of Germany.

Sinclair's second-in-command was Colonel Claude Dansey, but in the early 1930s Sinclair appeared to fire him or forced him to retire. There were tales of financial impropriety, but nobody was quite certain what happened. Dansey was an unusual man, even by the standards of the motley crew who made up our intelligence organisations at that period. He had fought in Africa, in the Matabele campaign, and in the Boer War. He was large, overbearing, frightening to juniors, colleagues, and even to some of his superiors! He was regarded by many as devious and untrustworthy, yet he believed passionately in the rightness of his arguments, and was undoubtedly a patriot.

In 1936, Dansey reappeared in SIS circles and it seems more likely that his break away from the centre of things and the various stories were actually a cover-up whilst he created the 'Z' organisation. With Sinclair's knowledge, he set about instituting a separate network of SIS stations around the capitals of the world, that became known as the 'Z' network. He used commercial firms as cover in these places instead of the more usual Passport Offices, reporting back to Sinclair personally rather than to the SIS itself. At the outbreak of war, the Z Network was amalgamated with the PCOs. In Holland, both the Z Network and the PCOs were penetrated by the Germans, leading to the Venlo affair.

There was growing concern in Sinclair's mind at the way the German military was being reorganised, and in 1938 he formed 'Section D', under Major Laurence Grand, to consider the different ways in which a war could be fought. The result of his research into sabotage and operations by irregular forces was SO1 and SO2, which later became the famous and successful SOE.

In the run-up to the war, and particularly in 1938 and 1939, Sinclair had been under great pressure to reorganise SIS on a war footing. The stress of this task and his increasing ill health resulted in his death. Sinclair had been much admired by the men who had served him loyally through his difficult years in office. His memorial service held at St. Martins-in-the-Fields on Wednesday, 8 November, 1939, was attended by most of the senior personnel of the various units under his command.

Major-General Sir Stewart Graham Menzies who became head of MI6 following the death of Hugh Sinclair in 1939 – having previously been head of MI6 (Section II) its military wing. He continued as 'C' throughout the turbulent years of World War II – and until his retirement in 1952.

The suddenly vacant post as head of SIS was obviously of huge concern to the Cabinet, at the very time that war had finally broken out. Several names were put forward for the position including Dansey's, and that of the then Director of Naval Intelligence (DNI), Rear-Admiral John Godfrey. It was generally thought that there ought to be a change of services, since both the previous heads of SIS, Cummings and Sinclair, had come from the Navy.

It was on merit, and not on any concept of rotation amongst the services, that the name of Stewart Menzies was put forward to a meeting of an inner council of the War Cabinet, held on 28 November and attended by the prime minister, Neville Chamberlain, Winston Churchill and Lord Halifax. Menzies was appointed, after intense lobbying by Lord Halifax, the Foreign Secretary. This was regarded as a victory by the Foreign Office over the Admiralty, who had wanted the Senior Service to continue to 'rule the waves' as far as British intelligence was concerned.

Menzies held the vitally important position of 'C' throughout the momentous years 1939–1945, and though a milder man than either of his predecessors, he came hugely qualified for the post. All through the 1920s and 1930s, he had held various positions in intelligence organisations, from being the Military Representative of the War Office inside SIS to Head of MI6 Section II (Military), a post he still held at the outbreak of hostilities. It is from there that he was promoted to head the whole SIS organisation which he did brilliantly throughout the war.

Sir Stewart Menzies remained as 'C', head of SIS, until 1952. He died in May, 1968. I believe Britain was well served by the three heads of SIS in the period mainly covered by this book – Cummings, Sinclair, and Menzies.

Chapter 2

Pre-war Communications, Passport Officers and Broadway

Down the ages, secret messages have been hand-carried on land, by a runner or courier on horseback. Until the 1850s, diplomatic reports and intelligent dispatches were sent by diplomatic bag or, more recently, by post.

Napoleon created a semaphore system to speed the transfer of diplomatic and commercial intelligence, but this could only work with line-of-sight relayers. The same applies to semaphore with flags. Use was also made of the heliograph, especially during the Boer war but again it depended upon the sender being able to see the recipient.

Other means of communication developed during the latter part of the nineteenth century. The telephone came into increasing use after its invention by Alexander Graham Bell but it relied upon on land lines and so was vulnerable to attack and interception. Morse code by telegraphy was already established but was vulnerable in the same way, although the messages were not easily monitored by the unskilled, and of course, a message could be encoded.

The other development was the establishment of a submarine cable network and it gradually became possible to contact most parts of the extensive British Empire by means of telegraphic messages. With the British Navy controlling the seas, it was highly unlikely that any foreign power would dredge up the cable and listen in to the traffic so that communication by telegram was considered secure, so long as sender and receiver were known to each another. It became common practice to send For sensitive commercial messages in code and not *en clair*. Commercial codes were used primarily to cut down the number of words.

All these methods of communication had their limitations. Semaphore and the heliograph depended upon the right weather conditions, telephone depended upon the availability of a direct line linking the parties, cable was hugely expensive and, again, could only be applied to a limited recipients. Whilst these were great steps forward in communications, there was still no way of passing messages unless sender and receiver were linked by wire or cable.

It was at this point that a massive breakthrough occurred which was to have an enormous effect on communication – the creation of practical wireless telegraphy by the Italian inventor, Guglielmo Marconi. Marconis' primary aim was to provide a means of communicating with ships at sea which were completely out of touch once they had left port and were over the horizon.

Marconi's own government showed little interest in his invention so he took his apparatus to London where he had connections through his Irish mother's family. She was a member of the wealthy Jameson family, the Irish whiskey distillers, and they proved willing to help him. He demonstrated his wireless in a series of displays including one involving the Army and the Navy when messages were passed over a distance of one mile.

Marconi's gradual development of the device, enabling him to relay his wireless message over increasing distances is well documented, but perhaps it is not so well known that the apparatus was tried by the British Army during the Boer war without much success. The breakthrough had been made, however, and rapid communication was no longer restricted by the need for daylight (semaphore and heliograph) or by vulnerable wire and cable.

Although the British Army had not been impressed with the performance of wireless communication in South Africa – largely due to its own ineptitude – the British Navy certainly adopted the invention enthusiastically. It was used to intercept ships trying to break the blockade and the Admiralty was so pleased with the results of the trials conducted in South Africa that it ordered wireless equipment from Marconi to fit out over 20 ships. At the same time, he was given a contract to establish a number of coastal stations in Britain.

Marconi's later feat in sending wireless signals across the Atlantic from Cornwall to Newfoundland led to a rapid increase in wireless installations of all kinds. It was recognised that wireless traffic could not be controlled or censored in the same way as cable traffic and this led to the passing of the Wireless Telegraphy Act of 1904. The Act required anyone in the UK wishing to use a wireless transmitter to be licensed and this requirement lasted for the rest of the century.

During the years leading up to World War I, both Germany and France embraced wireless and developed a chain of stations. Germany, in particular, was glad to be free of reliance upon the British cable network and soon built wireless stations in their colonies around the world.

Whilst it was felt that submarine cables were secure, the land lines extending from the cable heads were clearly not, so the War Office set up a unit to censor both mail and cable traffic. At the end of the Boer War the unit continued in place and concentrated on setting up contingency plans for use in any future war situation.

It was obvious to everyone that although wireless was a step forward in terms of speed and flexibility of communications, the traffic could be freely intercepted and was dramatically less secure than cable telegraphy. With the outbreak of war in 1914, the intelligence services started to recruit interpreters but again it was the navy who led the way by setting up a listening post at Hunstanton to intercept German wireless traffic across the North Sea. The information was passed to associated offices known as Room 40 in the Admiralty Buildings.

There, three German-speaking interpreters, all provided by the navy, started to analyse the coded wireless messages passed on from Hunstanton. One of these men was A. G. Denniston who, in 1921, was involved in forming the Government Code and Cypher School (GC&CS) funded by the Foreign Office, which in turn went on to become Government Communications Headquarters (GCHQ) that continues today.

By the end of World War I, wireless had already become the prime method of communication across the world. The technology improved rapidly as the apparatus became cheaper and more readily available. This lead to an ever widening number of wireless stations of all kinds.

During the war, funding for the SIS had not been difficult to obtain, but in 1919 the organisation was transferred to the control of the Foreign Office. GC&CS was funded by the Foreign Office and SIS from the 'Secret Vote'. It is believed that the whole of SIS had to exist on £150,000 per annum in the early 1930s. So clearly, there were very tight financial constraints upon SIS which restricted staff recruitment and made its operations difficult. At the same time, and in the culture of appeasement that was being ardently pursued in diplomatic circles, the existence of SIS officers placed in embassies was felt to be counter productive to good relations with the host country. The ambassador might find it embarrassing if there were SIS men operating, notionally under his jurisdiction, but actually controlled from SIS HQ. It must be understood that SIS was in effect licensed to engage in criminal activity abroad, i.e. espionage.

It was recognised that, as the appointed representative of his Britannic Majesty King George, an ambassador could not possibly condone what at times was regarded as rather bad behaviour by the agents working from his embassy. The agents were thus not to be granted diplomatic status since they might jeopardise Foreign Office policy – but how else could they operate within the system?

It was decided to set up a seemingly separate organisation within the Foreign Office called the Passport Control Office whose members would operate alongside each major embassy but would not enjoy diplomatic immunity. This relieved the Ambassador of any possible embarrassment if the agent should be found out in any indiscretion or compromising situation.

The scheme had another major bonus. Since payments for British passports and visas around the world generated a considerable amount of cash and this helped subsidise local SIS operations. Indeed many argue that this was the prime reason for the move!

The senior SIS representative (or 'Head of Station') pre-war would have been the CPO (Chief Passport Officer). Staff answerable to him might be the Passport Control Officer (PCO), and further down the Assistant Passport Officer. A distinction should be made between these intelligence officers, who were on the SIS establishment in a country and based in the embassy's passport office, and the local agents they recruited.

The passport office was housed in a separate building to the embassy. Again, this put the SIS agents at one further remove from the Foreign Office officials – all with the intention of ensuring good relations with the host country.

It was soon realised, of course, by the intelligence services of other countries that the Chief Passport Officer was the local representative of the British secret service. He and his staff were thus almost branded by their titles and their association with the Passport Office at the British Embassy, thus removing most of the anonymity that was so important to much of what they were trying to achieve. It also meant that if a passport officer were questioned by the local police or agents, he could not plead diplomatic immunity. This was an obvious consequence of the Foreign Office's parsimonious approach but there was another.

Wireless equipment had not been supplied to most embassies as part of the cost-cutting approach, so the local passport officer had to rely on mail drops, or even open telegram systems. Letters containing secret or sensitive information were usually hand-delivered by messenger across the Continent, with the inevitable delays that caused.

In the increasing tension arising from Hitler's actions in the mid-1930s, the British government made larger grants to many departments and SIS was one of the beneficiaries of this belated largesse. At the same time, Admiral Sinclair, the head of SIS, had finally realised the deficiencies that existed within his organisation and set about dealing with them energetically.

By this time, it was clear that the communication system was one of the department's greatest weaknesses and the lack of speedy interchange of intelligence arose from the absence of wireless expertise within the organisation. France and Germany both had extensive wireless networks and most other countries could handle their worldwide intelligence gathering much more speedily than could Britain.

Fortunately, as it turned out, Sinclair was put in touch with Richard Gambier-Parry who was then Sales Manager of the UK factory of the giant American radio company Philco. Gambier-Parry's story is told in the next chapter.

From time to time, reference will be made to 'Broadway' (also known as Broadway Buildings). This was the headquarters of SIS. The building was actually number 54, Broadway, in the City of Westminster, neatly adjacent to many government departments. It was also close to SIS offices in Queen Anne's Gate and The Passport Office in Petty France.

Since throughout the book reference will be made to the different divisions of SIS, it might be helpful to readers to have an idea of its various component parts, as reorganised by Sinclair just before the Second World War.

The more familiar title, 'MI6', may be used to describe the Secret Intelligence Services – especially when referring to its Section VIII – rather than 'SIS,'but they are merely different names of the same organisation, and both names will appear in the chapters that follow.

By the time war broke out in September 1939, the HQ of MI6 was divided into ten sections and these are shown below. Each Section number is written in Roman numerals; this was then standard practice within the SIS.

The SIS Organisation at the Start of World War II – 1939

Chief of the SIS (CSS), Admiral Sir Hugh Sinclair, known as 'C'.
(Sinclair died in November 1939, and was succeeded by Colonel Stewart Menzies).

I. Political Section.

II. Military Section, under Colonel Stewart Menzies.

III. Naval Section.

IV. Air Section, under Wing-Commander Fred Winterbotham.

V. Counter-espionage.

VI. Industrial.

VII. Finance, under Commander Percy Sykes R. N.

VIII. Communications, under Colonel Richard Gambier-Parry

IX. Cypher.

X. Press.

This book is about MI6 (Section VIII), its head, Richard Gambier-Parry, its association with various premises, and in particular with Whaddon Hall. The story mainly refers to the various sections highlighted above and it will be seen that Section VIII is concerned with Communications.

Chapter 3

Enter – Richard Gambier-Parry

It has been shown how, in many respects, SIS communications were inadequate in the 1930s but fortunately this was belatedly recognised. No doubt the constant scrimping and saving by the Treasury was partially responsible for the poor state of affairs that had existed. Foreign Office communications was headed by Harold Eastwood but in 1938, Admiral Sinclair, the head of SIS, realising the transmission of intelligence was too slow and lacked security, recruited Richard Gambier-Parry to head up MI6 (Section VIII), SIS Communications.

Richard Gambier-Parry was born on 20 January, 1894, the son of Sidney Gambier-Parry, an architect with a practice in London. He was sent to Eton but was only there for one term. Eton College library records report that he started in September 1907 but left in December, only a few months later. They cannot now say why; it may have been shortage of funds but there is a more likely explanation.

Gambier-Parry was in the Lower IV, a low form for his age. It may have been realised early on that Eton's strong emphasis on the classics was not for him. At that time, the main test of ability was an aptitude for Latin and Greek and even very able boys whose talents lay elsewhere were badly served. If Gambier-Parry had continued at Eton and become a classics scholar we might well have been deprived of one of the most dynamic leaders our intelligence services ever had.

In the First World War he followed his older brother Michael into the Royal Welch Fusiliers and was commissioned 2nd Lieutenant to the 1st Battalion on 13th April 1915. Michael Gambier-Parry had been at Eton earlier, then went on to the Royal Military College at Sandhurst where he had graduated and joined the Royal Welch Fusiliers as a 2nd Lieutenant back in 1911. Thus Richard followed into his brother's regiment where Michael was by that time, a Captain and adjutant of the 8th Battalion.

Brigadier Sir Richard Gambier-Parry KCMG
Head of SIS Communications
(MI6 Section VIII) 1938 – 1946.

Richard was fighting with the 1st Battalion at Festubert when he was wounded on 16 May, 1915. As a result of his bravery, he was mentioned in despatches by General French. After recovering from his wounds, Gambier-Parry returned to France where he was attached to the 2nd Battalion. He was promoted to full Lieutenant on 3rd March, 1916 and then wounded again at High Wood in July, 1916. He continued to serve in the Royal Welch Fusiliers in both the 1st and 2nd Battalions. On 29 August, 1918, he was seconded to the Royal Air Force, where he stayed until the end of hostilities.

After the war, Gambier-Parry is believed to have travelled extensively in the United States but he later returned to England and became the BBC's public relations officer from 1926 to 1931.

He was a licensed 'experimental' (i.e. amateur) radio operator who used the call sign 2DV, which shows that he was active during the period 1920–1926. From 1926, an international prefix was added ('G' for Great Britain), which would have made his call sign G-2DV. This suggests that he may have given up his amateur wireless operator activities around the time he joined the BBC.

In 1931, Gambier-Parry left the BBC and joined Philco, the American radio manufacturer, as their British sales manager, eventually becoming General Sales Manager for the United Kingdom.

It is not entirely clear how Richard Gambier-Parry came to the notice of Admiral Sinclair. Both were old Etonians, as was Stewart Menzies who was by this time head of the Military Section of SIS – MI6 (II). It is said that this was the connection that brought them together but they were years apart at Eton and, since Gambier-Parry was only there for only one term, it is rather unlikely that they met at school.

Stewart Menzies had an estate in Gloucestershire and the Gambier-Parry family lived at Highnam, in the same county. Like Richard Gambier-Parry, Menzies was a keen horseman and regularly hunted, so it is possible that they met socially between the wars. Stewart Menzies eventually became head of the whole of MI6, after the sudden death of Admiral Hugh Sinclair.

Whatever the background, Gambier-Parry was invited to run the proposed new communications section of SIS for a number of reasons. His organisational skills, knowledge of the wireless world of the time, his extensive connections in the industry and earlier experience as an amateur wireless operator were the perfect combination. To cap it all, he had great charm, an essential attribute, as the overhaul of the worldwide SIS communications network needed tact and understanding.

Gambier-Parry features in a group photo taken at the leaving party he was given by Philco, which was reproduced in the company's house magazine *Philco News* of Thursday, 24 March, 1938. In the picture he is shown holding a glass of champagne and the caption reads:

'This picture is an opposite illustration to what we should be seeing. We, the editorial staff, and you, the readers, should have glasses in our hands bidding Captain Parry goodbye and the best of luck but space forbids it. So perhaps we can take it as a toast to Philco prosperity from Captain Parry before he left us.

It is with regret that we announce his resignation from the post of General Sales Manager. Captain Parry leaves Philco to take up and important post in the War Office and he will be succeeded by 'Jimmy' Noble who has been Distribution Manager.'

He is described here as 'Captain Parry' and not by his full name of Gambier-Parry. Also, their use in a civilian context, of a military rank below that of Colonel, is unusual for ex-service personnel. We know however, that he did not leave to take up a post at the War Office but instead it was with SIS.

In his new role, Richard Gambier-Parry was an outstanding success, becoming a major figure in secret wartime wireless communications, by completely reorganising the weak SIS network he had inherited. His achievement was due in part to his knack of picking the right man for the job and he had no qualms about blatant headhunting – even from his own firm, Philco.

A Farewell Toast

This picture is an opposite illustration to what we should be using. We, the editorial staff, and you the readers, should have glasses in our hands, bidding Captain Parry goodbye and the best of luck, but space forbids it.

So perhaps we can take it as a toast to Philco prosperity from Captain Parry before he left us.

It is with regret that we announce his resignation from the post of General Sales Manager. Capt. Parry leaves Philco to take up an important position in the War Office. He will be succeeded as General Sales Manager by J. G. G. " Jimmy " Noble, who has been the Distribution Manager.

Champagne toast at Philco's farewell party to 'Captain Parry' in March 1938 as he '...leaves Philco to take up an important post in the War Office.'

Thus MI6 (Section VIII) was formed which went on to play such a vital role in the Secret Wireless War. Under Gambier-Parry, the section expanded rapidly, not only in numbers of staff, but in the many additional responsibilities that he was always eager to accept, being regarded at Broadway Buildings as a bit of a 'pirate.' He would probably have gloried in the title, being totally convinced of the quality of the unit he had created, the ability of the men within it, and their ability to master the wireless technology of the time.

The functions of Section VIII were many and various, and included creating Main Line links with SIS offices in neutral countries, secret links with SIS agents and governments-in-exile, maintaining the HF (high frequency) network to supply ULTRA information based on the Enigma decrypts, the provision of transmitters for Black Broadcasting including the 600kW Aspidistra and provision of VHF/RT equipment for the Special Duties Home Guard (part of the Auxiliary Units) who were manning 'stay-behind' intelligence posts, in preparation for a German invasion.

In 1941, Gambier-Parry took over the Radio Security Service (RSS) 'lock, stock and barrel'. It had been run hitherto by MI5. A year earlier, Churchill had appointed Lord Swinton, a Conservative peer, as head of the Security Executive which assumed responsibility for MI5. Until then, the director had been General Vernon Kell. Churchill thus effectively secured closer political control of at least one of the two secret services. In January 1941, Swinton recommended that RSS be handed over to SIS, but this met with fierce opposition throughout the upper echelons of MI5, resulting in a battle that reached the highest levels.

No doubt, the already proven success of Section VIII's organisation, and its wireless technology, enabled SIS to prevail. Gambier-Parry became responsible for RSS, and its whole operation, from May 1941.

He then appointed his deputy, Ted Maltby, as Controller of RSS, which became responsible for setting up the special RSS intercept stations such as Hanslope Park and Forfar, and running its D/F network under Dick Keen and Louis Varney.

One of the hardest tasks for SIS and Bletchley Park was dealing with the continuing jealousy of the Admiralty. The Navy had been hugely successful in wireless interception during World War I and there is no doubt their Room 40 organisation was a model for some of the subsequent work of Bletchley Park. Add to that the thought that perhaps Admiral John Godfrey, Director of Naval Intelligence, might have become 'C' on the death of Admiral Sinclair, and it can be seen why there was simmering resentment on the part of the Admiralty and a reluctance to take part in anything to do with Bletchley Park, even accepting Ultra traffic with a lack of enthusiasm.

It is difficult to say now whether Gambier-Parry's great personality helped to break that barrier down, but he certainly developed a good relationship with the NID and especially with Commander 'Joe' Loehnis who worked very closely with Admiral Godfrey. Commander Loehnis was the NID liaison with MI6, its connections to Bletchley Park, and to Section VIII at Whaddon.

Gambier-Parry's relationship with the Navy perhaps showed results at a lower level, where NID were prepared to work closely with Section VIII in fitting wireless gear to MTBs, Slocum's Navy, MFVs and so on.

Yet even Gambier-Parry was defeated in his efforts to retain complete control of all secret radio communications. In July 1940, the Special Operations Executive had been established 'to set Europe ablaze'. It consisted largely of Section D of the SIS, although it was an independent unit. SOE was soon seeking to establish its own radio links, but for some time Gambier-Parry argued successfully that Section VIII was the only organisation with sufficient experience. There followed a year to eighteen months of growing discord, with SOE seeking to develop its own wireless sets, arguing that Whaddon was unable to supply all its needs and adding (*sotto voce*), that the design of Whaddon's wireless equipment was already out-of-date.

Since SOE was formed later, and its initial wireless links and equipment were still the responsibility of Whaddon, not surprisingly, it was in a good position to see how improvements to wireless sets and procedures could be introduced.

By late 1941, SOE had established its own design team and arranged for some of its early suitcase sets to be manufactured by Marconi. In 1942, SOE was able to set up its own Signals Directorate. Gambier-Parry, however, retained control of the links with France for de Gaulle's BCRA groups.

As a footnote, it should be added that through its handling of its wireless links, SOE suffered the Allies' worst failure in the Secret Wireless War.

The German control of the entire Dutch SOE network over a period of almost two years had devastating results. It was responsible for the deaths of 47 out of 51 agents sent into Holland over a period of almost two years, the arrest and often death of some 400 members of the Dutch Underground, the loss of twelve RAF aircraft and some 84 aircrew on Special Missions, and the delivery of agents, funds and supplies directly into the hands of German organised reception parties. A sorry tale. At one stage, the Germans controlled over a dozen W/T links between Holland and the UK. All this did little to help the already poor working relationship between SIS and SOE.

From the start of his involvement with MI6 in 1938, and throughout the war, Richard Gambier-Parry confirmed the faith placed in him by Admiral Sinclair. Starting with just a handful of men, and initially with limited resources, he created a unit whose importance is only now being appreciated.

Gambier-Parry was a distinguished looking man, always immaculately dressed whether in uniform or in civilian clothes. When in uniform, although no longer serving with his regiment, he always wore the Royal Welch Fusiliers 'flash'. I remember wondering what it was when I first saw him during one of his frequent visits to the workshops at Whaddon.

The 'flash' consists of five pieces of black ribbon folded over and fitted below the back of the collar. It is a direct link with the time when soldiers wore pigtails. These were powdered and greased and in order to protect the red coatee, they were contained in a pouch, which over time, developed into a protective patch. The Regiment continues to wear the 'flash' and is unique in this regard, having had it endorsed in 1900 as their right by Queen Victoria. Clearly, Gambier-Parry continued to feel a close affinity with the regiment in which he served in the trenches, during World War I.

Gambier-Parry was a gifted man, highly regarded – even loved – by his staff and respected by all who came into contact with him. He always signed himself as 'Pop' in his annual letter to the Editor of *Stable Gossip*, the wartime Whaddon house magazine, and in private correspondence to many of his staff.

Those of us under his command, referred to him amongst ourselves as 'The Boss', but in the nicest possible way. We had great respect for him; we knew that he was completely in control of the operation of Section VIII, but at all times he also had our welfare in mind.

All the nations participating in the 1939–1945 conflict were involved in The Secret Wireless War to some extent. In Great Britain, however, virtually all aspects of it had connections with, or were controlled by, Richard Gambier-Parry.

Chapter 4

The Early Days – Barnes Station X and Funny Neuk

I have assembled this chapter from a number of sources, some of whom are listed in the acknowledgements, references in the various diaries that follow in the book, and my own personal knowledge of Funny Neuk.

In the late 1920s, Foreign Office wireless communications were run from a tiny station in the Passport Office, under the control of Harold Eastwood. He was also in charge of its station run jointly with SIS at Barnes – along with Henry Maine and Harold Kenworthy, who later went on to run the Foreign Office wireless intercept station at Knockholt, in Kent.

The Barnes station was near the River Thames and in the 1930s its operation was actually being funded by GC&CS (via the Foreign Office), rather than by SIS from the Secret Vote. That was probably because it was carrying out intercept work as well as handling wireless traffic. That traffic included outstations such as Peking (Beijing), and later from Spain during the Civil War, where Charles Emary and Don Lee were working as agents. Certainly, this changed when Gambier-Parry took over in 1938 and Barnes then came fully under the wing of his Section VIII.

The Barnes station was only a receiving control station so did not have transmitters of its own. During its lifetime it worked to the overseas stations via transmitters provided by the GPO, probably at Daventry. It is also possible that transmitters run by the services in Hampshire were used at times. Edgar Harrison was an operator working for the Foreign Office in Peking in the 1930s and worked two way traffic from there to Barnes. (See his story in chapter 31).

One might ask why the quiet suburb of Barnes in south west London was chosen as the Foreign Office wireless station? The simple response might be – *well it had to be somewhere* – but I think there may be a more positive answer that has eluded those of us puzzled by the location.

The postcard photograph of Barnes High Street was taken in the 1920s and is reproduced by kind permission of the Barnes and Mortlake History Society. The picture was taken from the northern end of the High Street where it meets the Thames embankment and branches left and right as Lonsdale Road. A tall wireless aerial can be seen in the background. It existed until 1935 – 1936 in the grounds of an old property called Threlkeld House. This property occupied the space between 61 and 67 Barnes High Street. The building was pulled down around 1935 – 1936, and redeveloped into a block of flats named Seaforth Lodge that still stands today. Kelly's Directory 1923 to 1929 lists 62 – 65 High Street as being occupied by Radio Communication Company Ltd., which may have been a subsidiary of Marconi Wireless. From 1930 through to 1934 the occupant is listed as being Marconi International Marine Communication Limited and the large scale map from the period indicates it as being a 'Wireless Factory'.

Kelly's for 1935 and 1936 shows no entry for 62 – 65 Barnes High Street. However, in the Directory for 1937 the name Seaforth Lodge appears on the site with 15 flats and number 64 listed as occupied by a Ladies Hairdresser and the only other shop on the block number 65 occupied by Norjon Radio Services Ltd., Radio and Electrical Engineers.

Although I have nothing whatever to support my suggestion, it is my view that the choice of Barnes as the Foreign Office/SIS wireless station was due to the close and long-standing connection between Marconi and SIS – that had existed since before World War I.

The reader will, no doubt, recall the case of Peter Wright who caused world-wide media interest, when he published 'Spy Catcher' in 1987. In the book, he refers to his father Maurice Wright who had joined Marconi in 1912, and later worked as Engineer-in-Chief for Marconi. In World War I, Maurice Wright was released by Marconi and employed by Naval Intelligence Department on wireless research. He even went to Norway to run a clandestine station for MI6. He was known, from those days spent in NID, as 'GM' – the same initials as Guglielmo Marconi. The close link formed between Marconi and MI6, continued after World War I and Maurice Wright rejoined Marconi as soon as he was released by NID. He was still working with Marconi as a consultant in the 1930s.

Now to the actual location of the Barnes station. In telephone conversations with Pat Hawker after the war, Steve Dorman referred to our pre-war wireless station at Barnes being located in a 'brick building near the Thames' with aerials on it.

Now to the actual location of the Barnes station. In telephone conversations with Pat Hawker after the war, Steve Dorman referred to our pre-war wireless station at Barnes being located in a 'brick building near the Thames' with aerials on it.

Barnes High Street is full of shops, restaurants and pubs and meets the Thames as a long arm of a 'T' junction on Lonsdale Road. To the west, the houses on Lonsdale Road are mostly examples of fine Georgian buildings, right up to the Barnes railway bridge, so there is no chance the station could have been there. To the east, there are one or two shops on the corner of Barnes High Street and Lonsdale Road, then a pub on Lonsdale Road called the Bull's Head, next to the Barnes Police Station. This also housed the local River Police.

Behind the Police Station was a very extensive complex of buildings, used as the area vehicle repair depot for Police cars, and motor cycles. We know, from John Darwin's diary entry for 6th June 1939 (see chapter 21), that he visited Florence House (Barnes) with Ted Maltby on that day and so Station X was in that building by that time. However, we also know that the SIS wireless station was located in Barnes long before that building became available to them.

Exhaustive research by the Foreign Office Archives has not been able to pinpoint the site of the Barnes station, nor can the Police Museum at Charlton in East London, confirm if the station was inside the Police property.

It seems clear to me, that the wireless station was firstly in the Police complex – a view shared by Pat Hawker. Both buildings – the Police station and then in Florence House. Both fit in with Steve Dorman's description of a *'brick fronted building by the Thames.'* The site, previously the police station and the complex behind, is being redeveloped as flats at the time of writing. However, under the planning consent given, the developer has to keep the brick facade of the Police station itself facing the Thames, but not the brick walls of its main premises down the Barnes High Street frontage.

Further along Lonsdale Road to the east, there are several houses and then one comes to Florence House. In the early 1930's, this was occupied by the International Art Company, publishers of post cards, including the comic sea-side ones by Donald McGill. Later, in 1936-37 it was used by Lennard Inter-Art company and in 1938 and in 1939 listed as being empty. Perfect for rehousing the SIS station from its existing site behind the

Police complex just along the road. That must have taken place during the early days of Richard Gambier-Parry's control of the newly formed Section VIII.

The view is reinforced by small details shown in the post card of Barnes High Street. On the left, above the police depot flank wall, one can just make out a faint aerial array. This has now been discussed at length by David White and Pat Hawker. Their conclusion is as follows:

Postcard of Barnes High Street of the late 1920s, taken from the top end near the River Thames. The tall wireless mast at the rear, is in the grounds of the Wireless Factory at 62 – 65 High Street. To the left of the picture, is the brick wall surrounding the Police station complex with a wireless aerial just visible above it.

The connection between the Foreign Office and the police is well documented. For many years they had cooperated in the use of the Metropolitan Police station at Denmark Hill, Camberwell in London, to monitor embassy wireless traffic. That took on more significance after the German embassy installed wireless equipment. Denmark Hill was used as the transmission station for the early wireless vehicles but also undertook work on interception. One good example of police interception was the tracing of an unknown wireless station, during the General strike of 1926. It was apparently found to be coming from the offices of the Daily Mail who were concerned that the GPO might go off the air during the strike. Another interception by the police led to SIS listening to the wireless of a Russian subversive organisation in Wimbledon – and so on.

Another instance of cooperation between the Home Office and our secret services, is the provision of cells in Wormwood Scrubs Prison for the infant RSS (Radio Security Service) of MI5. The RSS was later absorbed into MI6 (See chapter 15).

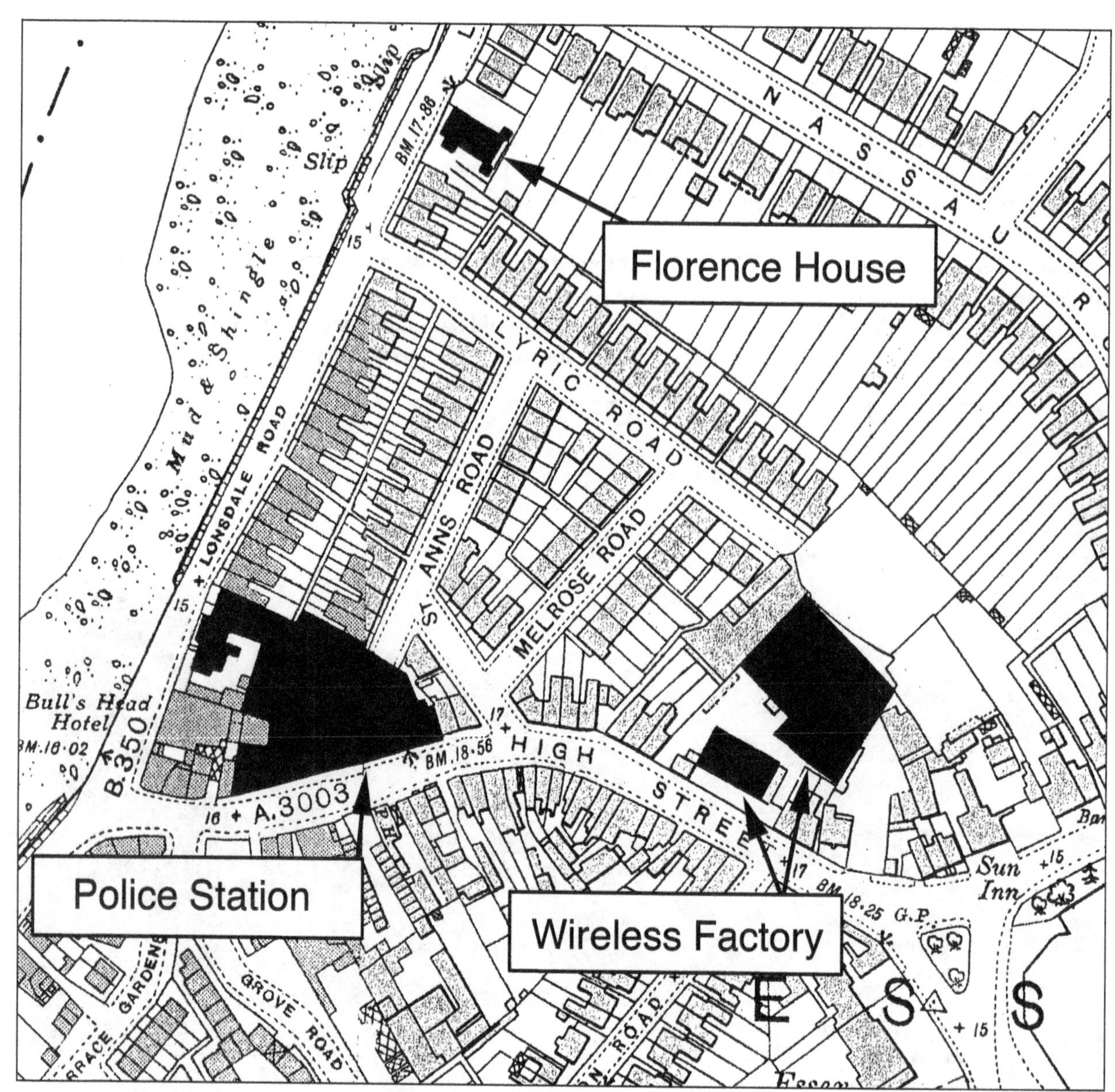

A map of Barnes in the late 1920s, showing the two possible sites for Station X on Lonsdale Road, facing the Thames. Firstly, Florence House. Secondly, the Metropolitan Police station which faced the river, with an extensive complex behind it, and down one side of Barnes High Street. The map also shows the 'Wireless factory' believed to be owned by Marconi.

We know that the Barnes station was small, with insufficient space for a meaningful workshop, so a shop was purchased elsewhere in Barnes and run as a retail wireless shop. The MI6 workshop there was used by Bob Hornby and others, for the building of our early agent's sets and for research. The chief operator at Barnes was K. Secretan who, we believe, also ran the shop. However, according to John Darwin (Chapter 18), a further workshop was opened in Barnes in 1938 but I am presently unable to find its location.

As Gambier-Parry began the reorganisation of the SIS wireless unit that was to become the wartime Section VIII, he decided that the Barnes unit would not be able to cope with the necessary expansion of traffic in the event of war, he also wanted to have his own transmitters, so that he did not have to rely upon other facilities.

These transmitters were to be located some twenty miles away in Woldingham, near Caterham in Surrey. The actual site was a large bungalow with the strange name of 'Funny Neuk' in an area known as Woldingham Garden Village. In addition to having sufficient space for the transmitters and aerial array, there was also room inside the bungalow for a properly equipped wireless workshop for Section VIII.

Until then, SIS had its miniscule station at Broadway in Westminster, and the Barnes premises over on the west side of Surrey. Both are a substantial distance from Woldingham. This had always seemed to me to be a strange location to choose for their transmitters. Whilst the distance between a receiving station and its transmitters is not of great importance, the choice of Woldingham added considerably to the travelling necessary for the growing workshop staff. Later on, two of the wireless engineers, Bob Chennells and Wilf Lilburn, tired of the journey to and from Barnes, decided to reside locally in nearby Caterham. Incidentally, that is how my family became involved with Section VIII (see my story chapter 38).

The front gate leading into the grounds of Funny Neuk with its name visible above the sign: 'War Department – No Admittance'.

However, in researching the history of Funny Neuk, I was advised to look at the electoral registers for the years leading up to World War II and I made an extraordinary discovery. In 1937, the road and the house were not listed but in 1938, the name on the electoral role for Funny Neuk (Ref: CC802/55/2), is none other than 'Hugh Sinclair' – Admiral Sir Hugh Sinclair – 'C' – Chief of the Secret Intelligence Services!

It may well be that 'C' actually owned the property himself, and for some time had used it as a 'retreat' from the cares of his office at 54 Broadway in Westminster. The rear of his official residence in Queen Anne's Gate, leads directly through to the SIS offices in Broadway. In a real sense, Sinclair was living 'over the shop' and a mere thirty minute drive to the peace of Woldingham must have provided him with a welcome break.

In 1939, the register shows those on the electoral role were Leonard George Morris, Violet Agnes Morris (whom I do not know) and then Wilfred Lilburn and again – Hugh Sinclair. Wilf Lilburn is a name that appears many times in this story of MI6 Section VIII. No electoral registers exist for the years 1940 – 1945.

Florence House in Lonsdale Road, facing the Thames.

The fascinating story of 'Funny Neuk' itself starts in World War I. Until its demolition in 1998, it was changed very little from when it was originally built in 1914, as part of the quarters to house the Public Schools Battalion of the Middlesex Regiment.

Immediately after the declaration of war in August 1914, patriotic feelings ran high, and many of the new battalions being created were assembled from men having some common interest. One such was the Public

A view of the bungalow called Funny Neuk

A view of Woldingham Camp in 1915

Schools Battalion. This was organised by a number of enthusiastic and obviously well-off serving or retired officers from various regiments of the regular army.

They were to be responsible for the feeding, clothing and accommodation of the unit at specified rates pending its official takeover by the War Office. One of the prime movers appears to have been Lieutenant J. J. Mackay, formerly of the Westminster Dragoons, who was appointed chairman of the committee formed to undertake this formidable task. He was promoted to the rank of Major in the new battalion.

On 4th September 1914, advertisements were placed in the principal daily newspapers seeking volunteers and these produced many recruits. Only a few days later successful applicants received a telegram requesting them to parade at Waterloo Station on Tuesday, 15 September bringing 'enough kit for ten days and one blanket'. The officers and 700 men dressed in civilian clothes and cadet uniforms boarded a train for Kempton Park racecourse where they were to be temporarily based, until more permanent accommodation could be provided.

How Woldingham came to be selected for this purpose we may never know, but it is likely that Mr. William Buttle who owned Muscombes Field where the camp was built had personal or business associations with members of the committee. He was a partner in a firm of solicitors in the City.

The following description of the Battalion appeared at this time in 'The Standard'.

'The men come from almost every Public School in the country. There are enough internationals to assemble two Rugby teams and at least one Association. Among the athletes is Private H. E. Holding, the treble Oxford blue who ran for England in the Olympic Games.'

The battalion strength was now twenty eight officers and 918 other ranks. There was a continual loss of men being appointed on temporary commissions with other units but this was being made up with a continual supply of new recruits.

By the end of December 1914 the camp was ready for occupation, and on the 29th they marched from their most recent temporary accommodation at Sutton, heading for Woldingham. Amazingly, in less than three months, a camp of some sixty or so buildings with gas and water services and cesspool drainage had been planned and built, on an awkward site – and in adverse weather conditions.

All the huts were of timber frame construction, those for the other ranks being clad with timber and those for the officers and the various services with corrugated galvanised steel sheets. Thirty men (other ranks) occupied each hut, sleeping fifteen down each side, on beds formed of three planks laid on two trestles. During the day these were propped against the wall of the hut. Heating was by one 'Tortoise' type stove to each hut.

The level area at the top of the field was used as the parade ground. The huts of E and F lines were shortly to be built on part of this but even so there was plenty of room. Training consisted of squad, platoon and company tended order and eventually rifle drill, but no firing.

The Battalion was involved in much of the fighting on the Western Front and in the mud of the Somme. In April 1919 the camp finally closed. The properties were sold off and developed as a so-called 'Garden Village' in 1921.

The bungalow was larger than most other buildings on the site and stood in about a half acre of ground. The plan shows a building marked 'Cook House' as being the largest, and this became 'Torres' with its first occupant being a Mr. Tolmie. In 1927, the new owners were Mr. & Mrs. Wignall and they gave it the new name of 'Funny Neuk.' Apparently 'Neuk' is a Scottish name for an outlying or remote place.

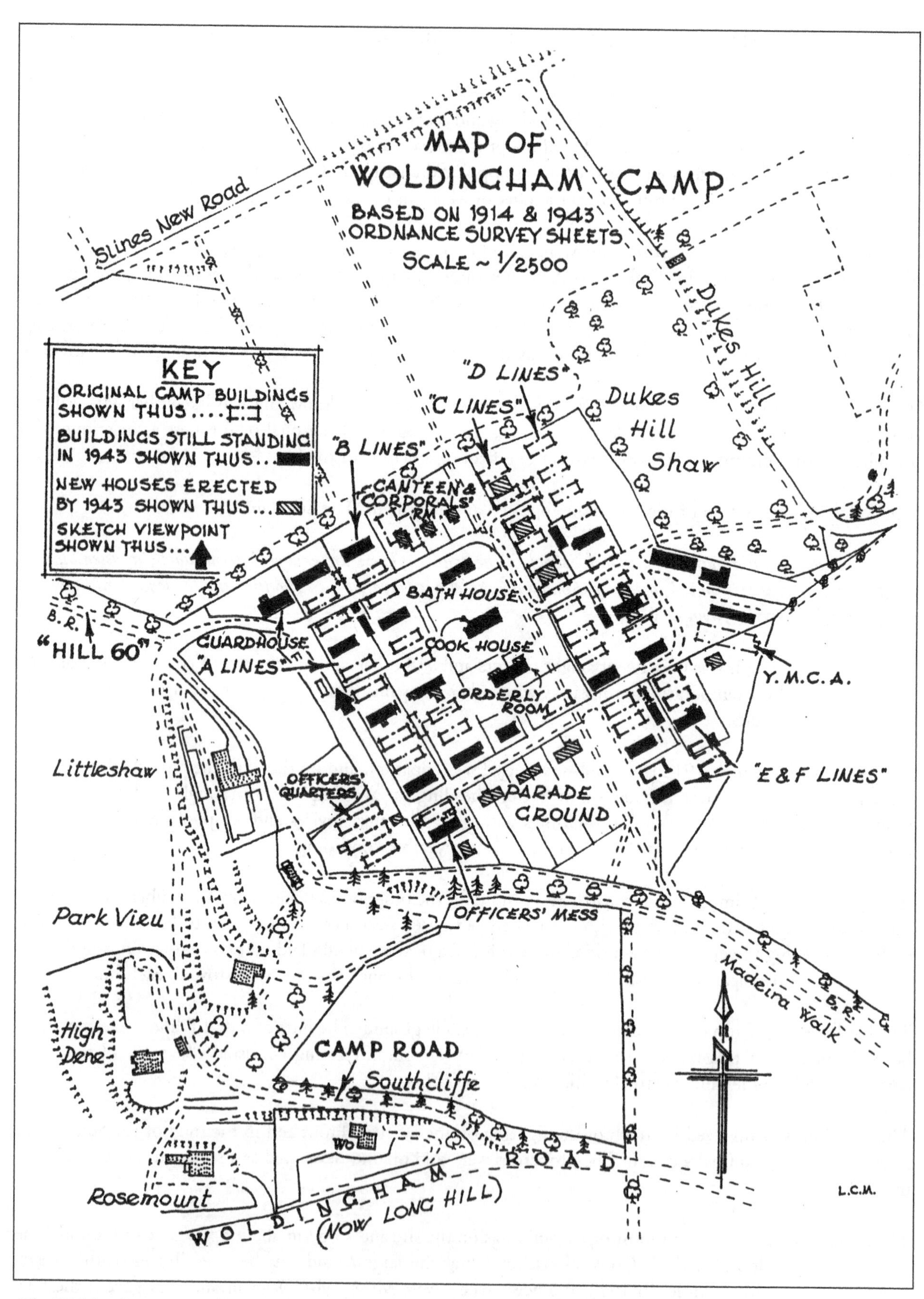

The Woldingham Camp layout during World War I. The building shown as the Cook House near the centre became Funny Neuk.

We know for certain that many of Gambier-Parry's newly-formed team of engineers worked at Funny Neuk. Over a period, these included Bob Hornby, Spuggy Newton, Jack Saunders, Bob Chennells, and Wilf Lilburn.

However, what might have been suitable as a wireless transmitting station just for SIS in 1938 or early 1939, was clearly going to be inadequate for all the many new and varied tasks arising for Section VIII, with the outbreak of war. By the time war began, parts of the Section had already moved into Bletchley Park, and its overall plans changed dramatically.

The inside of Funny Neuk as an operational wireless Station. Although built by MI6 Section VIII this was possibly taken during its later use by the Czech Intelligence Services.

The Funny Neuk station was completed well before the end of 1939, but I cannot be sure if it was ever fully operational – as intended – as the main transmitter for Station X at Barnes. By that time of course, Section VIII had already established new wireless stations at Bletchley Park in Hut 1, and in the tower of the mansion. Before the end of 1939, work was well under way to create the major wireless stations at Whaddon Hall.

In May 1940, Funny Neuk was handed over to Czech Intelligence as their wireless communication centre. It went on to play an important part in the conduct of the war. Amongst its many operations, it was directly involved with the Czech agents who were parachuted into German occupied Bohemia and assassinated Heydrich – known as the 'Butcher of Prague'. Incidentally that operation was run in conjunction with SOE, but the wireless sets used by SOE at that time, were supplied from the Section VIII workshops at Whaddon.

Chapter 5

The Cast Assembles

Richard Gambier-Parry now had the enormous responsibility of creating a world-wide communications network for SIS – almost from scratch – and was mindful of the urgency of his task. The Foreign Office's own communications setup, under the control of Harold Eastwood, operated on a shoestring. During his stint as Sales Manager of Philco, Gambier-Parry had been in a unique position to judge the quality of wireless engineers, not only those working for Philco, but those employed by other firms with whom he came in contact. His time at the BBC must have expanded his circle of contacts, and he was a former amateur wireless operator so he had the basic knowledge required for selecting those suitable for his infant organisation.

Not surprisingly, Gambier-Parry found his first engineers at his old company. The first recruit was **F. R. (Bob) Hornby** (see chapter 22) who was soon followed by the latter's recommendation, **Arthur 'Spuggy' Newton** (see chapter 23), with whom Bob had worked so closely at Philco.

Not long after came another of the brilliant wireless engineers to be employed by Section VIII – **Harold Kilner Robin**. His part in the story is told in chapter 18 – Black Propaganda. Harold Robin had worked for Philco but left in 1937 to open a wireless station in Liechtenstein that operated in a similar way to the hugely successful commercial station, Radio Luxembourg. Advertising was banned, of course, on British wireless stations, all of which were controlled by the BBC. Robin was recalled and arrived back in England at the outbreak of hostilities. He joined the unit but, curiously, remained a civilian throughout the war.

Robert 'Bob' Hornby

Arthur 'Spuggy' Newton

While he was recruiting engineers, Gambier-Parry was also seeking senior personnel to run the organisation with him. Again, he turned to the wireless world. Here he found another old Etonian, Edward ('Ted') Maltby, who was to eventually run SCU3 at Hanslope Park.

Edward 'Ted' Maltby

Edward Frank Maltby was born in 1904 and educated at West Downs, Winchester and Eton. He took a keen interest in wireless from an early age. When he was only 12, he built a receiver that was able to receive broadcasts from Finland. Apparently, this was a considerable achievement because he demonstrated it to several leading figures in the industry. While still at Eton, a military car collected the adolescent once a week to enable him to use his radio expertise for the Army!

After leaving Eton, Maltby attended the Royal Military Academy at Woolwich, then joined the Royal Signals. While with the Signals, he spent some time in charge of communications for the Army's annual exercises on Salisbury Plain. At about this time, the Army was considering mechanisation and Maltby participated in the testing of various types of vehicle. At some point, the Army sent him to Clare College, Cambridge, for a year.

Unfortunately, most of Maltby's father's money was lost in a Lloyds' syndicate that crashed and he suddenly had to rely entirely on his own resources. The main effect of this was that he had to leave the Army, because, at that time, it was just not possible to survive as a junior officer without a substantial private income.

Ted Maltby then became chief engineer at EMG, one of the leading manufacturers of what is now termed hi fidelity equipment. Many of EMG's customers were wealthy city financiers and one of them was having difficulty in providing wireless reception for the flats in one of his new developments. EFM was able to solve this problem and, subsequently, others. Eventually, Maltby set up his own consulting firm, where he specialised in advising financiers on the technical validity of inventions requiring finance, saving several of them considerable sums of money in the process. He said that this was very easy work as many of them were basically 'perpetual motion' machines and could be dismissed accordingly.

One of Maltby's tasks was to help the Royal Navy who were having trouble at one of their long-wave shore transmitting stations (this turned out to be due to deterioration of the insulation on the high voltage cable to the aerials). It is then that he may well have met Captain Frank Slocum RN (who later ran 'Slocum's Navy') and this may have been the start of Maltby's involvement with the SIS. Certainly he was brought to Gambier-Parry's attention at that point. Along with Micky Jourdain and John Darwin, Maltby became one of the 'founding fathers' of the organisation.

Colonel 'Micky' Jourdain was an officer who has sometimes been referred to as the original deputy to Gambier-Parry. Very little is known about him, but he was certainly in all the discussions in 1938/1939 that led to the organisation of the wartime unit, as shown in John Darwin's diary of the period (see chapter 21).

The late Norman Walton worked for Jourdain at Whaddon Hall before going off to North Africa with our 'A' Detachment as its administration officer. Walton also told me that he had met him subsequently near the Riverside transmission station outside Cairo but had no idea what his job entailed. Certainly, it was not with either our SIS work, or with our 'A Detachment' that was dedicated to Ultra traffic.

Brian Sall, Alec Durbin and 'Jan' Ware

Norman Walton related the odd tale of the constant quarrelling between 'the wives' at Whaddon and in particular the hostility between Mrs. Phyllis Gambier-Parry and Mrs. Jourdain. This situation is confirmed in John Darwin's diary entry for Tuesday 12 December, 1939: 'Is there such a thing as justifiable femincide?' And again on Friday, 22 December: 'Bleeding women still fighting over something or the other'. This unpleasant situation at the top was clearly disruptive. Unfortunately, it eventually became common knowledge forcing Jourdain to leave the unit around the middle of 1940. Before his departure, he gave Norman Walton his SIS Station Code Book of which the main list is reproduced as Appendix 1: Station Code numbers. These SIS code numbers were a closely guarded secret at the time but they are mentioned in some detail in Nigel West's book *MI6*. The last entry in his notebook is for July 1940, so I suspect he left soon after that date.

In his lightening recruiting drive, Gambier-Parry blatantly poached talent from Philco his old firm. It was lucky that this was one of the largest wireless manufacturers in the country at the time, with a substantial staff! He also trawled through other areas for the first-class engineers and telegraphists his expanding unit needed. He found that the Merchant Navy and Royal Navy were good sources, especially the latter.

Gambier-Parry's first 'finds' in the Royal Navy included **Jack Saunders, Charlie Bradford, Claude Herbert, Jan Ware, Syd Cole and Harry Tricker**, all of whom became key personnel in the operational side of SCUs. They all eventually reached the rank of Major in the Royal Corps of Signals, whilst remaining Section VIII personnel, NPAF.

The story of how they became involved with Section VIII is interesting. They were all Chief Petty Officers at the Navy's Flowerdown wireless station, wireless telegraphy operators with naval experience of wireless engineering.

Several RN shore establishments had such W/T staff who were already involved in interception and some limited covert work. All Navy W/T operators needed to be both telegraphists and wireless engineers, the perfect combination for the tasks they would be required to perform in MI6, Section VIII.

Claude Herbert was apparently the first out of the Navy. According to his son Dennis, he obtained a civilian post as an operator at the Admiralty. Naval W/T personnel were often used by the Foreign Office to construct and maintain their wireless gear (See Edgar Harrison's story chapter 31).

Claude Herbert

Charles Bradford

It seems that Claude was sent to Prague on loan and was later recruited into Section VIII while it was still in its infancy. It should be noted that Bob Hornby visited Prague to install wireless equipment and commented on the high quality of the ex-Royal Navy operator working there. Bob Hornby may have been the link.

Claude in turn, spoke for his friends who were still in the Navy as Chief Petty Officers in the W/T branch. They were at the end of a long period of service and were hesitant about signing on again. To be offered a job in such a position must have seemed like manna from heaven and they all accepted. The commanding officer, HMS Flowerdown, lost many of his senior and most qualified telegraphist/engineers at a stroke.

Shortly before his tragic death in 1944, Jack Saunders was awarded the MBE. At the investiture, he was naturally wearing his uniform as a Major in the Royal Signals, but it is reported that King George VI spotted Saunder's Long Service Navy Medal, and commented 'What's the matter, Saunders? Wasn't my Navy good enough for you?'

Section VIII personnel encountered other problems to do with their military status. Claude Herbert was stopped several times by keen-eyed Military Police whilst in London, who had spotted him wearing army officer's uniform with naval medal ribbons. They thought, quite reasonably, he was either a spy or someone masquerading as an officer and choosing the wrong ribbons! One of those whom Gambier-Parry recruited from the Merchant Navy was **Gerry 'Jack' Gerrish**. Gerrish had connections with the Merchant Navy training schools which proved to be a rich source of skilled operators. Others came from the General Post Office. Soon after the beginning of the war, Gambier-Parry developed 'insider connections' with the leading army signals units, including formally through the Royal Corps of Signals, or in individual regimental signals sections.

Charlie West

Charlie West worked for Philco where he specialised in the manufacture of radio chassis. He was a very kindly man, and incidentally, my first boss when I joined the unit in 1942. He had designed all of the chassis used in the production of our major sets. His workshops manufactured many of the components, including spacers, Morse keys for agent's sets and the fronts and casings of the sets.

Wilf Lilburn, like Spuggy Newton, had been in the merchant navy as a wireless operator and later acquired a reputation as a wireless engineer, especially with short wave. He had been responsible for the installation of

Wilf Lilburn

wireless cars for the Glasgow Police whilst he was the Glasgow Service Manager for Philco. He was a life-long pal of Spuggy Newton as they had grown up together in the same small town in County Durham. They remained close friends until Wilf's untimely death. He played a big part throughout the story of Section VIII and his name appears in a number of the chapters that follow.

Wilf Lilburn was recruited alongside **Bob Chennells** who soon decided against joining the family farm and became an outstanding engineer. **Percy Cooper, RNVR** (Royal Navy Volunteer Reserve, Engineering Branch) became another member of the team, and more were recruited by the usual and safe method of personal recommendation. That procedure continued for years. These men, and many more like them, were drawn through head-hunting and careful selection into Section VIII by Gambier-Parry. The infant unit soon possessed many of the leading brains of the wireless world.

Alec Pollard came through yet another route. He was a representative of a wireless component manufacturer and he brought in **Ewart Holden** who owned a wireless shop in Twickenham. Thus the cast was gradually assembled. At the top, it was always by 'knowing someone'. This was vital because there was a need to have someone to vouch for each man, in view of the great secrecy surrounding the work that the Section was about to undertake.

Ewart Holden

Jack Saunders

Miss 'Monty' Montgomery came to the unit via the Passport Office in London where, for a long time she had handled cyphers, until the Passport Office organisation was absorbed back into SIS. She was moved to the 'War Station' at Bletchley Park and was in charge of the SIS cypher room, working on traffic for the SIS wireless room (Station X), in the tower of the mansion.

When Whaddon Hall was purchased, 'Monty' was billeted in The Chase, a large house beside Whaddon Hall, within the security zone of SCU1. She remained there throughout the war and worked at BP, first in the mansion itself, and later in Hut 10, where she had an area under her control, devoted solely to SIS traffic, emanating from Main Line at Whaddon, Nash, Weald and Forfar.

Alf Willis came from Philco, where he had worked as Service Manager for Birmingham. In view of the number of staff lured away from the company, it is a measure of its size that it was able to keep going.

It has been said that most of the Philco men were 'domestic' wireless engineers but that is rather unfair. Bob Hornby had been the Chief Engineer of this large and important company, Alfie Willis was a research and development engineer, and the work of Spuggy Newton on miniaturising wireless sets – in view of the clumsy components of the time – speaks for itself.

J. M. C. 'Mike' Vivian was the son of Lt. Col. Valentine Vivian, who was for a time Deputy Chief of SIS (DCSS). This was under Colonel Stewart Menzies, who took over control of SIS following the death of Admiral Sinclair. In 1940, Mike Vivian was a W/T operator in Cairo, SIS number 89952, and later was the officer commanding SCU11/12 in Calcutta. (See chapter 38).

Brian Sall was a wireless engineer and constructed the early North Cerney station along with 'Spud' Murphy. Unfortunately, I never learned 'Spud's' real Christian name but his good reputation as an engineer means I will not leave him out of this list just because of that minor detail!

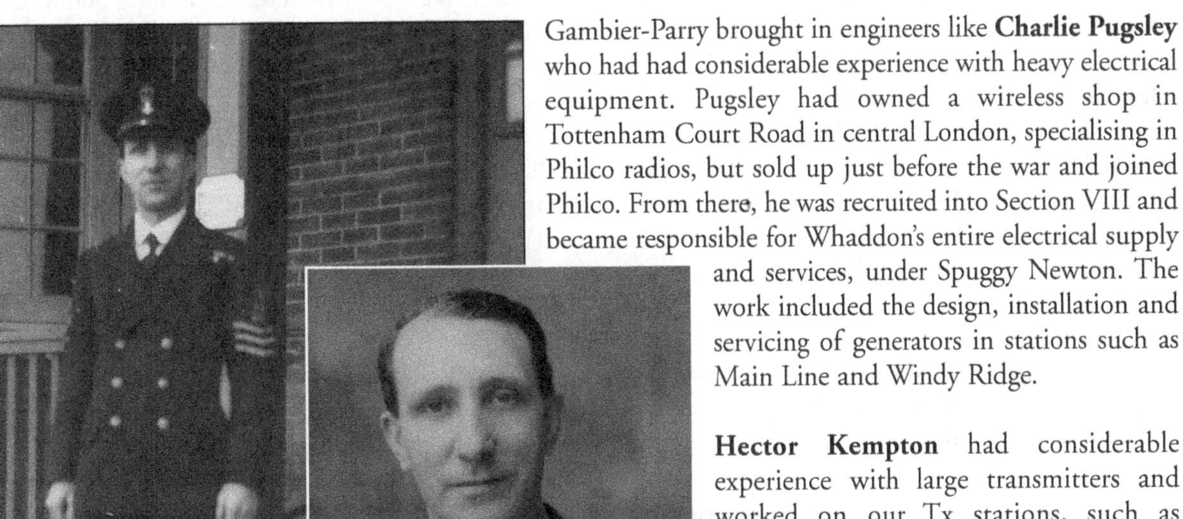

Gambier-Parry brought in engineers like **Charlie Pugsley** who had had considerable experience with heavy electrical equipment. Pugsley had owned a wireless shop in Tottenham Court Road in central London, specialising in Philco radios, but sold up just before the war and joined Philco. From there, he was recruited into Section VIII and became responsible for Whaddon's entire electrical supply and services, under Spuggy Newton. The work included the design, installation and servicing of generators in stations such as Main Line and Windy Ridge.

Hector Kempton had considerable experience with large transmitters and worked on our Tx stations, such as Tattenhoe Bare. He later ran the new transmitting station at Creslow.

Robin Addie was a well known wireless engineer who had worked for Metropolitan Vickers. He was involved in designing some of the first aerial arrays and

Harry Tricker

the wireless transmitters at Hanslope. Much later, I worked for him in Calcutta where he had designed the Dum Dum relay wireless station.

Don Lee was already a member of MI6 having been in Abyssinia with Adrian Trapman and then operating as an agent in Spain. He came home and joined Section VIII where he played numerous roles.

Ann Trapman who joined at the outbreak of war. Her husband was Adrian Trapman, Head of Station at Addis Ababa who died in a motor accident on their honeymoon in 1937. Although she married Don Lee in the mid-forties, she was usually called Ann Trapman in the office. Her maiden name was Hill see chapter 24.

John Darwin had already been a member of SIS whilst in industry and was a personal friend of Admiral Sinclair. He was well-connected in Whitehall circles and must have been a particular asset to the infant unit. His fascinating story is told in chapter 21.

Don Lee and Ann Trapman

Charles Emary was an agent in Spain before the start of World War II but was recalled by Gambier-Parry. On the way home from Gibraltar, however, his ship struck a mine and Charles was injured in the explosion. Instead of the more demanding active role intended for him, he was given the task of running Station X at Bletchley Park, and subsequently the SIS station in Hut 1 when Station X moved to Whaddon.

Major Kenneth MacFarlane R.A. was already a member of MI6 and apparently joined Gambier-Parry in the earlier stages of Section VIII. He was involved in negotiations with the Poles over the Enigma cypher machine and in May 1940, represented SIS in discussions with the French and the Poles at the HQ of the French decryption service. Later that year, he headed our 'A' Detachment, sent out from Whaddon to North Africa. In Daily Orders Part 1 for 18 September, 1945 MacFarlane is shown as being a Lieut. Colonel and 'Commanding No. 1, Special Communications Unit (To which No. 7. SCU was affiliated)'. By all accounts, a talented man.

In describing the quality of the staff Gambier-Parry assembled, I have tried not to exaggerate the engineering skills of the team. Apart from anything else, I am certainly not sufficiently qualified myself to judge their capabilities. However, they came with the highest credentials from the British wireless industry, the wireless engineering division of the Royal Navy, and its equivalent in the Merchant Navy, so it is difficult to believe they would have been bettered by many in their field. More importantly, the results that were achieved speak for themselves.

Part II

War Comes to Whaddon Village

Chapter 6

Whaddon Village

It is perhaps fitting that an operation of such importance as the transmission of Ultra should have been located in a village with as colourful a history as Whaddon. Whaddon lies just beyond the southern boundary of present day Milton Keynes, within a few miles of the former code-breaking station at Bletchley Park. Its name is derived from the Saxon words for 'Wheat Hill' and there is evidence of Roman occupation.

Just before the Norman conquest, the village was owned by a one of the courtiers of Edward the Confessor. It spent the following six or seven centuries 'in the gift of the Crown', the reigning monarchs granting it to a succession of families in return for services rendered. This process was started by William the Conqueror; Whaddon was one of batch of Buckinghamshire villages he gave to the French nobleman, his kinsman Walter Giffard, Lord of Longueville, who had commanded his army at the Battle of Hastings. Every so often, Whaddon reverted to the Crown because the latest recipient had either died without leaving a male heir, or committed some misdemeanor.

The value of this Royal gift owed much to the attraction of Whaddon Chase, a title granted by Henry III in the thirteenth century. This was one of Henry VIII's favourite hunting grounds and Elizabeth I also expressed great satisfaction with the sport 'in such a magnificent amphitheatre of wooded scenery'. On a somewhat gentler note, it was at Whaddon that the poet Edmund Spenser is said to have written his masterpiece *Faerie Queene.*

Other notables associated with the Manor of Whaddon during the last millennium include Richard, Duke of York and Ulster (slain during the War of the Roses), Jane Seymour (it was part of Henry VIII's dowry to her), Lord Grey de Wilton, one time Lord Deputy of Ireland, Sir George Villiers, Duke of Buckingham, and Browne Willis, the noted eighteenth century historian. During the eighteenth century Villiers' spendthrift son, then Lord of the Manor, ran up huge debts. As a result, the estates were sold to James Selby, Sergeant-at-Law, and Browne Willis's father, who was a famous physician of the time.

There was a further complication one generation further on when Selby's son, now sole owner of the estates, died without naming a 'right and lawful heir'. Many claimants came forward and it was ten years before the High Court ruled in favour of William Lowndes, Lord of the Manor at nearby Winslow, on condition that he adopted the Selby name.

Of course, none of this had any bearing on why MI6 set up its new SIS communications unit at Whaddon. That decision was made for the reasons explained in the following chapters.

So between 1939 and 1945, this quiet, sleepy village, in the North of Buckinghamshire, played a crucial part in the Secret Wireless War, an aspect of the war effort that has been largely overlooked by historians, many of whom simply never even knew of the existence of Whaddon Hall, or the numerous roles it played in the wartime story.

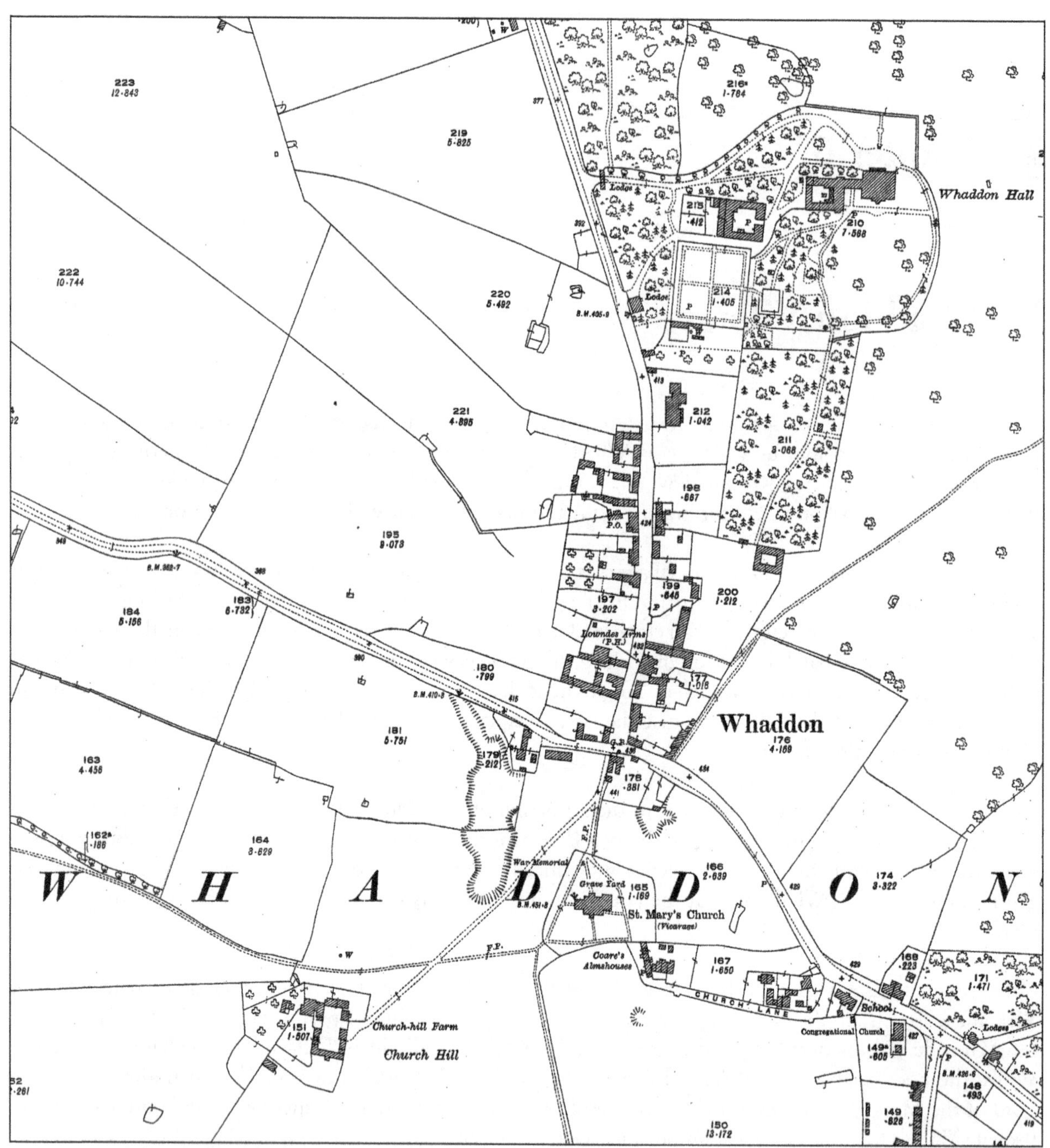

The 1925 Ordnance survey map of Whaddon Village. It was almost identical when I was there in the 1940s, and is remarkably little changed today.

Chapter 7

Whaddon Hall – the House on the Hill

'A fair old mansion place builded of brick and stone and covered with tiles.'

'A miserable gloomy place, though seated on an exceedingly beautiful knoll of a hill.'

'A magnificent country house built in classical style with its facade dominated by superb Grecian-style pillars.'

Those are all descriptions of Whaddon Hall which is situated at the northern end of Whaddon village. They differ, because the village's premier residence had been much altered, as well as pulled down and rebuilt three times, during its chequered history.

The first description comes from a surveyor's report dated 1541; the second is the opinion the Revered William Cole during the eighteenth century; and the third from a 1975 brochure for the ill-starred Whaddon Hall Country Club which went up in flames only a year after it was opened.

The third 'Whaddon Hall' was built by the Lowndes family, who later took on the name of Selby-Lowndes. The work was completed in 1820.

The Whaddon Hall mentioned in this story of secret wireless communications during World War II, was a quite magnificent building. Shades of its grandeur, as described above are still apparent today, despite the fire which completely gutted it during the attempt to turn it into 'The Whaddon Hall Country Club' after the war. The rebuilding that took place after the fire has left the building truly recognisable as the wartime HQ of our unit. It now forms four substantial properties with the section described as 'Whaddon Hall' containing the magnificent frontage.

The selection by Gambier-Parry and his colleagues of Whaddon Hall as the home for MI6 (Section VIII) was due a number of factors. These were mainly its proximity to the SIS War Station at Bletchley Park, where Section VIII presently had its own wireless unit – Station X, its elevation at nearly 500 ft above sea level at its highest point, the availability of eminently suitable accommodation in Whaddon Hall itself, and in its many outbuildings. There can also be little doubt that its rural location – and thus the relative ease with which it could be made secure militarily – were contributing factors.

This book is about the work of Gambier-Parry's MI6 (Section VIII) but it is also about its association with the Whaddon Hall on its 'knoll of a hill' that exists today.

The frontage of Whaddon Hall in 1939

The dining room of Whaddon Hall in its pre-war days

The library steps at the back of Whaddon Hall

The entrance lodge to Whaddon Hall

Chapter 8

Bletchley Park – Admiral Sinclair's 'War Station'

Bletchley Park mansion

This chapter is in several separate but linked parts. It starts with the history of Bletchley Park itself, then its purchase by Admiral Sir Hugh Sinclair as his 'War Station.' It goes on to describe, the work done by the codebreakers, and their dependence upon the wireless interception service or 'Y' stations for the messages, and the SCU/SLUs for the dissemination of the intelligence.

So far as the actual codebreaking work of Bletchley Park is concerned there are innumerable excellent books on the subject, and there is little I could add.

1. Bletchley Park and the Leons by Ted Enever

After the Romans left Britain about AD 400, various settlements grew in the region we now know as North Buckinghamshire. Among these was a clearing within an area later to be known as Whaddon Chase and made by a man named Blecca. It is from Blecca that Bletchley gets its name, 'ley' being a clearing or grassed land. So, over the centuries, 'Blecca's ley' has become 'Bletchley.'

Part of Etone Manor in medieval times, the first reference to Bletchley as a separate manor is found in 1499. Prior to this we know that the estate, to use contemporary wording, was given by William the Conqueror to one of his notable commanders at the Battle of Hastings, Bishop Geoffrey, of Constance, in Normandy. The estate was subsequently won and lost by several families. A certain Walter Gifford was made Lord of the Manor in 1092 by the then king, William Rufus, but died childless. His principal relative, Richard de Clare, took over and called himself the Earl of Buckingham, which apparently seemed not to annoy Richard the Lionheart who, in 1189, formally gave the manor, or estate, to de Clare.

In 1211 de Clare's daughter married Sir John de Grey and the manor passed into this family line, de Grey becoming Baron Grey de Wilton, from where the current entrance to Bletchley Park, Wilton Avenue, gets its name. At this time records show that the present Bletchley Park was the deer park of Sir John's estate and in 1563 mention is made of a moated keeper's lodge in the middle of this deer park. The manor was to remain in the hands of the Wilton family until 1614.

In 1616, after the Crown had confiscated the de Grey lands following charges of treason against Lord Thomas Grey, who died in prison, the estate was given to Sir George Villiers who was created Earl of Buckingham and later, in 1623, Duke of Buckingham. He was followed by the second Duke, also George, but in 1674 he sold the estate to Dr. Thomas Willis, a famous physician in the reign of Charles II. Following the death of Dr. Willis in 1699, the estate passed to his son, Dr. Browne Willis, whose wife, by coincidence, was a direct descendant of Walter Gifford, the Lord of the Manor in 1092.

It was Browne Willis who, in 1711, first built a house close to the site of the present Mansion in Bletchley Park, which he named Water Hall, but by 1798 the house and adjoining lands had been sold and the house pulled down by the new owner, Thomas Harrison, a steward to the Northampton Spencers, the family of Diana, the late Princess of Wales. In the late 1870s, a descendant of the Harrison family sold the estate to a Mr. Coleman who built the first part of the present day Mansion. By 1881 the property had come into the ownership of Samuel Seckham who enlarged the building before selling it on, a year or so later, to Herbert Samuel Leon, one of Bletchley's greatest benefactors.

Herbert (Sammy) Leon was born on 11 February 1850, the second son of a Jewish financier and founder of the stock exchange firm of Leon Brothers, which Sammy entered at the age of twenty-four, rising to become its senior partner. Just before he entered the firm he married, but sadly lost his wife, Esther Beddington, after only two years, leaving him with two small children, a son, George and daughter, Kitty. Some years later, in 1880, he was to marry again, his bride being Fanny Hyam. The marriage was to last for forty-six years, until his sudden death in 1926.

An ardent Liberal, Sammy Leon was a long standing friend of Lloyd George, who was a frequent visitor to the Mansion, the family's new home at Bletchley Park. In 1891 Sammy Leon became Liberal Member of Parliament for North Bucks, a seat he held for four years before losing it to the Tory candidate, Sir Walter Carlisle. He was destined never to return to Parliament again but his loyalties to the Liberal Party did not go unrewarded and in 1911 he was made a baronet. Immediately, he dropped the name Sammy and became known as Sir Herbert Leon, but change elsewhere was not noticeable, particularly locally.

He continued to support Bletchley in many ways, and the current Leon recreation ground, the building of Leon Avenue, the Leon cottages in Church Green Road and many other property projects remain as evidence of his largesse. One of the town's current major comprehensive schools is also named after him. He was a good employer who rewarded loyalty and every Christmas a bullock from the estate was slaughtered and the meat distributed to his estate staff.

Sir Herbert did not find favour with members of the suffragette movement, who chained themselves to the Park's main gates, or with the nearby rector of St Mary's church, Revd. William Bennitt. Sir Herbert hated the sound of the bells, apparently, which he could not help but hear, the Mansion being only yards from St Mary's. Many times he tried to stop the bells being rung but the rector stuck to his guns, pointing out that the church had been there for 700 years before Sir Herbert came to Bletchley. Only as Sir Herbert lay on his deathbed in 1926 did Revd. Bennitt grant his request. But once the funeral had been held, the bells pealed out again from St Mary's ancient tower.

Sir Herbert was sorely missed. He had established Bletchley Park and its grand house, the Mansion, and nearby Home Farm – now the site of Home Close, a small cul-de-sac of residential houses off Whalley Drive – as a thriving estate. Lady Fanny Leon, continued his good works, not least by allowing the Park to be used for the annual Bletchley Park Show, which was the town's major event during the years between the two wars and for another score or more years following victory in 1945.

Lady Fanny died in January 1937 and on her death the Bletchley District Gazette reported: 'Lady Leon can, without dispute, be described as Bletchley's greatest benefactress. Her generosity towards and interest in many of the town's organisations are well known. What will never be fully known are those many acts of kindness which have so endeared her to the people of Bletchley'.

During the Leons' lifetime at Bletchley the Mansion was enlarged considerably. From the small house purchased by Sir Herbert in the nineteenth century, the 1937 building had grown in size to the building that now takes pride of place in the Park today. As soon as he acquired the property, Sir Herbert added a domestic and servant wing. In this wing is the ice-house, used for cold storage of meats and produce and which, from the outside, is often mistaken for either a large, brick built dove-cote because of its shaped roof, or a private chapel, owing to its stained glass windows.

After 1883 Sir Herbert extended the property by further stages, taking it forward and adding the drawing room, dining room and main entrance hall and lounge hall that can be seen today. The lounge hall has a stunning glass painted roof.

Architecturally, some will argue that the Mansion is a hotch-potch of styles, but to those living locally, it is a constant reminder of Sir Herbert and is a magnificent country house, boasting huge fireplaces, ornate ceilings and panelling.

By the time of Lady Fanny's death in 1937, storm clouds were already gathering in Europe as Nazi Germany steadily grew in power and influence. Sir Herbert's son George, the second Baronet, had made his own life and was ready to dispose of the Bletchley Park estate which was split into various lots and duly sold off. The Bletchley Park that we know today, bounded by the older Church Green Road and the newer Whalley Drive and Sherwood Drive, was one such lot, as was Home Farm and its surrounding land, running adjacent to the main London to northwest railway line.

The buyer of the Bletchley Park parcel was a local consortium of developers headed by Captain Hubert Faulkner. A keen horseman, Captain Faulkner would often appear on site dressed in riding or hunting wear and his plan was to break up the Park into smaller parcels of land for residential development. He intended to demolish the Mansion and build himself a new property slightly to the south, on the flat ground used as a croquet lawn by the Leons alongside the lake. He began his site clearance in the stable yard, and took down some stables at the eastern end which were flanked to the south by the Leons' apple, pear and plum store, a building which was destined to make its own impact on twentieth-century history, as the 'think-tank' for such as Turing.

2. Admiral Sinclair's 'War Station'

I question the widely accepted view of the purchase of Bletchley Park by the Government Code and Cypher School. Agents are said to have approached Captain Faulkner on behalf of GC&CS but in fact Admiral Sinclair, as Head of SIS, was not trying to find a wartime home merely for that section, but a safe haven (or 'War Station'), away from potential bombing for a number of sections of SIS. These included GC&CS, Section V (Counter espionage), and the new Section VIII (Communications), along with Station X from Barnes. We know from John Darwin's diary (see Chapter 21), that Bletchley Park was known throughout the SIS as its 'War Station'.

The headquarters itself and several departments of SIS were housed in just one or two buildings in London, but mainly at 54 Broadway, Westminster. Clearly, in the event of a war, and the air raids that would surely follow, this vital nerve centre of our Secret Service would be very vulnerable being in the heart of the city. Sinclair had come to accept that war with Germany was inevitable and only the timing was difficult to predict.

The devastation bombing could cause had been only too apparent in the Spanish civil war, when Hitler and Mussolini supported their fellow fascist dictator Franco, and used the war to test out much of their new military equipment, especially aircraft.

Admiral Sinclair believed Bletchley Park was quite excellent for his purpose. It was a large building, with extensive surrounding land for expansion, access by road to London was good via the A5, and a major railway line went through Bletchley station, only a short distance away. However, having located a suitable property, he found difficulty financing its purchase. The Foreign Office thought, as a War Station, it was the responsibility of the War Office; in turn, they suggested he approach the Admiralty – who referred him back to the Foreign Office! Frustrated by this 'passing-the-buck', and in some fear of losing the property by inaction, Sinclair paid the asking price of £7,500 out of his own pocket.

In his book, 'The Ultra Secret', Group Captain F. W. Winterbotham says that Denniston and the GC&CS had moved into Bletchley Park by August 1939, which he states, Sinclair had purchased earlier as a 'wartime hideout'. Clearly this refers to a 'wartime hideout' for SIS as a whole, not just GC&CS.

Further on, Winterbotham says – *'In September, 1939, the SIS were also evacuated to Bletchley Park, some fifty miles north of London, near the main road and the railway to the north-west'.* He refers here to the whole of the SIS organisation but I should record that Admiral Sinclair and his immediate subordinates remained at Broadway. After his death in November 1939 his successor Stewart Menzies continued to be based at Broadway where they had a teleprinter room, and small wireless facility, in the basement.

GC&CS and Gambier-Parry's team, were amongst the first to move in to Bletchley Park, and indeed, representatives of both had worked in the property since its purchase in 1938. Sinclair intended to move other sections there too on the outbreak of hostilities, but the rapid growth of the cypher work in the early days of the war, arising from the breaking of the Enigma codes, meant space was suddenly at a premium.

Until recently, it was thought that Gambier-Parry only authorised the building of a wireless telegraphy station in Bletchley Park in mid-1939, but it now appears that it may have been started towards the end of 1938. Dennis Herbert relates that his father, Claud, having been engaged by MI6, moved his family to Bletchley in 1938. He then started work on installing a wireless station in the tower and aerials in the grounds. Gambier-Parry initially intended Bletchley Park to be a twin of the Foreign Office/SIS wireless Station X at Barnes, as part of the 'War Station' concept of being able to evacuate sensitive units to Bletchley Park.

Whatever the start date, we know the station was operating in 1939, in a hut now called 'Hut 1' but at the time was unnumbered – simply because it was the only hut in the grounds! At that period, the GC&CS organisation was housed within the Mansion or its original outbuildings.

The wireless station in 'Hut 1' was used for diplomatic traffic for the Foreign Office and some covert traffic to embassies. The SIS traffic was handled only in the new wireless room built into the tower of the Mansion. The result was that, between the station in the tower, and the station in the hut, they completely duplicated Barnes.

One of our agents in Spain was Charles Emary. He was recalled to relieve Spuggy Newton with the BEF in France but on the way home his ship was mined, and he was injured. However, he finally arrived home safely and after recovering, was given the relatively light task of running of the SIS station in the tower – supervising it until 15th November 1939 – when it was transferred to Whaddon Hall.

By the end of November 1939, Section VIII began to move out of Bletchley Park to its new home at Whaddon and the transfer was completed in early 1940. Section V moved to Ryder Street in London; in 1943 with the major shake-up at Bletchley Park, GC&CS moved out to Berkeley Street in London to distribute diplomatic, commercial and RSS material, with Travis taking over at Bletchley Park.

On page 22 of his book on Bletchley Park – *'Britain's Best Kept Secret'* – Ted Enever mentions Station X being in the tower at BP, and he goes on to say:-

'The radio room, small and cramped, was given the code name 'Station X'. The radio's aerial was slung between the finials of the Mansion's Victorian roofline, before running to a tall cedar tree....'

further down on that page, he goes on to say:-

'The result of this policy thinking was that Station X, and the aerials spanning the trees and roof, were quickly dismantled and a new base established some seven miles to the South West at Whaddon Hall.'

On the removal of station X – Room 29 of the mansion was changed to a teleprinter room to connect Whaddon Hall, SIS HQ in Broadway Buildings, and the Central Telegraph Office in London.

The original wireless station in 'Hut 1' continued to handle just diplomatic traffic until it too was closed down and transferred to the Main Line wireless station at Whaddon. A room on the floor in the North East wing of the Bletchley Park mansion (known today as the Board Room), was the SIS cyphering and decyphering unit for its own traffic, under its supervisor, Miss 'Monty' Montgomery. The team were later transferred to part of Hut 10, where they continued as an autonomous unit.

3. The 'Ultra Trilogy'

Much well deserved praise has been heaped upon those working in GC&CS, who collectively have become known as the 'Codebreakers,' but they did not exist in a vacuum. The work at Bletchley Park was dependent on the thousands of wireless messages coming in daily, since the codebreakers did not attempt to intercept messages themselves. So where did all these messages come from? Then, how was the intelligence gathered from them, as a result of the brilliance of the codebreakers, sent out to our military leaders? I believe this is best described as *The Ultra Trilogy* and I will explain it in the three parts that existed.

The first part of the Ultra Trilogy

Wireless Intercept – the 'Y' service, by Joan Nicholls.

Wireless Intercept became known as the Y service because of the Civil Service habit of abbreviating everything to initial letters – Wireless Intercept became WI, pronounced Y, and so this very secret organisation had its own code name. All sections of the Y service, Army, Navy, Air Force and civilian had special selection procedures when recruiting their operators.

They had to pass the IQ test above a certain score, they had to be assessed as being capable of working under pressure, have the patience to sit waiting for hours for a station to come to life and remain alert, to be mature enough not to gossip about what they did and not least maintain a very difficult work pattern with meals and sleep patterns disrupted. They were not told what they had been selected for and indeed did not find out until after they were trained.

Training, for the forces special operators, was 19 weeks. In that time, they were taught the Morse code, wireless procedures and electromagnetism all alongside the military training of marching, cleaning billets, inspections and drill, and not least route marches all over the countryside.

The Morse was keyed by an instructor through head phones and the speed of the delivery was increased and speed tests taken each day to assess the operators ability to read Morse accurately, at ever increasing speeds. In slower speeds progress was constant, but all reached a plateau, where they could not seem to attain their goal of 25 words a minute. It was always a very trying time, however, with perseverance, the block was overcome and speeds were a constant 25 and accurate.

Reading Morse in training was a relatively simple task, now the operators were introduced to the reality of listening to Morse from a wireless set, which had the added problems of static electricity crashing in their ears. Stations would drift and they would have to tune the set as they read and recorded the Morse.

The signals faded and had to be adjusted to give the best possible sound level. This sometimes meant being connected to a different aerial, all the time the operator would be reading the Morse, recording it, handling the wireless set, and all at quite high speeds. Now they were ready to be tested, not only in Morse but also electromagnetism, wireless procedures, and only those who met the standards would be allocated to a Y Station. They were then told they were to be Special Wireless Operators and their job was to intercept the enemy wireless traffic.

They were to be the ears of the nation, passing their intercepted messages to the centre, now known as Bletchley Park, where they would be decoded and the intelligence gained from them acted upon. Imagine how these bright young teenagers felt at being so close to the war zones but sitting at their wireless sets somewhere in England.

As time went on they accepted the routine and the importance was elusive as they were never told how effective their work was. Everything in those days was on a need to know basis, and the powers that be didn't consider they needed to know. Accuracy was of the essence and every operator strove to get an accurate record of the message which was transmitted in code.

The German Wireless Operators were possibly the best in the world. They had a very precise, almost staccato way of keying the letters. Their wireless procedure was faultless and had we only had the keying to worry about, intercepting the messages would have been easy. However, life is never that simple.

The German stations had ways of blocking reception by jamming the signals with loud noise, sending on a frequency that had other stations next to the one they were monitoring and with the added natural hazard of static electricity crashing and crackling in our ears, it all made reading the Morse quite a challenge.

As Y' service operators became experienced at intercepting Morse, they soon realised that they could identify the operator sending the Morse and also the sound of the transmitter. When the German station would suddenly change frequency in the middle of a message to shake them off, they would first search for the transmitter and then check the operator sending was in fact 'ours'. This was known as 'fist' – no two people key in the same rhythm. It is like a fingerprint, find a match and you have your man.

The call signs, frequencies and known stations the German operators worked to, helped to track them when they went AWOL. During and after the war there were many automatic radio systems that attempted to replicate this ability to recognise the senders but none could match the experienced intercept operator. Increasingly, our aerials were hard pressed to cover the long distances between the intercept stations and the enemy wireless operators.

The war zones grew, as Hitler marched through Europe and other countries and added to his empire. Therefore, it meant that the numbers of operators and wireless sets needed to cover them became a very real headache for MI8 who were responsible for controlling the stations that had to be monitored either as a priority, or only if capacity allowed.

Churchill worried that the growing array of aerials might be spotted from a plane and he had this checked by reconnaissance planes. The aerials could not be seen, but the concrete footings that held them could. Fortunately they were erected in a random fashion, not in neat rows and a casual observer would see only what looked like scattered stones in the fields below.

From the 1920s the government was asked to provide more sets and operators, as it was realised that in the event of another war, this service would be vital. None were forthcoming. Wireless sets suitable for intercept work from this country were not manufactured here as there was no real demand for them and what we had were built by the operators themselves. One set was made out of scrounged parts and the casing was an orange box.

At the beginning of World War II, the Americans were asked to supply us with sets. In 1937, they had 80,000 short wave amateurs which created a demand for good quality receivers and allowed us to tap into this source on lease-lend. The first of these sets, the HRO, was followed by the SX28, and the AR88. Later, 'Y' service had the Sky Pilot and the Sky Rider. The HRO was a small set with a range of 50 to 430kc/s and 480 to 30,000kc/s and 9 sets of coils covering these ranges. The coils were cased in lead and quite heavy to manhandle. Whenever a frequency changed it often meant changing the coil too. One great advantage of the HRO was it's ability to be fine tuned, which meant that a station that was drifting could be followed with the minimum of movement on the dial.

The Royal Signals Research and Development Section later produced the R206 in 1943 and by 1945 they had developed the Mark 11. Operators were being recruited in their hundreds, and at Beaumanor, the War Office's largest Y Station, at the peak of the war in 1943, there were 900 ATS operators, and 300 civilian male operators, manning five set rooms, and all the Direction Finding Stations around the clock.

It is not possible to establish the number of Y Stations in World War II, as there is no official record. Each of the services had their own, and also had small mobile units near enemy lines in all the war zones. Add to that, the civilian operators monitoring the Embassies etc., it would seem as if a large part of the British public had ended up in various branches of the Y service!

It is not sufficient to record wireless messages and decode them. It was essential to know where the signals were coming from and to whom they were being sent. This meant that not only had the new stations to be plotted, by the civilian intercept operators from Beaumanor working in the direction finding (D/F), section of the Y Service, they also had to keep checking those stations in the battle areas, to see if the enemy had moved.

Beaumanor D/F stations were located all over the country, from Scotland to Cornwall. They also carried out work for other stations and the Air Force and Royal Navy. They worked in tiny huts in isolated locations, usually alone. They would pinpoint an enemy station, pass the coordinates to Beaumanor Control room, who would plot them on a map then advise Bletchley Park of the exact geographic location of the station.

Bletchley Park would then pass the information on to the commanders of our armies, but as with the results of the decoded messages, the source was never revealed. Intelligence was, of course, received by the commanders from many different sources besides Ultra; for example from agents, air reconnaissance, and battlefield intelligence like Phantom.

By closely monitoring the frequencies and call signs used by the different German army groups and knowing who was the control and who the outstations, it was possible to track the whole of the enemy organisations such as the SS and the Panzer Divisions, the Luftwaffe, and the German Navy.

The Y service became, not only the ears, but the eyes of intelligence gathering, and all from the safety of this country. To maintain this immense advantage, was of course due to the operators keeping the secret of our intercept organisation, and maintaining it not only throughout the war, but until the government released the information in the nineteen seventies.

Much has been written about Bletchley Park and the decoding of the messages without ever mentioning where those messages came from. Without the Y Service, they would have had nothing to do and would have sunk without trace. The decoding was only a part of a whole system, the beginning of which was the German wireless operator. Then came the Y service intercepting the signals and tracking the locations and then the decoding and sending out the intelligence thus gleaned to our commanders in the war zones.

An Enigma machine

The now famous Enigma machine, which was invented and exhibited before the war, was originally designed for commerce, no one was interested at that time. The German High Command however, saw it had a potential for military purposes, and developed it into a very sophisticated machine. It was small, cheap to produce and completely mobile. The Germans had one in each U-boat, in the staff cars used by their generals, as well as in all wireless control stations.

It meant that messages from Rommel for example, could be processed on the machine and the coded messages sent off from wherever the General was located. He could call for Luftwaffe assistance, or supplies to be brought in by the German Navy, and all while visiting his troops and actually fighting the war. (See Siegfried Maruhn's story chapter 37).

The machine coded the text into a five letter code. In the preamble before the coded text was sent, would be the key to tell the receiving operator the setting of the machine, he could then set his Enigma machine to receive the text which would be converted back to German language. This preamble, and the first five blocks of five letter code, were vital in the work of the decoders to break the code.

It was also realised by the enemy that they must prevent us getting these signals and so they used jamming devises to drown the signals. The intercept operator had to strain to hear the message and if any letters were missed show exactly where that letter would have come in the text. Not only were they coping with the natural drift of the station, static electricity, and the howl of the jamming device, they had to read and record the Morse, and record the missing letters – all at fast speeds. The Germans could re-send any message that did not get through to their outstation, the intercept operator usually only had one chance. Just very occasionally, they would have to re-transmit their message and that gave them a second chance.

The second part of the Ultra Trilogy
The Bletchley Park codebreakers
By early 1940, BP were breaking the Enigma code more frequently. Their story has been told many times, sometimes by those who were actually there. I say this is the second part of the Trilogy, since obviously Bletchley Park had to first have messages from the Y Service to work on, before its codebreakers could start their work.

I entirely endorse the view that there were many remarkable – some quite brilliant – people working there. At the peak, there were upwards of 10,000 employed; they were billeted out in the many small villages around Bletchley Park, and some were in Stony Stratford where I lived.

One would often see them when off duty, in the many local village pubs, and I became 'drinking friends' with one or two. In Stony Stratford, for example, their favourite pubs were The Cock, The Bull, The George, and The Plough.

It was inevitable that they stood out since most were not in uniform, many were young, and their sheer numbers in a small local town or village, made them conspicuous. However, I think the 'locals' view was that some were quite a rum lot, with a few being described as 'snooty'. This was not surprising, since they included many of the country's leading intellingentsia. A discussion on Thales, Plato, and other Greek philosophers – *in Greek* – in a pub in a remote village, was somewhat daunting conversation for the locals, trying hard to concentrate on a game of darts. Nevertheless, some good friendships were formed and remained firm in later years.

It should also be remembered that, however clever their work, the codebreakers were helped enormously by the recovery of actual Enigma machines and codebooks. This is the subject of the excellent recently published book *Enigma – the Battle for the Code* by Hugh Sebag-Montefiore. Whilst wanting to take nothing away from the codebreakers, he does point out there was a mistaken belief that they broke Enigma entirely by themselves.

One other duty of the Y service was to provide Bletchley Park with the geographical location of the source through their D/F (Direction Finding) capability. Thus Bletchley Park were given, not only the messages, but where they came from, which was obviously of enormous help to the whole operation.

On the input side, we must remember the interception work done by the Radio Security Service both at its fixed stations and by the many hundreds of VIs listening in their homes – often through the night – to that most secret of the German wireless traffic, the Abwehr. They also listened to the traffic from the Reichssicherheitshauptamt (RSHA) run by the Gestapo (see chapter 15).

The third part of the Ultra Trilogy.
SCU / SLUs – the Voice of Ultra
Clearly having broken the codes, and analysed the intelligence it contained, the intelligence information gleaned had to be sent to the Admiralty, the War Office, RAF Commands, and the various military SLUs, mainly via Windy Ridge at Whaddon and certainly that for North Africa and Europe went from there.

The dissemination of Ultra was the third part of the Trilogy and is described in other chapters of this book, being just one of the many wartime roles of MI6 Section VIII. These included the interception of Abwehr traffic – one of the most important of all wireless interceptions – by Section VIII's own RSS / SCU3 unit based at Hanslope.

After the 1942 disasters in the Far East, it was decided that great priority be given to communications in that theatre of war. So in 1942, it was decided that Bletchley Park needed a high power station to handle this long distance communications. In conjunction with the large control station at Leighton Buzzard, the RAF was utilised for this overseas communications. A special station was built at the village of Stoke Hammond exclusively for the use of Bletchley Park. All signals were put through to Bletchley Park on landlines. RAF personnel were employed at Bletchley Park which was almost exclusively WAAF.

The termination station at BP was located in E block and it was fitted out specially for high speed work by using electric Morse senders and the printing undulator. Both these machines could send and receive Morse code at very high speeds, far faster than the best human operator.

By the beginning of 1943 E block was in full operation. Some of the high speed SLU's were sending and receiving from E block, but the major places being contacted were Melbourne, Delhi, Calcutta, Cairo, Colombo, Malta, Toronto and Alexandria. Over 100 girls were employed on each of the 4 shifts and initially the girls were housed at Wrest Park, a country mansion near Silsoe, in Bedfordshire. But it became so difficult to transport them every day that a large air force camp was set up in 1943 to accommodate them and it was adjacent to Bletchley Park and it became known as RAF Church Green.

However, almost all the military Ultra traffic was handled by Section VIII. Certainly that was the case for the North African campaign, Sicily, Italy, and the re-entry into Europe after D-Day. Gambier-Parry's organisation was also involved with the dissemination of Ultra to other parts of the world, through its Main Line and Windy Ridge stations. In addition, it provided back-up wireless facilities at a number of key SLU installations like the Admiralty and RAF commands.

In suggesting we look again at the established perception of Bletchley Park, I mean no disrespect to those who worked there. It is only that the earliest books on the subject were written by such people as Group Captain Frederick W. Winterbotham with little or no mention made in it of the others involved. It has been argued that this was due to the secrecy surrounding both interception and dissemination, but until relatively recently it was barely acknowledged that these components of the operation existed, or that they were important in the Ultra process.

After the publication of Winterbotham's books, good marketing by BP kept the emphasis on the codebreakers role, so there seemed little need to disturb the image so successfully created. Much however, has now been done to correct that perception, and the Wireless Museum at BP, now housed in Hut 1, is a small but important part of the change of direction towards a more balanced view of the Ultra story. The Bletchley Park Trust have also reproduced the 1939 Section VIII's SIS wireless room in the tower of the mansion, and make it clear that it was later removed to Whaddon.

Nothing whatever can diminish the importance of the work of the dedicated thousands who worked at Bletchley Park, in shifts around the clock. The staff varied from eminent scientists and mathematicians, to filing clerks. They were involved in designing the Bombe and Colossus, breaking the keys, interpretation, traffic analysis, evaluation, there were secretaries, typists, teleprinter operators sending Ultra to fixed SLUs, telephone operators, and the all-important filing clerks. Filing was a hugely important task keeping records of the location of various units, and making that available to the overall intelligence picture.

However, it is important to remember there were also many thousands of staff working in the various Y Services providing the fodder for Bletchley Park. All the armed services had interception stations of varying sizes and for different purposes. As well as Beaumanor where Joan Nicholls was working, there were intercept stations at Keddelston Hall, Chicksands, Forest Moor, amongst others, and they were home for these thousands of young operators.

They were mostly girls from the WAAF, ATS, and Wrens. They had been chosen for their high IQ and recruits had rapidly achieved the skill of reading Morse at high speed, before going on watch duty. Here, they had to concentrate hard for hours on end, trying to isolate sometimes faint Enigma messages through a cacophony of other Morse traffic, on over-crowded wireless bands.

Together with their male colleagues, they were working in equally unpleasant circumstances as those at BP, in draughty, partially-heated huts, keeping what today would certainly described as unsocial hours.

The SCU/SLU organisation was disseminating the Ultra military traffic, and in the UK, that would be from such places as Windy Ridge – its very name says it all. At the other end, were our SCU wireless teams in the field, often operating in most uncomfortable, and sometimes dangerous circumstances.

So, the real story of Ultra is that it was a superb combined effort. This involved the three parts of the trilogy, wireless interception, the brilliant work of the Bletchley Park codebreakers, and then the successful dissemination of the intelligence gathered. To these must be added contributions from other intelligence sources, and from such work as the recovery of actual Enigma machines and codebooks, mentioned in Hugh Sebag-Montefiore's book.

Chapter 9

Special Communication Units

The Special Communications Units (SCUs), were the para-military organisations set up by Gambier-Parry to cover the many and various operations of MI6 (Section VIII), including the dissemination of Ultra traffic. That particular aspect of Section VIII's work was handled by Special Liaison Units (SLUs) in conjunction with the SCUs.

With the acquisition of Whaddon Hall in November 1939, Section VIII senior personnel working at Broadway and elsewhere were gradually moved into the Hall or into local houses and hotels. Firstly, the staff from Funny Neuk in Woldingham, then those working at Barnes who had been brought together at Bletchley Park, now moved on to Whaddon, as soon space and accommodation was made available. The operational and engineering side was based in the Hall and its outbuildings, whilst other staff were billeted locally in houses or public houses in the many surrounding villages.

The extensive stables and coach houses around the stable yard were the obvious first choice for the engineering workshops that had been moved from Barnes and Funny Neuk, so they were set up there under Bob Hornby, Spuggy Newton and Charlie West. The stables also became the home of the rapidly expanding wireless stores under the guidance of Ewart Holden. These moves were taking place whilst wooden huts to house the workshops were being built in the grounds of the Hall between the stable yard and Stratford Road outside. The contractor for the work was Cowley and Son of nearby Stony Stratford, and they continue in business today.

The 'founding fathers' of the organisation – Gambier-Parry, 'Ted' Maltby, John Darwin and Micky Jourdain – were already in uniform. As serving MI6 officers, they automatically assumed their services rank. With the end of the 'phoney war' (approximately the first six months after the declaration of hostilities), it was realised that the rest of the staff of the rapidly expanding Section VIII could not continue to appear as civilians. Military intelligence at the War Office were approached and it was agreed that the Section VIII civilians should be enrolled ostensibly as members of the Royal Corps of Signals – although they were not to be paid from army funds.

Thus, the first of the '**N**ot **P**aid [from] **A**rmy **F**unds' (**NPAF**) personnel appeared in the routine 'Daily Orders' as a further part of the plan to disguise Section VIII's true function by giving it the appearance of an army unit. Later on, these staff were referred to in the unit's otherwise strictly military-looking Daily Orders just as 'Special Enlistments'.

In the early days of 1940, the unit was referred to as 'Special Signals Unit No. 1' (SSU1) but it was felt that the word 'Signals' might give the game away. In *The Story of the SLUs* (PRO HW3-145) presumably written by Winterbotham in 1945 as part of the History of Bletchley Park, it is suggested that the change of name was prompted by the assumption soon made by other units that SSU stood for Secret Service Unit! There was also the unpleasant link with 'SS' who by that time had already become a symbol of hatred and fear across Europe.

DAILY ORDERS PART I
by
Colonel R. Gambier-Parry.
Comd., S.S.U. No.1.

Issue 20 dated 25 Jul 40.

---

96. POSTINGS.

(a) The following personnel have been enlisted at Whaddon and are posted as shewn :-

N.Y.A.	Sigmn.	Holden. G.S.	Workshops.	18 Jul 40.
"	"	Sweet. C.F.	H.Q.(S.O(SS)).	"
"	"	Kempton. H.A.	Workshops.	"
"	"	Morrow. I.A.	H.Q. Tailor.	"
"	"	West. C.W.	Workshops.	20 Jul 40.
"	"	Hill. J.	"	"
"	"	Roberts. J.G.W.	"	"
"	"	Pidgeon.H.E.C.	H.Q.(S.O.Tech).	"
"	"	Castleman.H.J.	Workshops.	"
"	"	Ord. J.T.	"	"
"	"	Lax. D.	"	"
"	"	Bromley.K.A.	Trg. Group.	22 Jul 40.

(b) The undermentioned Officer arrived in the Unit :-

Lieut. C.M. Harrison. Signal Officer. 24 Jul 40.

(c) The undermentioned [illegible] posted to [illegible] todays date :-

2585609 Sigmn. Hooper. K.W.J.
2308091 " Thompson. C.

The undermentioned are posted as operators to Training Group :-

2091512 Sigmn. Parrish. J.C.
4742065 " Hodgson. A.
4445433 L/Cpl. Wilkinson. R.

The undermentioned are posted to Transport w.e.f. todays date :-

2337575 Sigmn. Dally. W.S.
2337797 " Calvert. W.

97. ATTACHMENTS.

The undermentioned are attached to the Scottish Command, w.e.f. todays date :-

2321782 A/U/L/Cpl. Nisbet. I.S.
2580926 Sigmn. Cullen. W.N.
T/131018 Dvr. Howie. W.

The undermentioned are attached to the Admiralty, w.e.f. 26 Jul 40 :-

2308091 Sigmn. Thompson. C.
2585609 " Hooper. K.W.J.
T/66188 Dvr. Broadfoot. J.

Daily orders Part 1. SSU No.1 Comd. Colonel R. Gambier-Parry. 25th July 1940

So, in the climate of the heightened concern with secrecy, in 1941, the word 'Signals' was changed to 'Communications' and the name 'Special Communication Unit' (SCU) was used until the end of the war. I believe that, when the war ended, there were fourteen SCUs around the world. An example of the Daily Orders Part 1, Issue 20 of the embryonic unit, dated 25th July 1940 is shown opposite, issued under the name of Colonel R. Gambier-Parry as Commander. This would mean that the organisation began to pose as an army unit in early July, 1940. As can be seen, the following are shown as being 'enlisted at Whaddon':

G. E. (Ewart) Holden, who had overall responsibility for all wireless stores, inwards and despatch, for the units world-wide;

C. W. (Charlie) West in charge of the metal workshops making agents' wireless sets. J. (Joe) Hill in the forge where he made frames for the larger transmitters; Hector Kempton, a clever engineer; Hughie Castleman who later ran the factory at nearby Little Horwood making the increasing numbers of agents' wireless sets needed before 'D-Day';

Douggie Lax and Tommy Ord, engineers selected for their outstanding skills, were also posted to 'Workshops'. Two other names on that list dated 25th July, 1940 are also worthy of special mention.

The first is Ken Bromley who was engaged as a draughtsman and became the first artist to create the cartoons that illustrated the unit's house magazine, *Stable Gossip* (See Chapter 20). He went on to join the workshops that produced agents' wireless sets.

Ken had been seconded from an army unit and thus did not actually belong to Section VIII, although I believe he was one of those in the army whose pay was later enhanced somewhat. This was done so as not too arouse animosity amongst men doing the same job, side-by-side, on very different pay scales, but even so the pay differential remained fairly large.

The second is that of my father, H. E. C. Pidgeon – 'Pidge' as he was known to everybody from Gambier-Parry down. His first Christian name was Horace which he did not much like, preferring 'Horrie' instead, but he became 'Pidge' at Whaddon and the name stuck throughout his time there from 1940 to 1945. Needless to say, when I joined the unit I was also called 'Pidge' or 'Young Pidge' – and probably other, rather less pleasant names on occasion!

Father's posting is shown rather cryptically as 'H.Q. (S.O. Tech). At the time, this was the name for the wireless stores housed in the stables – '**S**tores **O**ffice **Tech**nical'. Later, the wireless stores were known as VIIIS.

You will see the initials 'N. Y. A'. against each name and that indicates that their army number was '**N**ot **Y**et **A**llotted'. Their numbers would have been issued to them shortly thereafter and it was possible to tell if a man was a 'Special Enlistment' from the first three figures of his army number. Note the names at the bottom of the page under 'Attachments', **258**0926 Sigm. Cullen and, below that, **258**5609 Sigm. Hooper. The first three figures show immediately that they were Special Enlistments and not paid from Army funds. In other words, this shows them to be part of MI6 Section VIII – i.e. SIS Communications.

Shown overleaf are 'Daily Orders Part 1' issue No. 292 dated 9 December, 1943, issued under the name of Lt. Col. G. Rooker R. Signals. This shows him as Commanding Special Communication Unit No 1'. (SCU 1). Under Item 1582, Enlistments – O. Rs. [Other ranks], the second name is N.Y.A. Rct. [Recruit] G. Pidgeon. He is shown as being specially enlisted in this unit on 30 November 1943 and posted to No. 2 Technical Group w.e.f. [With effect from] same date.

Although I had joined the unit in 1942 as a civilian, my own formal enlistment into the army unit occurred on the day I became old enough to be entered in the army lists – at the age of seventeen and a half.

DAILY ORDERS PART I
BY
LIEUT-COL. G. ROOKER, R. SIGNALS
COMMANDING, SPECIAL COMMUNICATIONS UNIT NO. 1.

Issue No. 292. 9 Dec 43.

1576. DUTIES AND ROUTINE.

	10 Dec 43	11 Dec 43
Duty Officer of the Day	To be detailed by No.7 SCU	Lieut. H.J. Castleman
Duty Warrant Officer	RSM. V.W. Robinson	CQMS. J.F. Brash
Orderly Sergeant	Corpl.N. Sutherland	To be detailed by No.7 SCU
Camp Piquet	Lo.Corpl. C.F. Anderson Sigmn. J.H. Cook and 2 men to be detailed by No. 7 SCU.	Lo. L/Cpl. A.G. Dudley Sigmn. E.H. Tyler and 2 men to be detailed by No. 7 SCU.

Night Fire Piquet - Period 13 Dec 43 to 19 Dec 43.
N.C.O. i/c to be detailed by Depot Group.
M.T. Group and Depot Group will each detail two men.
Two men to be detailed by No.7 SCU.

Reveille	0630	1st Works Parade	0900	Tea	1715-1800
Check Parade	0655	Dinner	1200-1215	Piquet Mounting	1800
Breakfast	0645-0800	Dinner	1230-1315	Tattoo	2200
Sick Parade.	0815	2nd Works Parade	1355	Roll Call	2230
P.T. & Drill	0830-0900	Tea	1630-1645	Lights Out	2245

Black-out times for 9 Dec 43 are from 1721 to 0824.

1577. BARRACK INSPECTION.
There will be a barrack inspection, (including Canteen, Cookhouse, Serjeants Mess and Tailor's Shop) at noon on Saturday, 11 Dec 43.

1578. POSTINGS - INTERNAL.
2344513 Serjt. (Loc. CQMS.) G. Laidler
is posted from Depot Group to No. 1 Special Operations Group w.e.f. 10 Dec 43.

1579. COMPULSORY EVACUATION OF HOMES IN S.W. ENGLAND.
Personnel whose families are being compelled to evacuate from S.W. England are entitled to compassionate leave with a free railway warrant in addition to normal leave and free warrant entitlement. In cases where personnel have paid their own fares refunds may be claimed at military rates.
Officers are entitled to additional free warrants for the above purpose up to and including the rank of Captain.

1580. AFRICA STAR RIBBON, EMBLEM AND CLASP.
All personnel who claim to hold the necessary qualifications to enable them to receive the Africa Star will hand in their number, rank, name and initials - Officers to the Adjutant and O.Rs. to the Orderly Room - not later than noon on Friday 10 Dec 43. This is in addition to their official application on AFB.2063.
Part I Order No. 1542 dated 2 Dec 43 refers.

1581. CANCELLATION - COURT OF INQUIRY.
Part I Order No. 1545 dated 2 Dec 43 is cancelled.

1582. ENLISTMENTS - O.Rs.
2602899 Sigmn. S.H. Cole
specially enlisted in this Unit on 26 Nov 43 and is posted to No.1 Operations Group w.e.f. same date.
N.Y.A. Rct. G. Pidgeon
specially enlisted in this Unit on 30 Nov 43 and is posted to No.2 Technical Group w.e.f. same date.

1583. DISCIPLINE - OUT OF BOUNDS.
No. 7 S.C.U. Lines are out of bounds to all O.Rs. of No. 1 S.C.U. except when on duty.

To Sheet Two............

Daily Orders Part 1. Special Communications Unit No.1. 9th December 1943 showing 'Special Enlistments' including my own.

'1 – Special Enlistments'
N.Y.A. Rct. Pidgeon G. [then my birth date] 27 May, 1926 – [followed by my medical category] A.1. **Not Paid from Army Funds.** It goes on to read: 'Enlisted and joined this unit at Whaddon, Bucks. on 30 Nov 43. A.F.E. 531 to Records herewith'. To further illustrate the point, my own army number, issued a few days later, was **260**2902, the **260** indicating my status as a Special Enlistment '**NPAF**'.

Incidentally, Lt. Col. George Rooker was a 'real' soldier, a regular army officer brought in by Gambier-Parry to ensure – so far as possible and without hindering our work – that correct army procedures were followed and that the unit appeared as closely as he could manage, to be a military establishment.

Those of us 'Not Paid Army Funds' were issued with the standard Army Pay Book (known as AB 64 Part 1) as proof of identity. It contained a record of service, listed training, the inoculations we had been given, names of relatives and one's last will and testament.

Most of us – and certainly those who had to travel to other units – had an insert in the book stating 'This man is on special duty and has permission to wear civilian clothes'. Overleaf there is a copy of two pages from my father's AB 64 Part 1. The insert is on the left-hand side, over the signature of George Rooker who may not have approved of granting yet another departure from army protocol, but was probably under orders to do so from the 'Boss', Richard Gambier-Parry.

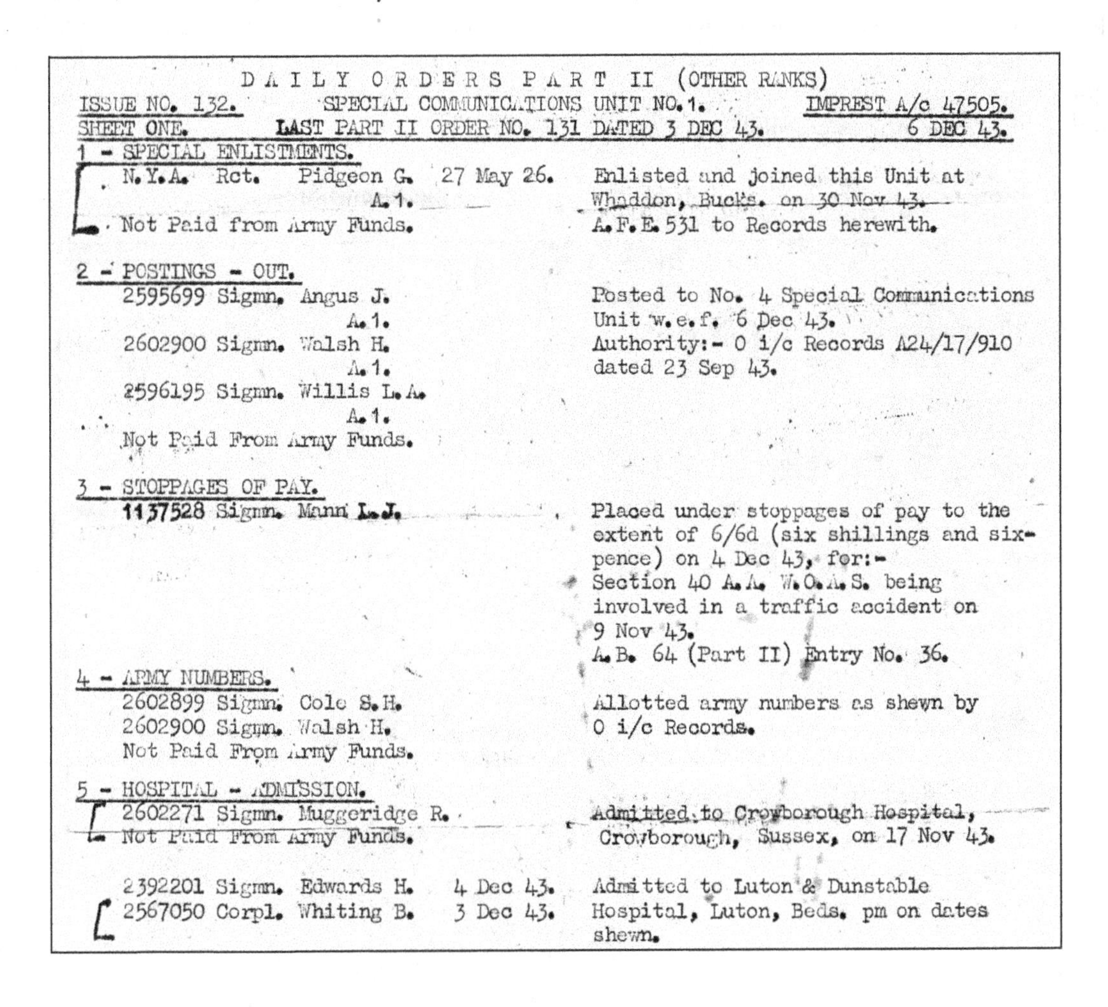
DAILY ORDERS PART II (OTHER RANKS)
ISSUE NO. 132. SPECIAL COMMUNICATIONS UNIT NO.1. IMPREST A/c 47505.
SHEET ONE. LAST PART II ORDER NO. 131 DATED 3 DEC 43. 6 DEC 43.

1 - SPECIAL ENLISTMENTS.
N.Y.A. Rct. Pidgeon G. 27 May 26. A.1. Not Paid from Army Funds. — Enlisted and joined this Unit at Whaddon, Bucks. on 30 Nov 43. A.F.E. 531 to Records herewith.

2 - POSTINGS - OUT.
2595699 Sigmn. Angus J. A.1.
2602900 Sigmn. Walsh H. A.1.
2596195 Sigmn. Willis L.A. A.1.
Not Paid From Army Funds.
— Posted to No. 4 Special Communications Unit w.e.f. 6 Dec 43. Authority:- O i/c Records A24/17/910 dated 23 Sep 43.

3 - STOPPAGES OF PAY.
1137528 Sigmn. Mann L.J. — Placed under stoppages of pay to the extent of 6/6d (six shillings and sixpence) on 4 Dec 43, for:- Section 40 A.A. W.O.A.S. being involved in a traffic accident on 9 Nov 43. A.B. 64 (Part II) Entry No. 36.

4 - ARMY NUMBERS.
2602899 Sigmn. Cole S.H.
2602900 Sigmn. Walsh H.
Not Paid From Army Funds.
— Allotted army numbers as shewn by O i/c Records.

5 - HOSPITAL - ADMISSION.
2602271 Sigmn. Muggeridge R. Not Paid From Army Funds. — Admitted to Crowborough Hospital, Crowborough, Sussex, on 17 Nov 43.

2392201 Sigmn. Edwards H. 4 Dec 43.
2567050 Corpl. Whiting B. 3 Dec 43.
— Admitted to Luton & Dunstable Hospital, Luton, Beds. pm on dates shewn.

Daily Orders Part II issue 132 dated 6th December 1943, showing my enlistment 'Not Paid army Funds'. The letters N.Y.A. refer to my number as being Not Yet Allotted. It was given shortly afterwards and was 2602902.

My AB64 had the same insert but unfortunately was lost whilst I was in Calcutta in 1945, celebrating my father's birthday on December 4th – perhaps rather too well – with a bunch of friends. That is another story and is certainly classified!

All real army personnel had another section called an AB64 Part 2, and this actually showed the owners the rates of pay and pay received. None of the Section VIII personnel were issued with AB64 Part 2.

Many of us working 'inside' Whaddon were Special Enlistments and we wore civilian clothes whenever possible. Sports jackets and flannels were fairly common amongst the Section VIII workshop staff, especially if we were working on Saturday mornings. Although allowed to wear civilian clothes, we usually wore uniform when away from Whaddon, especially when working with service units.

The casual approach to dress ended around the time of D-Day when the unit was put on full alert and everyone was required to carry firearms. This made us seem a little more like a military unit, with the exception of those whose uniforms simply did not fit, and were quite determined not to do anything about it! Of course, there were many who tried very hard to play the part properly throughout – some even revelled in it – and the occasional sight of polished boots and Sam Brown belts was not entirely uncommon.

However, the dress of many of the 'special enlisted' personnel was casual in the extreme and not at all military in appearance. One, I recall, often insisted on wearing grey suede gloves and matching shoes with his khaki uniform. His mode of dress looked particularly incongruous when we were later required to carry firearms. Try and picture this untidy soldier with long hair, grey suede shoes and gloves, carrying a rifle that seemed to have a mind of its own, as it pointed wildly in all directions!

As a silent protest against the low standard of military dress around him, my father seemed to go out of his

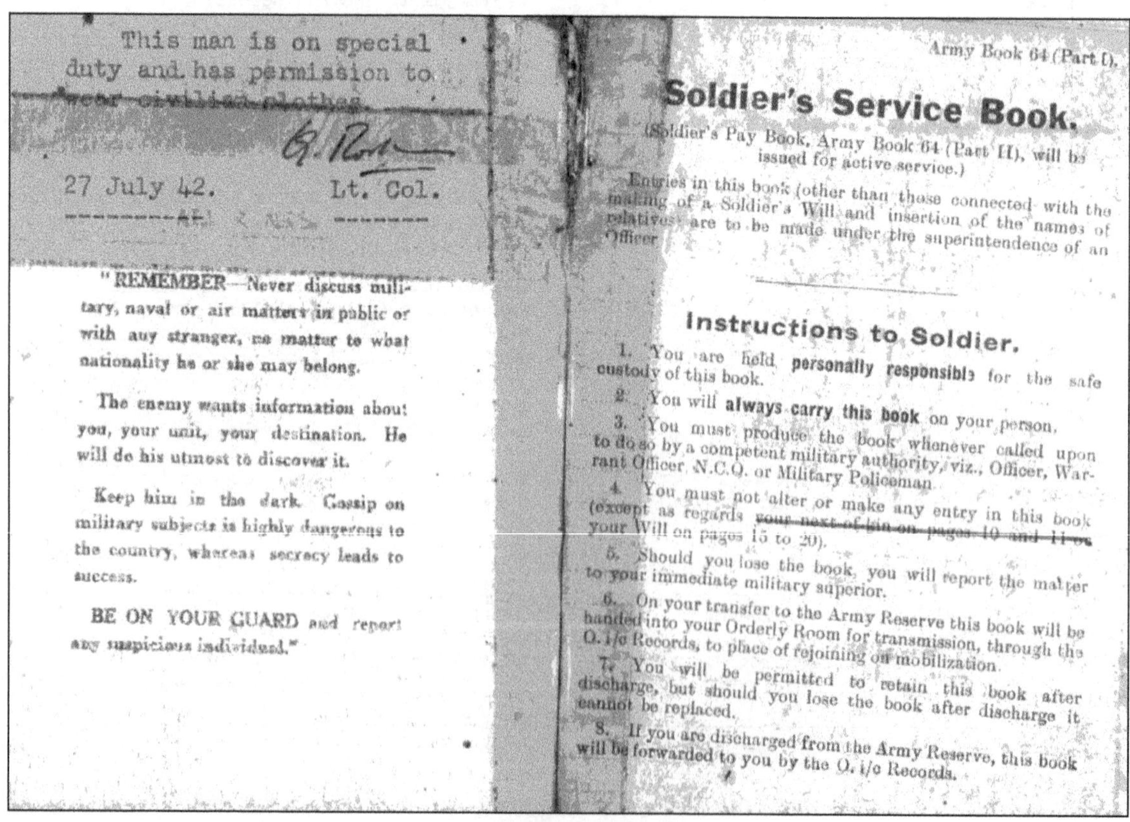

This man is on special duty and has permission to wear civilian clothes.

27 July 42. Lt. Col.

"REMEMBER—Never discuss military, naval or air matters in public or with any stranger, no matter to what nationality he or she may belong.

The enemy wants information about you, your unit, your destination. He will do his utmost to discover it.

Keep him in the dark. Gossip on military subjects is highly dangerous to the country, whereas secrecy leads to success.

BE ON YOUR GUARD and report any suspicious individual."

Army Book 64 (Part I).

Soldier's Service Book.

(Soldier's Pay Book, Army Book 64 (Part II), will be issued for active service.)

Entries in this book (other than those connected with the making of a Soldier's Will and insertion of the names of relatives) are to be made under the superintendence of an Officer

Instructions to Soldier.

1. You are held **personally responsible** for the safe custody of this book.
2. You will **always carry this book** on your person.
3. You must produce the book whenever called upon to do so by a competent military authority, viz., Officer, Warrant Officer, N.C.O. or Military Policeman.
4. You must not alter or make any entry in this book (except as regards ~~your next-of-kin on pages 10 and 11 or~~ your Will on pages 15 to 20).
5. Should you lose the book, you will report the matter to your immediate military superior.
6. On your transfer to the Army Reserve this book will be handed into your Orderly Room for transmission, through the O. i/c Records, to place of rejoining on mobilization.
7. You will be permitted to retain this book after discharge, but should you lose the book after discharge it cannot be replaced.
8. If you are discharged from the Army Reserve, this book will be forwarded to you by the O. i/c Records.

My father's AB64 Army Pay Book showing dispensation to allow him to wear civilian clothes due to his being 'on special duty'. Most of us in Section VIII had this prerogative.

way to appear as soldierly as possible and in his old-fashioned army peaked cap he certainly looked the part of the trained military man – even if he was not.

It is important to point out that in contrast to this seemingly laisser faire approach at Whaddon, the personnel entering army service in the 'real army' unit at Gees, whether as drivers, despatch riders, cooks, car mechanics or trainee wireless operators, all lived under a quite different regime. Gees was a military camp in every sense of the word, with the discipline and military order that goes with such a unit.

This later caused some resentment, as those entering the service in that way saw the easy lifestyle of those privileged to work inside the SCU units, and more importantly when a few discovered the large difference in pay scales. The tight discipline for those at Gees did not change, but the pay scales were adjusted with a small 'top up' to a few who worked alongside Section VIII men, doing the same job. However, this was certainly not universally applied. Due to the great secrecy surrounding us all, only a very few people found out that they were being paid less than the 'civilians'.

There is some confusion over the numbering of SCUs and SLUs. A SCU of one designation might well be working with a SLU with a different number. Before going further, it might be timely to explain the link between an SCU and an SLU.

The name 'Special Liaison Unit' (SLU), was given by Group Captain Fred. W. Winterbotham, to units handling Ultra traffic coming out of Bletchley Park. He had been charged by 'C' (then Colonel Stewart Menzies), to ensure the total secrecy of the information to be disseminated by Bletchley Park.

There were a number of fixed SLUs in the UK, at such strategic locations as the Admiralty, the War Office, the Air Ministry and RAF Fighter Command. These static SLUs usually received their information directly from Hut 3 in Bletchley Park via the teleprinter operators. However, most also had a mobile support team from SCU as a wireless backup, in case the ministry was put out of action or had to move in the event of an invasion.

The information from Hut 3 would, in that event, go out via a BP teleprinter to Section VIII's W/T station at Windy Ridge, and then to the SCU team at the designated SLU. The SCU team in the 'Dugout' in St. James' Park, supporting the War Office SLU, is an example of where this backup existed.

It is known that in 1940, a mobile SCU unit was sent from Whaddon to the Admiralty, as a mobile emergency support W/T station to receive traffic from Bletchley Park. (SSU1 Daily Orders Part 1. dated 25 July, 1940, Item 97 Attachments). The wireless communications at SLUs further afield were almost always run by SCU personnel.

Things were quite different for military commands abroad where the SLU clearly had to be mobile, moving with the army or air force commanders. When a mobile SLU was attached to a command, such as in North Africa, Sicily, Italy and, later, in the re-occupation of Europe, the unit consisted of two parts. The SLU cypher crew in one vehicle would be coupled with an SCU wireless van that actually handled the traffic. The cypher element was manned by RAF personnel and the SCU (wireless telegraphy) element by Royal Corps of Signals staff supplied exclusively from Whaddon or Little Horwood.

The first four SCUs (SCU1, 2, 3, 4) were not set up until after March, 1941 when Gambier-Parry took over the RSS (Radio Security Service), on behalf of SIS, 'lock stock and barrel' (formally on 7th March, 1941). Previously the pseudo-military units of Section VIII had been known as 'SSU' (Special Signals Units).

SSUl apparently became SCU1 and SSU2 (mobile installations, etc) became SCU2. The two Whaddon mobile units sent to France before May 1940 may have been designated SSU2, and there is evidence that

SCU2 existed throughout the period 1941-1946. In Germany, in 1945, SCU9 was discontinued, presumably becoming part of SCU2.

So, Whaddon was the first of the Special Communication Units but a whole series of them followed up to SCU14. The numbers and functions are, so far as can now be ascertained, as follows:

SCU 1 This was based in Whaddon Hall and was the HQ of the whole MI6 Section VIII operation during the war, handling all the communication requirements of SIS. One of its most important tasks was to forward the Ultra traffic from Bletchley Park to commanders in the field, and to bases that had no teleprinter connections. Most of this work was done through mobile units at local commander level.

Windy Ridge was set up in the autumn of 1939, specifically to handle Bletchley Park communications. It handled the traffic originally known as 'Boniface', known as 'Ultra' during the war. In the later stages of hostilities, the traffic was eighty per cent outgoing and twenty per cent incoming. It was destined mainly for Egypt and North Africa and later for the British return into Europe, Sicily, Italy and then D-Day and its aftermath.

Mobile SLU unit of SCU1 'A' Detachment with its wireless gear fitted in a Packard.

The first major unit sent abroad by Gambier-Parry specifically to handle Ultra traffic was known as 'A' Detachment. It was formed early in 1941 under the command of Major Kenneth Macfarlane, R.A. His second in command was Captain A C 'Jan' Ware. The engineer-in-charge was Lieutenant Brian Sall and the Admin. Officer was Norman Walton. They were based at Abassia Barracks, Cairo, and the majority of the operating personnel were recruited locally from Royal Signals and RAF.

All equipment for the detachment was supplied from Whaddon, including several Packard mobile units. These quickly proved unsuitable for off-road work and were replaced by stripped-down army wireless 'Gin Palaces' as offices, and Humber Super Snipe Estate cars, into which the Whaddon transmitting and receiving gear was fitted.

I believe the Abassia base station worked solely to the Special Operations Group at Windy Ridge, Whaddon and the mobile stations worked with Abassia. Most of 'A' Detachment returned to the UK, after the end of hostilities in North Africa. This detachment was clearly the forerunner of larger SCU/SLU mobile units in the field that culminated in the formation of SCU8, the large organisation required for the invasion of Europe.

There was an Agents' station that worked with the agents in North Africa, Asia Minor, Crete, Greece, Italy and Yugoslavia. It was located at Riverside adjacent to the British Embassy. Jock Adamson took it over from Charles Emary in late 1941 and he in turn was replaced by Dick Pott in late 1943.

SCU2 Pat Hawker and I decided that all we knew about SCU2 was that it was, *'An early mobile outstation of Whaddon sent to France before Dunkirk consisting of two vehicles'* and those words were entered into the final proof of this chapter. However, just in time, I have noticed the details of my 1945 movement order to New Delhi (see chapter 38) headed 'To whom it may concern' – shows that we were to report to **ISLD (SCU2 'B' Group) at GHQ New Delhi.** We went to this hugely impressive place then, I recall, on to the Palace of the Maharaja of Baroda, where we waited for a while. We were then transported to the SCU11/12 wireless station in the cantonment to the east of the city. So, seemingly without knowing it, I may have passed through this mysterious unit. I can only guess it was by that time a generic term for SCUs overseas, and they all belonged to it when abroad – perhaps for administration or finance reasons. I have no other explanation.

Humber Estate mobile SLU attached to Eighth Army HQ, and Tactical Air Force.

SCU3 Under Gambier-Parry, the military side of RSS (Radio Security Service), became SCU3 and SCU4, with Lt. Col. Ted Maltby as commander. The newly formed SCU3 was based partly at Hanslope Park but was largely controlled from Box 25. RSS Engineering under Major Keen was based at Hanslope Park and later, with an influx of REME personnel, it became concerned with the building of Rockex cypher machines, based on the work of Canadian Professor 'Pat' Bayly. Incidentally, Alan Turing worked at Hanslope Park in 1944, developing an on-line speech coding system (Delilah) which never became operational.

SCU3 soon controlled intercept stations at Forfar, Thurso, St Erth (Cornwall), Gilnakirk (near Belfast); fixed D/F stations at Thurso, Forfar, Bridgewater, Gilnakirk, St. Erth, Wymondham and Hanslope, and could draw on D/F facilities at Gateshead.

SCU4 This comprised the mobile D/F facilities in the UK and both intercept and D/F operations overseas, including a large base in Cairo.

SCU5 and **SCU6** remain obscure, although there is a reference to SCU5 in *The SLU Story* and it may have been part of SLU5/SCU5. SCU6 may have been the SIS unit originally established in Algiers after Operation Torch (the invasion of North Africa in 1942) and later moving to Bari, Italy, to handle SIS operations (setting the pattern for SCU9 for the Normandy campaign). There is, as yet, no firm evidence on this point.

SCU7 The training wing at Little Horwood was set up in late 1943 or early 1944. It was based originally in the stables and outbuildings of the country home of Mr. Gee of Gee, Walker, Slater Limited, major building contractors in pre-war London. The unit trained Morse operators to supply the increasing number of units and the stations that would be required in the invasion of Europe. Later in the war, with the number required still increasing, the recruits did their basic training at Ashbys, a large house on the London Road in Stony Stratford. They then moved on to Little Horwood for intensive Morse training that was given in huts in the 'SCU7 Lines', part of the established camp known universally as 'Gees'.

This was already the billet of all the army support staff for Whaddon, the many drivers, motor mechanics, up to fifty despatch-riders, cooks, tailors, medics, etc., needed to make the unit function like a real military establishment. Everything at 'Gees' however, was basically under SCU1 except SCU7, the training wing.

SCU8 The operational unit handling Ultra for the invasion of Europe from 'D' Day onwards. It was originally formed at Little Horwood, and one by one, the SLUs went to join major army and air force commands. All Ultra military traffic, intended for commanders in the field from Montgomery to Patton, and emanating from Bletchley Park, went through these SCU/SLUs, via the Whaddon wireless station at Windy Ridge.

SCU9 This was a small Section VIIIP unit (initially about 20 men of whom 12 were operators) headed by Major Tricker. Most of the operators were young recruits from the Weald or Nash stations. SCU9 was set up in April/May 1944 in readiness for the Normandy campaign. The unit was attached to No 2 Intelligence (Underground) Section [21(U) Sect] who provided the coding liaison with 21st Army Group, administration, etc.

The function of this unit was to provide a link between Whaddon (Weald) and the 21st Army Group to forward intelligence received from agents in France. It was very much involved in the large joint SIS/OSS/BCRA 'Sussex' operation which involved dropping some 50 two-man teams of French agents in a wide sweep from Brittany to the Belgium border.

The unit had two signals vehicles. The first, a large six-position vehicle (QL 4x4), was fitted out by Mobile Construction at Whaddon and equipped with HROs, MkIIIs and one MkX. The second was a single-position (15cwt Guy) vehicle with HRO and MkIII, also built at Whaddon, as a forward unit.

The unit operated from three places in Normandy after the D-Day landings, but by the end of August it had a forward unit in Paris, where it remained for six weeks. In early September, the main unit moved from Normandy to Brussels where it remained until May 1945 when it moved into Germany. It soon established

Inter Service Liaison Department (ISLD) mobile unit in the desert of North Africa. Standing second from right, Wing Commander 'Jock' Adamson, second left is Edgar Harrison. Kneeling far right is Norman Walton who was actually visiting from our 'A Detachment' where they handled Ultra traffic.

itself at Bad Salzuflen with single operator outstations in various parts of West Germany.

In October 1944, a forward unit was established in Eindhoven and single operators were allotted to intelligence officers in various places. In January, 1945 two operators (Lawley and Pat Hawker) were loaned to the Dutch Bureau of Intelligence (BI) at Eindhoven and remained there until the end of the war in Europe (May 1945). In about Autumn 1945, SCU9 and 21(U) Sect. became respectively SCU2 and 5 and 7CCU, and were employed on post-war SIS intelligence and counter-intelligence operations in Germany.

SCU11 I believe SCU11 was formed in Palestine and then sent to India as ISLD (SIS). It handled SIS traffic to India and beyond, relay work and the like.

Watson 'Bill' Peat working with SCU9 at 21st Army Group HQ after D-Day. He is using an HRO receiver and a MkII transmitter.

SCU12 This joined with SCU11 as **SCU11/12** with its HQ based in Calcutta, under Bill Sharpe. Its major role was working to agents across the broad sweep of Burma and south east China, as well as to the ISLD station at Kunming, being run by Tom Kennerley. Dave Williams and the late Bill Peat were at Calcutta where we were also working to SIS agents in Malaya. These two were preparing an SCU/SLU wireless van to accompany the full-scale invasion of Singapore which was aborted as the Japs surrendered.

I was in SCU11/12, first in Delhi and then, by the time the atomic bombs were dropped on Japan, I was down in Calcutta.

SCU13 and **SCU14** These are obscure, but I believe they were originally intended to be RSS and the SLU in Singapore from 1945 onwards, after its re-occupation. In the event, the SCU unit from Calcutta moved down to Singapore basically as the same organisation that had existed there, and in Delhi, under our boss Lieut. Col. Bill Sharpe.

I was given the task of organising the packing of most of our Calcutta station wireless equipment as it was closed down in 1946. Some of it went back to Whaddon but the bulk was taken to a warehouse on the quayside ready for shipment to Singapore. It arrived there soon after a party of us had settled into our quarters in the city, following our journey down from Calcutta. I had it unloaded at the docks, with the help of Jap prisoners of war and put into store. I am therefore unsure whether 13 & 14 ever functioned as such.

In Singapore, SIS had offices in the Cathay Building on its 8th and 9th floors, whilst the SCU living quarters and wireless receivers were in 12 and 13 Adam Park, off the Bukit Timah Road which lead to Johore Bahru. For a while, I believe we used the transmitters at the Far Eastern Broadcasting station on the Caldecott Estate, near the McRitchie reservoir although we later had MkXs which worked to our station at Delhi for onward transmission. The station at Delhi had remained open even though we had closed down Calcutta.

Finally, I must record there was an SCU/SLU wireless facility provided on General Eisenhower's train in his travels around the various military commands across Britain, shortly before D-Day. The coaches were apparently supplied by Great Western Railway (GWR) and after the war I believe were used as a reserve coaches for the Royal Train. Edgar Harrison, who travelled as Churchill's SCU operator on several journeys abroad, used one of the Whaddon agents suitcase sets to provide a wireless service for Churchill. Although I cannot be certain, I suggest that an agent's set might have been sufficient to provide General Eisenhower with a wireless facility for that short period within the UK. Any wireless operator would certainly have been from SCU.

F. W. Winterbotham describes in his book 'The Ultra Secret' how he was asked to furnish a secure Ultra connection for Churchill during his travels abroad. Apparently he received a polite message from Downing Street 'Pray make the necessary arrangements'. Thus Churchill always had a secure channel for Ultra, and for his own messages, provided by Gambier-Parry's Section VIII. Using only a low powered agent's set would mean that his wireless messages in the Middle East for example, would have been relayed on from our station at Cairo, through to Main Line at Whaddon.

General Montgomery also travelled extensively by train across Britain in the weeks before the invasion of Europe. He visited the British, Canadian and US forces that were to come under his command on D-Day. His four coach train was known as 'Rapier' and was effectively a travelling HQ and had a flat car with it to carry his Rolls Royce. Under the circumstances, it is highly likely that his HQ staff would have had a secure wireless channel on the train for him.

In 1946, Gambier-Parry's Section VIII/SCU organisation, that had served SIS and the country so well, was largely reformed by him as the Diplomatic Wireless Service, and initially took on most of its functions.

These subsequent changes, the engineering side of HMGCC at Hanslope, the building and activities of GCHQ at Cheltenham, and the other changes to the war time sections, are outside the story I have tried to tell here, which is largely confined to the war years – 1939/1945.

Chapter 10

SCU/SLUs – The Mobile Voice of Ultra

I originally started to write this chapter on SCU/SLUs myself, since I assisted in building a number of them. However, when John Lloyd sent me some notes on his work with the SLUs during the war, I recognised that here was the definitive story, both on the concept and on the operation of SCU/SLUs in the field. Nobody could do the job better so here it is with only minor alterations arising from later information.

The SCUs were the para-military organisations set up by Gambier-Parry to cover the many and various operations of MI6 (Section VIII), including the dissemination of Ultra traffic. That particular aspect of Section VIII's work was handled by Special Liaison Units (SLUs).

When Norway, Denmark, Holland and finally France fell with lightening speed to the German invasions of the spring of 1940, it became obvious that it was a waste of time examining intelligence gathered from the various sources available, once a battle had been lost. If the war was to be won, intelligence had to be gathered, collated, analysed and disseminated at great speed to commanders, so they could act upon it prior to and during the battle, whether on land, sea or in the air.

With the arrival of Ultra information through Bletchley Park's increasing success in decyphering the Enigma codes, it was necessary to form SLUs to enable the information obtained to be disseminated directly to the commander to whom it was relevant. It was decided from the start, that the information thus provided should be distributed only at the very highest level to protect the Ultra secret.

Such a unit had to be totally autonomous, protected by complete secrecy, supplied with the finest communications equipment possible and, above all, with highly qualified staff to operate it. Most importantly, it had to be beyond the reach of any officer, regardless of rank and status, other than the authorised recipient of the intelligence.

The code-breakers at Bletchley Park were in receipt of a mass of information from all directions. Intercepted German military traffic via the 'Y' service, Abwehr via the RSS at SCU3, information back from SIS agents in the field and from diplomatic links. That made BP the right place in which to evaluate the intelligence gleaned and decide on its usefulness to commanders before sending out the messages as Ultra traffic.

The sort of information that the commander in the field needed to know was the strength and disposition of the enemy, its logistical support and, at all times, the pre-battle movements of enemy forces, the battle instructions and aims of the various sub-commanders, their identity and tasks their forces were required to perform.

Intelligence-gathering had many other facets, including:

1. *Bletchley Park decrypts arising from 'Y' service interception.*

2. *Foreign Office and Ministry of Economic Warfare (Underground, SOE, MI6).*

3. *Battlefield radio interceptions: W/T forwarded to Bletchley Park by wireless links to RAF Church Green, MSR Station and R/T to local intelligence appreciation.*

4. *Intelligence – interrogation of prisoners.*

5. *Intelligence provided by army battlefield specialist units equipped with good communications rear link, in the shape of Phantom, Special Air Service (SAS) penetration, Intelligence and Reconnaissance Corps groups. Such information was either to be used in local appreciation, or once again rear-linked to Bletchley.*

All such intelligence was ultimately, and with great urgency, to be sent to Bletchley so that the wider picture could emerge.

An SLU consisted of two parts, the wireless service provided entirely by Section VIII at Whaddon and the cypher aspect provided by RAF personnel. Each SLU had one officer (with sometimes a deputy), who was totally responsible for passing the Ultra messages to the only person specified to receive them. According to circumstances prevailing at the time, commanders might be removed or added to the list of recipients. For instance, in the Ardennes Battle of the Bulge in late December, 1944 and early January 1945, the US General George Patton, commander of the Third Army, had a SLU attached to him during his famous hinge movement at Metz. This SLU was removed from him for a while, much to his disgust, by General Eisenhower, a move for which Patton probably blamed General Bradley at the time.

Each SLU was, with the exception of the one attached to SHAFE (Eisenhower's Headquarters), based upon a two-part unit which was required to be highly mobile within an army corps's operational theatre.

Of course, the SLUs varied in size, so that a large one might consist of twelve jeeps and trailers, one or more 15 cwt Guy wireless trucks and a cypher van staffed by between 35 and 50 men. (In the US theatre of operations the wireless vehicles were converted Dodge ambulances). The numbers of men involved might fall at times to 25 or 35 men, according to losses caused by sickness, postings for various reasons or enemy action.

A Dodge 'Ambulance' fitted out at Whaddon as a wireless van for use in American army areas

Each SLU was fully self-sufficient, being authorised to draw food, equipment, fuel and, most importantly, pay from any source.

The officer commanding an SLU was known as the Special Liaison Officer (SLO) usually with the rank of major. He was often in the Intelligence Corps, as was his second-in-command, a captain. His staff of specialists, officers, NCOs and men were naturally of the highest calibre. They included a number of translators of German, together with clerks involved in analysis and battlefield appreciation, and of course cypher and code experts. The cypher machines used for transmissions to

Bletchley Park, via Windy Ridge, were the keyboard-operated mechanical TYPEX, the British version of Enigma! The equipment also included a 110 Volt AC power supply in the form of the American Onan generator which was transported in the 15 cwt signals wagons.

Being a composite unit, the senior signals officer was usually a major in the Royal Corps of Signals, who together with his second-in-command, a captain, commanded what can only be described as a signal troop with every trade represented:

Drivers for all SLU vehicles
Motor-cycle despatch riders
(the Matchless 350 cc G3 L and the BSA 500 cc M20 were the motor-cycles used)
Motor transport mechanics

Wireless operators – telegraphy and radio telephony
Keyboard operators for TYPEX machines
Instrument mechanics to service the equipment
Stores clerks, technical and quartermaster
Clerks – signals
Wireless mechanics
Field telephone linesmen
Line mechanics
Cooks
Clerks – administration

A field telephone exchange was carried to the interface with normal Signal Corps, divisional and brigade telephone links if required. British Army No.19 sets and No.24 sets were used with various nets for radio, telephony, telegraphy links in similar nets plus the nets involving battlefield intelligence units.

A Guy 15cwt wireless van for use in British and Canadian operational zones, installed with Whaddon equipment.

Personal weapons carried were the Lee-Enfield .303 rifle and bayonet, the British Webley or US Smith and Wesson .38 calibre revolver and a 9 mm Sten machine carbine. For all round unit protection, the jeeps were fitted with a mixture of Potts 20 mm cannon, .303 Vickers 'K' machine guns on twin mountings and, in some cases, captured German MG 42 'Spandau' machine guns with their ammunition.

The wireless telegraphy equipment was the then 'Rolls Royce' of receivers, the US National HRO. The transmitters were the Whaddon-made MkIII, and Mk33 special, which were based on the time-honoured, amateur 6V6 Crystal Circuit with an 6L6 Doubler Circuit and 807 amplifier output. The Morse code was transmitted using a Marconi key. Power supplies were either from 110 volt Onan AC generators or batteries charged by the Tiny Tim engines generating 12 – 24 volts DC.

Each jeep and trailer carried its specified load together with a part-load in duplication of other vehicles, in case of loss of the first vehicle and load. Self-destruction explosive devices were fitted to both sensitive equipment and vehicles.

An ex-signaller who worked at SHAFE Headquarters reveals that the standard equipment included Creed Morse tape perforators, associated tape-head transmitters and Morse slip reader equipment for high-speed operation. All signals, with their decoding keys, were incinerated after use in a special high-temperature chamber.

John's experience was probably a little different to other SCU8 units. He was late to cross to France and he found himself attached to 2nd SAS brigade – a unit with a more 'frontline' role than the others – hence his reference to the important subject of personal weapons. See John's story in chapter 36.

Life in the SCUs was almost monastic. Fraternisation with other units, especially Signals, was strictly forbidden, for obvious reasons. Encampments, both temporary and semi-permanent, always had to be at a distance, hence the field telephones and field telephone exchange equipment. This gave the members of the SCUs the unfortunate reputation of being aloof and superior to others, although to young men in their late teens, such as John Lloyd, it was perhaps the most exciting period in their lives.

The crew of an SCU/SLU at Bad Kissingen showing the SCU8 number on its wing

Chapter 11

Whaddon Hall, its Layout, and The Chase

Bearing in mind the importance of Whaddon Hall to the story of MI6 Section VIII in wartime, I think it right to provide the reader with a plan of the grounds, the original buildings, the huts built in the grounds and their uses, and to mention some of the people who worked in them.

At my first meeting with David White, when he persuaded me to write about SCU1 and Whaddon Hall, he asked for a drawing of the layout. I did that from memory, and it is quite surprising how a man who now forgets quite simple daily things, could so accurately recall details of something that happened sixty years ago.

When, with the blessing of Richard and Pauline Windward, the present owners of Whaddon Hall, I was able to walk round its grounds, I found little wrong with my earlier sketch. One can only suppose the importance of our work, and the impact of the characters I worked for, made an indelible imprint on my mind. Nevertheless, I cannot claim 100% accuracy. This is the best I can now manage, with apologies for any errors that might come to light later.

The chapter is divided into two parts. The first is about the Hall itself and the second about 'The Chase' which lay within the Whaddon Hall security zone.

Whaddon Hall and its layout

Our roving electrician at the Hall was Percy Unwin whom I remember well. He rode a 1000 cc BSA motorcycle with sidecar, and was usually dressed in a full-length, leather overcoat, flying helmet and goggles. His tools were in the sidecar. On the pillion seat, hanging on tight, was his son Ron, who continued to work at Hanslope until his retirement a few years ago. As our general electrician, Percy was called in for anything from fitting a plug to rewiring a room, so he had more access than most of us to the various parts of the buildings.

Ron Unwin has been most helpful in confirming my sketches and I much appreciate his support. I think this is an appropriate place (see over) to show a picture of his father Percy astride his motorcycle – unfortunately not in the famous coat, so it must have been a very hot day.

The original buildings of Whaddon Hall are identified by letters, and the temporary huts in the grounds by numbers. Although most were of similar construction to the wooden huts of Bletchley Park, we also had some brick-built additions that have since been removed. The largest of these was alongside the wireless stores run by my father based in the stables. The new store was two storeys high and eventually quadrupled the capacity for the great expansion programme that started in late 1942.

The expansion included the workshops producing much larger quantities of agents' sets, fitting out mobile wireless units, and more and larger wireless stations at home and abroad. This required a stock of large quantities of wireless receivers and components of all kinds, from tiny resistors to screws, and including wireless valves of all shapes and sizes, generators, tools and much more besides.

The wooden huts that stood between the stable buildings and the outside road, were the workshops. Because of the slope from the rear arch of the stable yard (now closed off) to the main road, the huts were built on stilts that increased in height, the tallest being at the end of the second R&D (Research & Development) hut that incorporated Percy Cooper's office (16).

Percy Unwin, the unit's general electrician, on his 1000cc BSA motorcycle and side-car.

Aerial view of Whaddon Hall. All the buildings shown on the plan are identifiable.

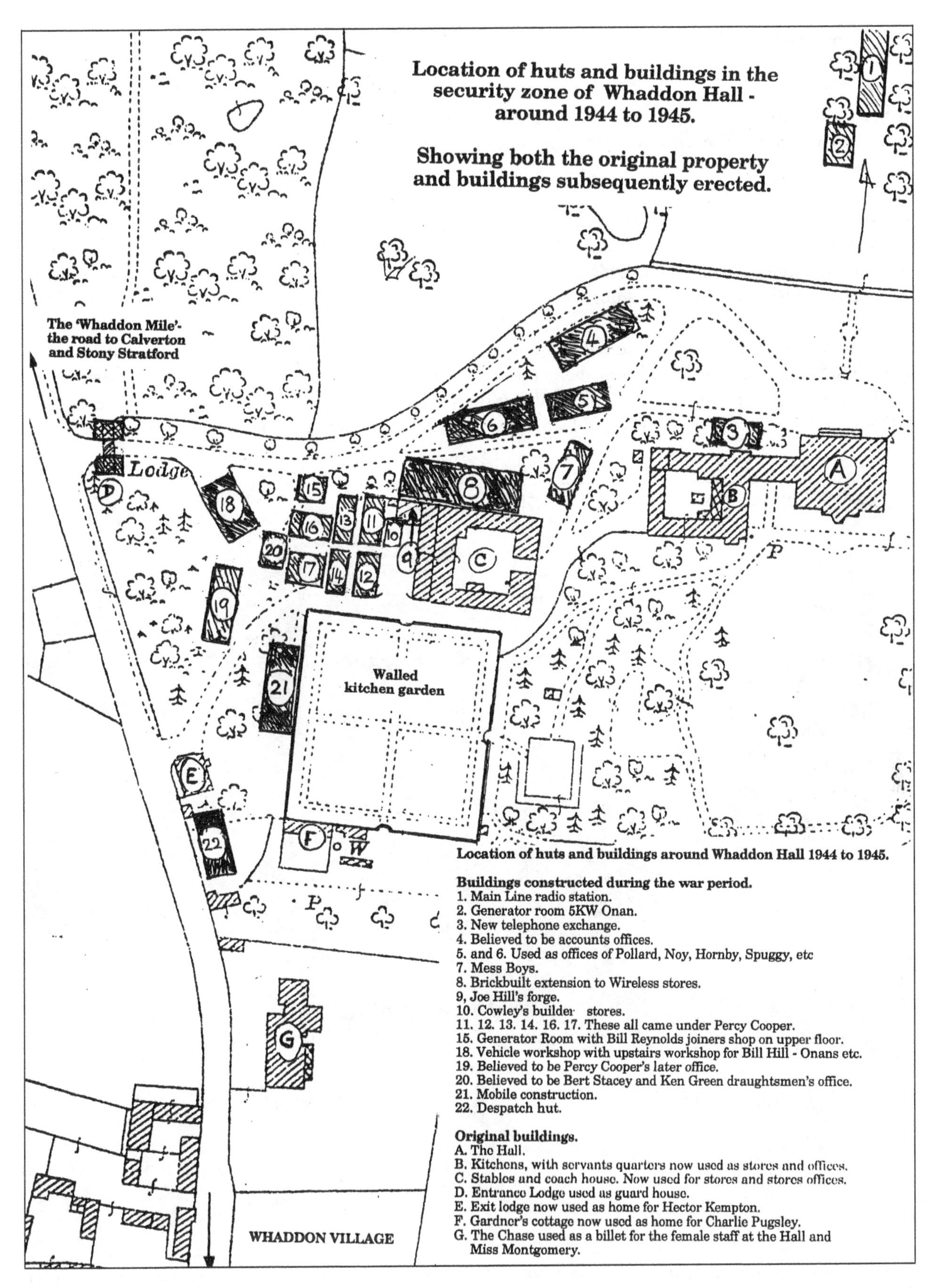

Location of huts and buildings in the security zone of Whaddon Hall around 1943 – 1944. Showing original property and buildings subsequently erected.

The Permanent Buildings

A. The Hall. This was the HQ of MI6 (Section VIII) – (SIS Communications), as well as the offices of SCU1. It housed the offices of, among others, Brigadier Gambier-Parry, Major Bill Sharp, Lt. Col. Russell and Lt. Col. Lord Sandhurst. The wireless receivers in some of the upper rooms were used for security work and minor tasks.

B. Kitchen and Servants' Quarters. The old servants quarters beside the kitchen, off the courtyard, were partially used as offices. The courtyard contained a central building (the former game store), which now housed our battery-charging unit. Offices surrounded the yard when I joined in 1942, but most of their staff were relocated when new brick buildings, such as huts 4, 5 and 6, were erected alongside the drive to the Hall.

C. Stables and Coach House. The stables housed the 'technical stores' that were run by my father. Again, brick buildings were later erected near the main drive to house the immense stock of wireless material. The stable yard contained a wooden building on the western side which housed the offices of Ewart Holden, my father's boss, who was in overall charge of stores and despatch. I am told that the actual stables, to the south

The library steps – one of the most used photographs of Whaddon but the names are seldom correctly shown. They are, left to right: Major 'Bill' Sharpe, Captain 'Pedlar' Palmer, Lt. Col. Lord Sandhurst, Ann Trapman on the steps, and Evelyn Watts. The dog is 'Whusker' and belonged to Lord Sandhurst.

side of the yard, had been used for the construction of wireless sets in the very earliest days, but by the time I joined in 1942, they were certainly being used for stores.

D. Entrance Lodge. This was used as the main guard house by armed military policemen. When I was there, the lodge had an ivy-covered arch between the two sections.

E. Exit Lodge. Occupied as his home by Hector Kempton.

F. Gardener's cottage. Now used as a home for Charlie Pugsley.

G. The Chase. The billet for most of the female secretarial staff at the Hall. Miss 'Monty' Montgomery, who ran the autonomous SIS cypher unit in Hut 10 at Bletchley Park, was also billeted there.

Jeannie Gorringe, Spuggy Newton and 'Fran' Seward at the back of The Chase.

Wartime buildings

1 Main Line wireless station. This was reached via steps directly in front of Whaddon Hall and handled 'Prodrome' traffic (Foreign Office Diplomatic) and 'Medal' traffic (MI6), amongst others.

2. Main Line generator building. With a 5kW Onan generator.

3. New brick-built telephone exchange, replacing one in the courtyard complex.

4. Wooden hut, believed to be the new accounts office.

5. & 6. Wooden huts, used as offices by Alec Pollard, Bob Hornby, Spuggy Newton, etc., with secretarial staff.

7. Wooden hut. Mess boys' accommodation.

8. Brick-built extension to the wireless store. Access was through the existing stable buildings.

9. Brick-built Forge. This housed Joe Hill and his forge and dealt with steel work for larger chassis and aerial equipment.

10. Wooden hut, used by Cowley & Son as their builders' hut.

11/12/13/14/16/17. A wooden hut complex known as 'The workshops'.

11. Metal workshops containing lathes, drills, presses, milling machines, a guillotine and workbenches along the side. Staffed by eight engineers. When I started work there in 1942, the benches from left to right were used by, Jack Bell, John Harding, me, Bert Norman, 'Nobby' Clarke, 'Polly' Perkins and Frank Franklin.

12. R&D, manned by Alfie Willis, Wilf Lilburn, Bob Chennells, Dennis Smith (until he formed his own unit called Mobile Construction), etc.

13. Contained Charlie West's office, engraving machines, drills, etc. I learned to use the engraving machine and cut the operating information labels (such as on/off) attached to the faces of various agents' sets such as the MkIII, MkV and the early versions of the MkVII. At the far end of the hut and with a separate entrance, there was the carpenter's workshop, under Bill Reynolds who was later succeeded by Bill Barnes.

14. Assembly of radio sets, radio engineering work of all kinds, under Hughie Castleman (until he took charge of the new Little Horwood 'factory') and Douggie Lax.

15. Brick-built generator room. Bill Reynolds had a new carpenter's shop on the floor above, entered via an external wooden staircase.

16. Lieut. Cdr. Percy Cooper's office (with his secretary Miss Marjorie Mainwaring) and a second R&D laboratory whose staff included Alec Durbin, Ted Turner, 'Mac' Hawkins, Steve Dorman, John (Tommy) Tucker, Bert Mason, etc.

17. Assembly of agent's sets before this operation was moved to the new workshops at Little Horwood. I think this was to give more space for R&D.

18. Brick-built vehicle workshop. Gossip has it that this was mainly used to enhance the condition of Bob Hornby and Spuggy Newton's Rover motor cars but I think other vehicles went on the ramp – occasionally! Upstairs, Bill Hill ran a workshop involved with Onan maintenance and it was also used by Jack Buckley on wireless repair.

19. Wooden hut. Believed to contain Percy Cooper's later suite of offices.

20. Wooden hut. I think this was the drawing office used by Bert Stacey and Ken Green, our draughtsmen, who were also the artists who like Ken Bromley embellished '*Stable Gossip*', our House Magazine. (See chapter 20).

21. Wooden hut. This was the workshop of Mobile Construction and stood near the exit gate. The hut contained an office for my boss, Dennis Smith, and was the workshop for his team of about eight. (I came back to Whaddon from the Little Horwood 'factory' to join this newly formed unit at the end of 1943).

The team included, Jock King, Tony Wheeler, Jock Denham, Norman Stanton, Wallace Harrison (brother of Edgar Harrison, see Chapter 31), me and two others, whose names I sadly cannot remember. We also had an excellent carpenter but again his name now escapes me.

We used this hut as our base for 'outside' work on aircraft, MTBs, and the like. We also created different types of SLUs (QL 4x4's, Guy 15 cwts, and Dodge ambulances), and this work took place in the immediate surrounding area, and alongside the building used for despatch (22).

There is no doubt that the creation of SLUs by Mobile Construction received the unit's highest priority, especially when work started on the SLUs for use in the invasion of Europe.

22. Wooden hut. This was the despatch hut which came under the control of Ewart Holden, who was in overall charge of all stores and despatch.

Entrance and exit gates.
These were guarded by the Military Police as was the security of the whole area of Whaddon Hall and its immediate grounds.

The Chase

This fine period house flanks Whaddon High Street, but opened at the rear into the grounds of Whaddon Hall. It was acquired as part of the arrangement with the Selby-Lowndes family and its wall helped to contain it within the security zone of the Hall itself. When first taken over, it was used by a number of senior personnel who had been in lodgings in Bletchley, whilst Section VIII was based at Bletchley Park. Now they came together in one place which saved travel and was convenient in view of the work being performed around the clock in the early days of 1940.

Evelyn Watts and Ann Trapman near the library steps of Whaddon Hall.

Later, I think by 1941, The Chase became the home of a number senior female staff working in the Hall itself. Most were officers in one of the services. Ann Trapman (later Ann Lee), Pru Clive and Fran Seward were all WAAF officers, as was 'Aunt Mary' who ran the place, and whose surname escapes me. Evelyn Watts and Jeannie Gorringe were ATS officers. An exception to the service appointments was Joyce Lilburn, who came from NID at the Admiralty, where she had been enlisted as Merchant Navy personnel (see Joyce Lilburn's story, Chapter 24). Another resident of The Chase was Miss 'Monty' Montgomery who ran the SIS cypher room at Bletchley Park which, by that time I believe, was already installed in part of Hut 10.

Those working in the workshops, or in the outside offices, would seldom see these ladies, as they went directly from the back gate of The Chase through the rear grounds of the Hall, and then into their offices.

Chapter 12

The Wireless Stations at Windy Ridge, Main Line, Nash, Weald and Forfar

This chapter tells of the outlying wireless stations in and around Whaddon. The two closest were Windy Ridge which handled Ultra military traffic and Main Line which handled Foreign Office, embassy and covert traffic.

The other SCU stations in the immediate neighbourhood were Nash and Weald, dealing solely with SIS agents. A number of colleagues have contributed to this chapter, notably Maurice Richardson, Stuart Hill, Ted Cooper and Alan Stuff who were at Windy Ridge; Ken Rymer and Dudley Bradford, who were at Main Line, and Pat Hawker and Jack Whitely, who were at Nash and Weald so this is an authentic record of the work of these vitally important wireless stations, written by the men who worked in them.

Windy Ridge

I must start by saying that whilst Windy Ridge had its 'finest hour' in providing our military commanders with Ultra information during the invasion of Europe in 1944 and 1945, it had been sending traffic of the very greatest importance and secrecy to many theatres of war, since 1940.

Windy Ridge became the wireless station of the SOG (Special Operations Group of Special Communications Unit 1). Its job, in 1944–1945, was to transmit the information provided by Bletchley Park to the British and American armies in the field after the D-Day landings. Each of the armies and air forces involved had a mobile wireless SCU/SLU (built at Whaddon) stationed at its headquarters; these were all manned by Royal Signals wireless operators, provided by SCU7 at Little Horwood.

Windy Ridge is the hill at the highest point in Whaddon, and the wireless huts were situated behind Whaddon church. Every hut had about twenty bays, each equipped with a National HRO receiver. The adjoining hut housed teleprinters connected to Bletchley Park. The receiving aerials were in the field near the huts, but the transmitters were in the Cotswolds, connected by landlines, and in fields at Tattenhoe Bare farm on the road to nearby Shenley. They were mainly 350 watt, with some 200 watt. A patchboard allowed each operating bay to be connected to any of the transmitters.

There were originally four watches. According to Maurice Richardson's recollection, each consisted of about twelve to fifteen signalmen, five or six airmen and a sailor. Most of them were posted there on completion of their army or air force signals training and consisted of those who performed best in the Morse code tests at the end of basic Morse training in the various army or air force units.

One assumes that Winston Churchill's determined support of the Ultra operation allowed the SCUs to cream off the best operators from wherever they could be found. The sailor, Peter Bishop, got into the unit because the entire crew of his ship were lost in the Mediterranean, whilst he was left behind in sick-bay. In a rare burst of compassion, he was posted to a non-naval job.

The watches were: (a) 0800 – 1300 and 1700 – 2359 (b) 1300 – 1700 and 2359 – 0800. At first the men had a long break every fourth day, but when the workload increased the watches were compressed into three, with no long break.

The unit took over the village hall in Whaddon, and most people slept in Nissen huts in the field behind. The concrete bases can still be seen behind the rear car park of the Lowndes Arms, Whaddon's only public house. Some of the foundations of the wireless huts are still visible up on Windy Ridge.

Ted Cooper who worked at Windy Ridge

The wireless traffic was always in cypher. The overwhelming majority was outward from Bletchley Park (BP), though some data was fed back from the armies, which helped to keep the operators skilled in receiving Morse. Maurice finds it difficult to remember the code names of the stations they contacted. Security in those days was very tight, one just didn't write down that sort of thing in diaries. The men knew the messages came from BP and that 'Intelligence' and possibly the Foreign Office were involved, but they never discussed the subject outside our circle, not even with the girlfriends they met at dances at BP. Many of them had passes to the BP Assembly Hall at Bletchley Park, to help to balance up the numbers in the predominantly female staff at BP.

Several operators can still recall the names of stations with which they worked. Just before the re-entry into Europe, there was traffic with names of chocolate – Fry, Cadbury, Terry and Rowntree, to name a few. The operators were not aware of the contents of the traffic, or where it was going, but it is now known to have been dummy traffic, going to the mobile SCU/SLUs taking up positions with the army and air force commands of the invasion force, as it prepared for D- Day. It was important that there was no sudden flare-up of activity to give the game away.

The operators were certainly aware that something was happening when some of the SCU/SLU stations closed down for a period, presumably at the point when they received orders to embark for the Normandy landings. The dummy 'chocolate' traffic ceased, and different call signs immediately came into use.

Maurice remembers Pack, Sybil, Atlas and Mermaid. Atlas and Mermaid were usually worked together, as most of their traffic was for both of them. This economised on operators, and only slightly slowed the process down by each having to ask in turn whether any repeats were required (usually after every 50 or 100 blocks of 5 letters).

Mermaid sticks in Maurice's mind because on Christmas Day, 1944, after his Christmas dinner, he sat down in the Atlas/Mermaid bay and was staggered to read in the log from the previous operator two Q codes which he had to look up. Their meaning was 'I am being bombed' and 'I have destroyed my cyphers'. This caused consternation, but the second one proved to be an error, even though the first was correct. He found out later that Mermaid was with the American First Army, which the Germans overran in their last major offensive, the Battle of the Bulge, during that Christmas.

Maurice recalls this as a largely happy time. 'All of us very worked very hard, and we were encouraged to use our skills in selecting transmit aerials and frequencies, so as to keep contact when conditions were bad. It was certainly difficult work; total concentration was essential in sending traffic that was clearly of great importance, even though its actual content was unknown for us.'

Maurice Richardson

In return for their dedication to the tiring work, the men were free of what the soldiers usually referred to as 'bull'. Dancing was a principal social activity for many of the men on just about every evening when they weren't on watch. Others, however, preferred the public houses in the local villages and so, on occasions, were not at their best when starting the night shift at midnight!

After the war in Europe ended, wireless traffic dwindled, and plans were made for a similar operation in the Far East. Kandy in Ceylon (now Sri Lanka) was to become the BP of the East. Six of the men from Windy Ridge set up a mobile station for the Fourteenth Army, ready for the assault on Malaya, but the Japanese surrender made it unnecessary. Maurice completed his service at Diplomatic Wireless Stations in Delhi and Singapore, where I first met him.

Alan Stuff's recollections included his own sketch of the layout and that provided by Stuart Hill has many similar features. I have therefore amalgamated them as best I could to produce a composite view. Whilst this is satisfactory I have found it difficult to line up the layout with the only known view of the place taken from above. This is a shot taken after the war, following an arson attack on the remaining buildings and some of the rubble and buildings have been cleared away. It is almost impossible to identify the various buildings but there at the back is the long Nissen hut that might have been the corridor shown in the drawings. The brick buildings in the foreground are not included in either sketch. The larger may have housed the generator, and the other possibly the battery room.

One ingenious device at Windy Ridge should be explained and recorded for posterity. There can be little doubt that this small wireless station probably handled the most important wireless traffic of World War II. Messages needed to be transferred between the two separate huts containing the operating bays and the teleprinters. They were placed on wheeled trolleys that were shot down covered, sloping wooden chutes. To make sure that no precious messages were not lost in the chute, a house brick was placed on top of the bits of paper to ensure their safe passage through the chute. A hi-tech solution!

Alan Stuff also recalls the lonely walk on dark winter nights past the cemetery of Whaddon's ancient church, into the field leading to Windy Ridge station. It was described as feeling 'quite spooky' by one old hand and Alan felt the same way.

The Windy Ridge huts were heated by cast iron stoves whose chimneys pierced the corrugated roof. The places were cold in winter and unbearably hot under the steel roof in summer. The cold and the damp are impressed on the memory of Stuart Hill. He worked there with other T/P operators – such as Norman Stephenson, known as 'Rocket', and W/T operators like Dave Peddie from Scotland, John Eccles and Stuart Jones, among others. Stuart has fond memories of the NAAFI behind the Lowndes Arms in Whaddon Village, presided over by a landlord called Harry. Stuart had come into the unit the hard way, via various army postings, until someone spotted his Morse ability. He was then whisked away secretly, like so many others, and sent on a journey to Bletchley railway station, 'Where you will be met'.

Main Line Station

This was based at Whaddon Hall from January 1940. The station was a collection of buildings directly in front of the Hall and reached by a flight of steps down near the drive, then via a path across the field.

Primarily, Main Line handled 'Prodrome' traffic (to embassies and certain of our overseas missions) and 'Medal' traffic, to covert stations abroad, often under the aegis of Passport Control Officers (PCOs).

Windy Ridge taken from the church tower after the war when the main buildings had sadly been vandalised. However, the layout can be seen in conjunction with the adjoining plan.

Exceptionally, Main Line worked to military missions overseas and to a few selected agents. What is not widely known is that it also handled some Ultra traffic to specific stations, such as Delhi, Kandy, Calcutta, the SLU at Brisbane, etc.

Major Jack Saunders was in charge of the Main Line, Captain C F (Charlie) Bradford was his No. 2 and Lieut. Claude Herbert his No 3. After Jack Saunders' tragic death, Charlie Bradford took over and was promoted to major.

Main Line wireless station opposite Whaddon Hall

The transmitters for Main Line were initially in a small building called the Dower House, which stood in the fields. There is now a group of houses called Briary View there, near the junction with the road to Little Horwood. The aerial arrays were in the fields behind and leading towards Whaddon Hall. Later, the aerials were resited at Creslow and a much larger transmitting station was built, completely replacing the Dower House.

Weald and Nash

These stations were both within a two or three miles of Whaddon Hall, and only handled SIS own agents' traffic.

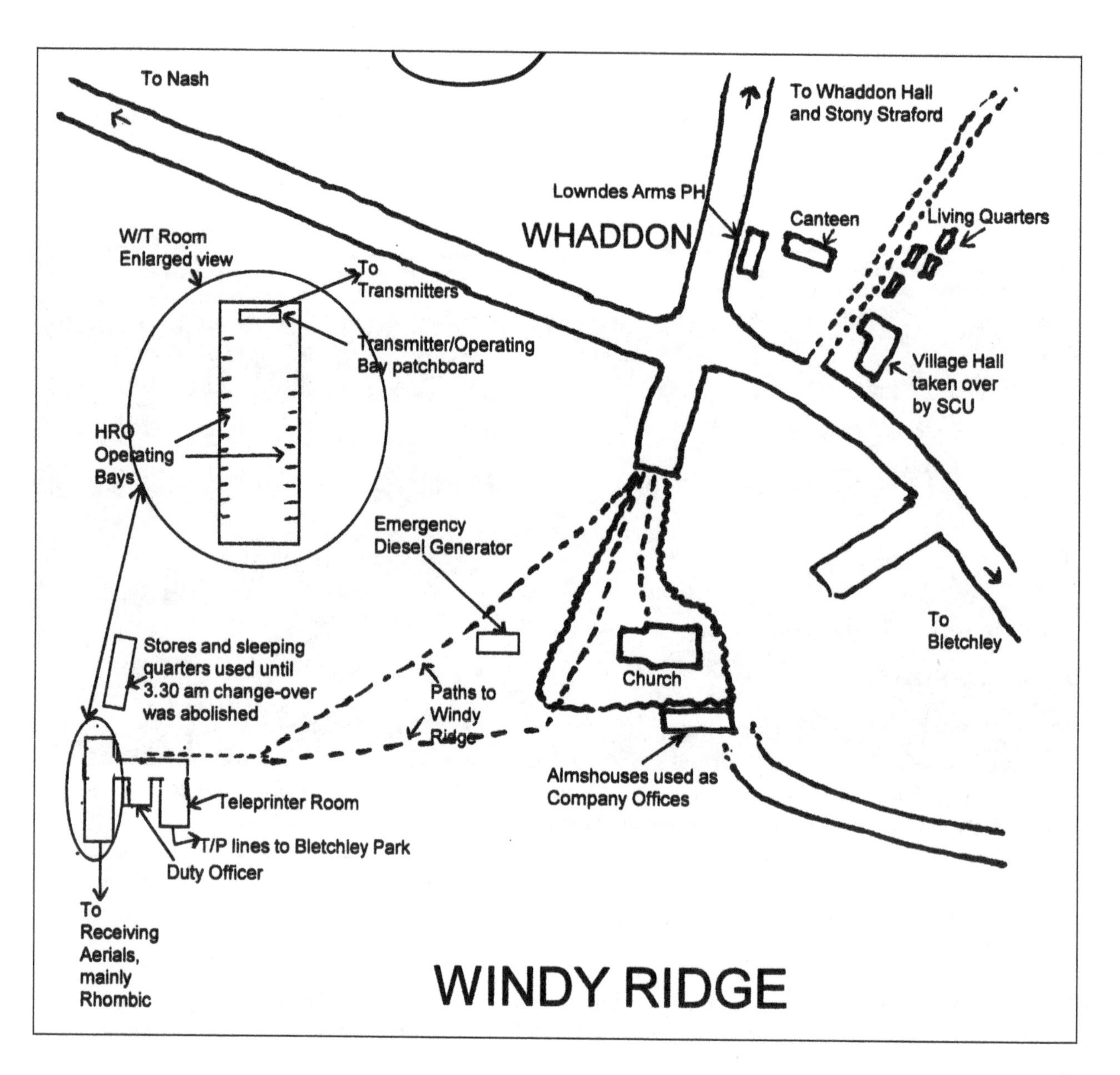

The layout of Windy Ridge showing its internal detail and the relationship to the village. Drawn by Maurice Richardson, an operator there.

Weald Station was at Upper Weald, Calverton, on the road between this small hamlet, and the then busy A5 Watling Street. The station was in a field to the south of the road and consisted of a pair of huts and a brick-built generator shed in which batteries were also charged. The operation of the station is described in some detail by Pat Hawker who spent a year there as an operator.(see his story in Chapter 24). The transmitters and aerial arrays were sited in the village of Calverton near Fountain's Farm, where there was another brick-built shed to house a generator. This transmitter site was run, in very great secrecy, by the US Civilian Technical Corps. The site also transmitted the outgoing traffic from the other SIS station at Nash.

When I lived in the district, I thought of Calverton as a village divided into three parts, whose names reflected their position and relative height. The largest part, nearest to Stony Stratford, was Calverton and it contained a pub,The Shoulder of Mutton, which was very popular with Whaddon personnel. It had a shop-cum-post office, a church whose vicar was the Reverend Ravenscroft, and Fountain's Farm in the centre of the village.

Away from the village, in the direction of Whaddon, stood a collection of houses called Middle Weald. Then, at a higher level, perhaps half a mile further on, one came to Upper Weald which was slightly larger collection of houses, and a farm owned by the Goss family. Middle Weald and Upper Weald were considered as extensions of Calverton, and their names were often combined, as in Calverton Middle Weald.

Nash Station was a short distance from the larger village of Nash which had two pubs, two churches, a school and one or two shops. The station was on high ground above the village and several of my colleagues were operators there. Nash station, the operating staff and the work done there, is well described in Jack Whitely's story, (see Chapter 33).

In addition to Weald and Nash, Section VIII had a station at Forfar, in Scotland, which was originally built as a sister station to Hanslope (SCU3) but was for interception only. The Norway agents' traffic was originally handled from Nash and/or Weald but the signal strength from the small agents' sets (MkVIIs) became a problem. It was therefore decided to install an entirely separate wireless station at Forfar, alongside the SCU3 unit there, to handle traffic from Scandinavia, which was entirely controlled by SCU1.

Inside Weald wireless station, Upper Weald Calverton

The generator building at Weald today

Nash wireless station handling European agents

Chapter 13

Whaddon's Multiple Roles

There is a misconception I have frequently heard expressed that Section VIII was a 'satellite' of Bletchley Park. Nothing could be further from the truth. Whaddon was a key player in the Secret Wireless World in its own right; the dissemination of Ultra by the 'Y' Service and the code-breakers at Bletchley Park – though vitally important – was by no means its only major undertaking.

Under Richard Gambier-Parry, Section VIII expanded rapidly not only in terms of personnel but in the many new responsibilities that he was always eager to accept, being regarded by some at Broadway as 'a bit of a pirate'.

The following is a list of just some of the tasks that Gambier-Parry agreed to perform on behalf of Whaddon, in a simplified form. I believe readers will be surprised at the sheer variety of the work he willingly accepted, and mostly performed successfully.

1. 'Main Line' wireless station was just a short distance in front Whaddon Hall, and it provided communication with our embassies and legations, throught the world. Main Line handled our diplomatic and covert traffic, including those arising in neutral counties. All woreless traffic for 'London" went through this station – including that for SIS HQ in Broadway – and wireless messages to and from Winston Churchill, when he was abroad.

2. The design, construction and operation of wireless stations at Nash, Weald and Forfar, to deal specifically with the increasing number of SIS agents in various regions of Europe entirely occupied by German forces.

3. The interception of Abwehr and other German Enigma networks initially by Arkley, then in October 1941 by Hanslope, (SCU3) then by main Hanslope (from May 1942). This work was also done by outlying stations such as Forfar, St Erth, Thurso and Gilnakirk.

4. The safe dissemination of 'Ultra' miltiary traffic via Windy Ridge at Whaddon, arising from the codebreakers at Bletchley Park. Something like 80% of Windy Ridge traffic was outgoing, with about only 20% incoming.

5. The wireless engineering division. This designed and built sets for use by agents and by British embassies. Its R & D departments created such special devices such as 'Ascension'.

6. The separate stores supply operation. This provided parts and equipment worldwide, either in 'raw' form as components for wireless sets, or as complete stations. It distributed completed wireless sets for reception and/ or transmission, as well as generators, chargers, etc. to a multiplicity of users.

7. The provision of skilled wireless operators to embassies and other strategic sites abroad.

8. The training of MI6 agents by Dave Bremner (later by Bert Gillies and others), in the proper use of our own wireless equipment, before being sent on operations. This took place at 23, Hans Place in Knightsbridge, west London.

9. The training (or retraining) of wireless operators to work on all the other wartime activities, such as Ultra traffic, Main Line and the SLUs being sent to North Africa and elsewhere. In particular, the training of the large numbers of operators eventually required for entry into occupied Europe as SCU8.

10. All messages transmitted to SIS agents were sent back to Whaddon to be examined en clair. This was in case any radio information had been leaked or breach of signal security had occurred. Once the messages had been checked, they were burned in the basement boiler of Whaddon Hall.

11. There was also the take-over in 1941 'lock, stock and barrel' of the Radio Security Service to which Gambier-Parry appointed his deputy, Ted Maltby, as Controller. Maltby then became responsible for the building of the special RSS intercept stations at Hanslope Park and Forfar, and its D/F network. Arkley, although nominally under Maltby, was actually run on a daily basis by Kenneth Moreton-Evans.

12. The making of W/T Plans for SIS agents to follow under VIIIP.

13. Monitoring an agent's 'fist' on a regular basis.

14. The design and construction of the wireless stations used by the operation known as 'black propaganda'. The technical staff for these stations was also supplied by Section VIII.

15. Providing the early wireless sets for the 'Hidden Army' of trained radio operators to be left behind in the event of the Germans invading and occupying the British Isles.

16. The provision of secure communications for Winston Churchill in his travels during the war.

17. Special communications with the United States via the Oshawa wireless link, worked by Main Line at Whaddon and also from 'E' block at Bletchley Park. The Rockex cypher link was controlled by SCU3.

18. The attempt, code-named 'Silent Minute', to deflect the V2 rocket attack on London by jamming the rocket's guidance system by high-powered transmitters sited at Crowborough.

19. Installation of British equipment in MFVs, special duty MTBs and aircraft. The building of the fixed SLUs in the UK and abroad.

20. Design and construction of mobile SLUs, from the very first one which took the form of a Dodge car sent to the British Expeditionary Force (BEF) in France, through to the fleet of Packards, SLUs for North Africa and Italy, and eventually the Dodge Ambulances and Guy vans sent as SCU8 units into Europe after D-Day.

21. The design, building and operation of wireless stations for Ultra and/or covert work, in Delhi, Calcutta, Kandy (Ceylon), Brisbane, Melbourne, Cairo and other stations outside.

And much more besides.

I hope this list, though far from exhaustive, will ensure that Whaddon will no longer be referred to by some writers as a 'Satellite' of Bletchley Park. Instead MI6 Section VIII will be recognised as a major player in the whole Secret Wireless War, on its own account.

Chapter 14

Clandestine Wireless and Agents' Sets

This chapter was largely written by Pat Hawker, with some additional material by David White and myself.

Pat Hawker was a wartime member of several of Section VIII's Special Communications Units – SCU1, SCU2, SCU3 and SCU9, from 1941 to 1946. He was born-in Minehead, Somerset, in 1922. His a lifelong interest in wireless began as a schoolboy; he obtained an 'artificial aerial' licence (2BUH) in 1936 and a transmitting licence (GVA) in 1938.

Early in 1940, Pat was recruited into the Radio Security Service as a 'Voluntary Interceptor' (VI) leading, in 1941, to enlistment for Special Duties in the Special Communications Units (SCUs), nominally part of the Royal Corps of Signals, but in reality Section VIII of SIS/MI6. He worked at Hanslope Park intercept station as a radio operator until 1946, covering German military intelligence (Abwehr) networks. He also worked at Weald (control station for clandestine radio links with France, etc) during this period; and he was in Europe from 1944-46, including a spell on loan to the Dutch Bureau of Intelligence as the war in Europe ended.

After the war, Pat Hawker became a highly respected technical journalist on wireless affairs. He has authored and edited books on this subject and still writes for several radio journals. His story is told in chapter 34.

By 1939, both the USSR and Germany had pioneered or prepared for the use of covert intelligence and diplomatic high frequency (HF) wireless links. Despite the early establishment of Station X and, in 1938, the setting up of Section VIII (communications) of the Secret Intelligence Service (SIS) under Richard Gambier-Parry, Great Britain lagged behind in this field.

In 1940 and 1941, as European countries fell successively under German occupation, there arose an urgent need to establish and then to improve wireless links with allied intelligence agents and resistance groups, leading to the design and assembly in the UK of fully portable W/T and R/T equipment by Section VIII, and later by SOE. The Polish and Czech fighters with the allies needed to be trained in the use of this equipment, as did agents who would be infiltrated into occupied territory.

When wireless telegraphy was first introduced, however, not everyone believed that it was of any strategic use. In 1903, an American admiral is on record as saying: 'Wireless is totally unsuited for war. The enemy could either hear all conversations, or could jam transmissions so nothing can be heard'. As late as 1939, another admiral said: 'As far as sinking a ship with a bomb is concerned, you just can't do it.' And yet another, commenting on the atomic bomb project in 1945, said: 'This is the biggest fool thing we have ever done. The bomb will never go off, and I speak as an expert in explosives'.

There was, of course, more than a grain of truth in dismissing the use of wireless in war. Signals intelligence, based on the interception of enemy traffic, proved a major source of intelligence in World War I and undoubtedly became the major source of intelligence in World War II. It was the decisive factor, not only in gathering information, but also in counter-Intelligence and in monitoring the success or otherwise of deception through the use of double agents.

The idea that spies might be able to use secret transmitters first emerged more than ninety years ago. William Le Queux, a now-forgotten writer of thriller and espionage romances, as popular in his day as John Le Carré is now, became interested in wireless telegraphy and used his knowledge of the wireless in his many books. By 1908, the possibility of a German invasion of Britain began to be mooted. Le Queux campaigned in the popular newspapers, claiming that Germany was flooding Britain with spies who planned to use wireless to transmit secret information to the German invasion fleet from houses along the East Coast.

With hindsight, it can be seen that this was pure fiction, worthy of his more far-fetched novels. But the campaign was taken up by other writers and attracted much attention, to such an extent that it led directly to the setting up of a new Secret Service Bureau largely independent of the Intelligence sections of the Armed Forces. Intelligence was soon split into two sections, one dealing with Home Security, the other devoted to Foreign Intelligence. Thus, the scene was set for what became the Secret Intelligence Service (SIS). Eventually, the MI5 Security Service was set up to support the Special Branch of the police which had been established in the late nineteenth century to combat Irish terrorism.

During World War I, signals intelligence was under the control of the Director of Naval Intelligence who established the highly successful 'Room 40' code-breaking operation. An important intercept and direction-finding station was set up at Hunstanton, Norfolk after two prominent pre-war experimenters – Hippersley, a Somerset landowner and Russell Clarke, a barrister – convinced the Admiralty that it was possible to receive the traffic of German naval vessels at far greater distances than previously believed. Engineering of the intercept and D/F stations was by H.J. Round of the Marconi Company. Leslie Lambert, a temporary Naval Officer, later joined Hippersley at Hunstanton.

The value of signals intelligence was proved beyond doubt when Hunstanton and Room 40 warned of the sailing of the German fleet, resulting in the major naval battle of World War I, off Jutland, in 1915. But even before 1914, at least two countries in Europe had recognised the value of secretly listening to the radio transmissions of other countries – Austro-Hungary and France.

It was Germany, however, which, almost by chance, scored the first major military success that can be directly attributed to signals intelligence. In 1914, Germany had no wireless-intercept service as such but by August 1914 the High Command had instructed the military radio stations of their telegraph troop to monitor Russian wireless communications in order to be able to jam them in case of war.

In the opening days of the war, the Koenigsberg station was able to provide the German Eighth Army in East Prussia with details of the disposition and Order of Battle of the attacking Russian forces leading, at the end of August 1914, to a major German victory at the Battle of Tannenberg. The Russians quickly learned an important lesson and issued an order that, in future, all military orders were to be transmitted only in encoded form. By the end of 1915, they were even monitoring their own wireless transmissions to ensure good wireless security.

This was a lesson that the British were forced to learn after it was discovered how much information Rommel's wireless intercept team was giving him, aided by the B-Dienst intercepts of the voluminous reports sent by the American Military Attaché in Cairo in a code broken by the Germans.

In the current and fully justified praises of the work of the code-breakers at Bletchley Park, it is often forgotten how successful the B-Dienst was in reading British GBMS (British Merchant Navy codes), and GBXZ (Royal Navy Codes), in the early years of World War II – with disastrous results for our shipping.

The use of wireless as a means of deceiving the enemy began to be used in World War 1. The date of Allenby's advance on Jerusalem was disguised by arranging for two apparently 'chatty' wireless operators to suggest that he was going on leave a day or two before the date of the planned attack.

'Deception' has been defined by Dr. R. V. Jones as making the enemy: *'...think you are somewhere else. Your weapons are different. You intend to do it at a different time. You intend to do it in a different manner. Your knowledge is either greater or less than it really is. His operations are more or less successful than they really are'.*

Despite the forecasts of William Le Queux, there is little evidence of any significant use in World War I of covert wireless behind the lines. However, in his history of the Marconi Company, Bill Baker writes: *'By 1916 all three armed services were depending heavily upon wireless. Paratroops were first employed in 1916 when Belgian soldiers, who in civilian life had been marine wireless officers, were asked to volunteer for special duties. After parachute instruction and a period of training by Marconi's, these men were dropped into enemy territory with small sets manufactured by the Company strapped to their backs, their task being to transmit intelligence from behind the German lines'.*

No information exists on the design of the sets, nor how many of the 'paratroops' survived either the drop by the crude 'chutes of the time, or the work behind the lines. One famous World War I spy, Mata Hari (Gertrud Margarete Zelle), the Dutch exotic dancer, was convicted largely on the basis of intercepted wireless messages between Spain and Germany. By curious coincidence, one of the group of musicians with whom she performed was the father of the World War II SOE radio agent, Noor Inayat Khan ('Madeleine'), who tragically died in Dachau after being betrayed to the Germans by a jealous woman.

The possibility of German clandestine wireless was taken seriously. On 3 August, 1914, as a security measure, the British government called for the immediate closing of all amateur experimental wireless stations, mostly spark transmitters and crystal-detector receivers: 'Remove at once your aerial wires and dismantle your apparatus'. Soon the apparatus was sealed up, and in the majority of cases removed by Post Office officials – a process that happened again 25 years later in September 1939.

In 1914, there was even a prohibition on the publication of magazine articles which might encourage the illegal construction of wireless equipment – receiving as well as transmitting. The Defence of the Realm Acts (the infamous DORA regulations), in addition to introducing the licensing laws, stated: 'No person shall, without the written permission of the Postmaster General buy, sell or have in his possession or under his control any apparatus for the sending or receiving of messages by wireless telegraphy, or any apparatus intended to be used as a component part of such apparatus'.

The spy scare that swept Britain had unfortunate results for several amateur experimenters: suspicious neighbours, recalling the sounds of past wireless operation, denounced the innocent experimenters as members of the German Secret Service. Police searches followed, often with the result that small pieces of forgotten apparatus – sometimes having only the most tenuous connection with wireless telegraphy – would be uncovered, and prosecutions followed.

In the prevailing atmosphere of suspicion, little official attention was paid (for at least 25 years), to a letter that René Klein, Honorary Secretary of the Wireless Society of London, wrote to The Times in 1915, suggesting that the authorities should enroll the Society's members to watch out for illicit transmissions.

It was the discovery in the early 1920s by amateur experimenters, and independently by Marconi and his chief assistant Franklin, that long distances could be covered at low power by using the previously despised short waves (wavelengths below 200 metres), that opened the way to effective covert radio. One possibility was to provide emergency links to overseas embassies, taking advantage of their extra-territorial status to set up wireless stations with or without the formality of obtaining the authority to do so from the host country.

It was not until the Vienna Convention on Diplomatic Relations, 1961 (Paragraph 1 of Article 27) that diplomatic missions were formally authorised to install and use wireless transmitters, and even then, it was only to be with the consent of the host State. For years, many countries, including Britain, maintained transmitters in their embassies often without formal permission.

Throughout World War II, such stations were, in effect, broadcasting semi-clandestinely. In the case of Great Britain, they and their operators were supplied by the communications section – Section VIII – of the Special Intelligence Service. The wartime operator in the Dublin embassy had to double as a butler! At the same time, the British government put pressure on the Irish to close down the transmitter in the German embassy.

In 1919, Room 40 became the Government Code and Cypher School (GC&CS). It was initially part of the Admiralty, but in 1921 transferred to the Foreign Office, Alaistair Denniston remaining its head. It had the task of all British code-making and code-breaking and was funded from the Foreign Office vote. Denniston answered to Admiral Sinclair, the then Chief of the Secret Service, and its training school was located alongside the SIS at 54, Broadway Buildings, SW1.

The small staff included one wireless expert, Leslie Lambert. Service intercept stations in various parts of the world copied the service traffic of target countries, concentrating on the USSR and later on Italy and Germany. The Metropolitan Police Wireless Station in Camberwell, south London, monitored diplomatic traffic from embassies and other foreign missions based in London.

During the 1920s, GC&CS developed a reluctance to disclose its decrypts to its political masters. This followed revelations in the House of Commons during the mid-1920s that information had been uncovered that could have come only from breaking Soviet codes. The result was that the USSR adopted the unbreakable one-time-pad code and GC&CS lost the ability to read their most secret traffic.

Before World War II, the Foreign Office did not set up its own HF wireless network but there is evidence that by 1925, SIS, in collaboration with GC&CS, was experimenting with the use of transportable HF stations. A London schoolboy amateur named Stanley Lewer (callsign G6LJ) broadcast a 'TEST!' call ('CQ' was banned by the Postmaster General), at about 00.45 on 8 December,1925. He was called by a station 'GB1' and asked to forward an urgent encoded message to the Foreign Office coding office. He agreed, but having no telephone in the house, had to wait until morning, cycle to school and obtain permission from his headmaster to telephone the message!

The following morning he received a letter from H. E. Eastwood thanking him. This was followed by correspondence and eventually a visit and small gift from an 'A J Allenby' whose letters were posted from Clarendon Road, Notting Hill. For over 66 years, the incident remained a mystery to Stan Lewer. He eventually wrote an account of the events for the journal of the Radio Amateurs Old Timers' Association. When Pat Hawker read this, he was immediately able to identify 'A J Allenby' as Leslie Lambert, not only from his using the same initials as 'A J Alan' but also from his address in Notting Hill.

There can be do doubt that GB1 was an experimental Secret Service station in the Middle East, testing the use of BF wireless. The results must have been satisfactory since within a few years SIS had established a radio station at Barnes known as 'Station X', under the administration of Harold Eastwood and Henry Maine. A prominent DX-er of the day, a man called Secretan, was Chief Operator. Previously Secretan had traded in radio components, using the slogan SEC SELDOM SLEEPS.

The first outstation was in Peking, China. One of the operators at that station was Edgar Harrison In the autumn of 1933, Edgar had been selected to be a member of His Britannic Majesty's Foreign Office W/T station (F.O.W/T) at the British Legation, Peking (see Chapter 31). Later, there were links with Spain during the Spanish Civil War where the operator was Charles Emary who was to be the Section VIII chief instructor at the radio-agent training school at Grendon Underwood, before this became an entirely SOE operation.

In the early 1930s, SIS did not have the funds to build up a large network or to supply the overseas Passport Control Officers or agents with wirelesses. The USSR was soon using HF wireless, however, to keep in touch with the NKVD and GRU agents. There was even a special school outside Moscow where agents were given a thorough training not only in Morse but also on how to build their own transmitters from designs published in the amateur radio handbooks. The graduates of this school included Max Klausen, who later handled communications for the important Russian spy, Richard Sorge, in China and Japan.

The first organised production of portable and suitcase sets specifically for agents and intelligence out-stations, was in Germany. In 1936, the Abwehr (German Military Intelligence), began setting up a Secret Wireless Reporting Service (Geheimen Funkmeldedienst), to collect Intelligence from many countries. It was based at Berlin-Stahnsdorf and the building also had a wireless station used during the war to communicate with Abwehr offices in occupied and neutral countries. Wireless stations for communication with agents in Western Europe and North and South America and Africa were set up in Hamburg, and near Wiesbaden. After the Austrian Anschluss, another main control station was set up, in Vienna, which was linked by teleprinter to Berlin.

Even before the outbreak of war, the Abwehr had networks of clandestine wireless stations in Czechoslovakia, Poland and France. In peacetime, like the Russian and other European networks, they came on the air only for occasional tests. At least one transmitter/receiver was brought into the UK, probably in the diplomatic bag, and deposited at Victoria Station. It was intended for use by a Welshman named Arthur Owens.

Owens was an electrical engineer who had supplied information to the Abwehr and had been trained as a spy in Hamburg. He had also been in touch with British Intelligence. His true allegiance thus appeared doubtful and he was thrown into prison at the outbreak of war. His transmitter was retrieved and handed over to a Voluntary Interceptor for the Radio Security Service – a prison warder who had formerly been a Service operator – to make contact with Hamburg. This was the start of the long-lasting Double Cross operation which ran right through the war and accounted for all the Abwehr radio links with the UK.

The Double Cross operation would have been far less successful had not the RSS, controlled after 1941 by SIS, been intercepting Abwehr stations and passing the traffic on to Bletchley Park (BP). BP was soon reading the German Abwehr and SD traffic (ISOS, ISK and ISOSICLE) and passing the decrypts to SIS Section V. RSS was greatly expanded after its takeover by SIS and by late 1942, some 150 German stations were being monitored by RSS/SCU3.

By the late summer of 1940, three types of Abwehr spy radios were in the hands of the British. These were: a low-power battery-operated transmitter, used without a receiver by reconnaissance agents; a 15-watt transmitter-receiver; and a Philco broadcast receiver adapted as a transmitter. From 1936 onwards, the Abwehr developed about 100 different models, with transmitters ranging in power from about 2 watts up to 250 watts. Like the British, the Abwehr design team gradually took advantage of the smaller and more rugged valves and components.

When Gambier-Parry started in MI6 the decision was immediately made that Section VIII would build its own low-powered communications transmitters rather than rely on the industry, which might easily show too much curiosity about their intended use. The building by the Thames at Barnes proved too small to house an assembly line, so an empty shop in Barnes was obtained for what were practically hand-built units.

The MkIII 'Tinker Box' with wood casing

The first MkI transmitters were intended for the SIS outstations working under diplomatic cover in the European embassies. These were single-stage power crystal oscillators using the tritet arrangement, with a built-in AC power unit. They were based largely on designs to be found in the amateur radio journals and handbooks. RF output was about 25 watts. Since the case consisted of a wooden box, they became known as 'Tinker Boxes', later emerging as the MkIIIs that were so widely used by Section VIII throughout the war. The Section VIII versions mostly took the form of a two-stage 6V6-807 transmitter with a separately boxed power supply unit. These transmitters were generally used in conjunction with a high-performance American receiver, such as the National HRO.

In 1939, one of the early Barnes transmitters was supplied to the Czech Intelligence station that had been set up in West Dulwich, following the flight from Prague, organised by SIS, of a dozen senior Czech intelligence officers on 14 March, 1939, the day before the Germans marched into the city. Other equipment for this station was procured through Reg Adams, then at Webb's wireless shop in Soho.

In the first months of World War II, the Barnes station X moved, first to Bletchley Park and then early in 1940 to Whaddon Hall, the wartime HQ of Section VIII. SIS had entered World War II with no secret wireless agents in place and no lightweight equipment suitable for parachute drops. The first attempt to establish a secret wireless link with German 'dissidents' ended in disaster. This was not the fault of wireless operation but of an offshoot of what became known as the Venlo incident. The German Security Service, the Sicherheitsdienst (SD), had thoroughly penetrated the SIS Dutch operations – both the Passport Control

Office and the Z-Network under commercial cover – and was able to pass off their own agents as representing a faction of German Army officers opposed to the Hitler regime. At an early meeting; one of the new Section VIII transmitters was handed over. But at a further meeting at Venlo on the Dutch-German border, two SIS officers, Stevens and Best, were seized and a Dutch liaison officer killed in a shoot-out.

The SD not only obtained the transmitter but also extracted, or at least obtained, confirmation of much information about SIS. The effect of the Venlo incident was that henceforward that SIS thereafter mistrusted all attempts by the genuine German Resistance to establish contact, although valuable information was later obtained from from senior Abwehr officers before the organisation was absorbed into the Nazi RSHA.

Admiral Canaris, head of the Abwehr, was a fervent Roman Catholic and became strongly opposed to Hitler after personally witnessing atrocities committed in Poland, even supplying masses of information to the British. He made personal contact with the SIS Jade network in Paris, supplying most of his information through a Polish woman in Switzerland from where it was collected and regularly brought to London, by an SIS officer, Philip Keun, via the secret SIS landing strip near Paris. Canaris was executed by the SS on 9 April, 1945, only two weeks before the liberation of Flossenburg Concentration Camp where he was being held.

Almost as frustrating as the Venlo Incident was the attempt to provide 'stay behind' wireless links with the Belgian 'Clarence' network under Walter Dewe. One of a number of transmitters supplied to the Belgians – an AC/DC version – is on show at the Imperial War Museum in London although the accompanying description of its use is, according to Pat Hawker, open to doubt. The transmitter, in a box similar to those used for many Barnes and Whaddon transmitters, had no tuning controls and would seem to have been an early version of the aperiodic Pierce crystal oscillator technique that reappeared much later as the Mark 14 and Mark 24.

An extraordinary Signal Plan was provided that was intended to allow messages to be passed by an agent having only a modicum of Morse ability. The call-sign was a single letter to be repeated for four minutes, no less! The message in figure code was then to be sent blind. Acknowledgement of receipt was to be broadcast in a disguised fashion over a Section VIII transmitter – presumably the 10kW transmitter installed at North Cerney, near Cirencester which was ostensibly the 'Radio des Beaux-Arts', allegedly in Brussels. After the German occupation of Belgium, arrangements were made to ensure that when the electricity supply in Brussels was cut, as it often was, Radio des Beaux-Arts would go off the air.

It is unlikely that the early Clarence group ever succeeded in passing messages by means of this system, although trained radio-operators with two-way equipment were later dropped into Belgium and the Clarence group became an important intelligence asset. Unhappily, Walter Dewe, its leader, was eventually shot while resisting arrest.

The Whaddon MkIII transmitter and an early heavy suitcase transmitter-receiver, the Mark V, were used by SIS and SOE as, agents' sets in 1940–41. MkIIIs were supplied to the Polish–French Inter-Allied group via Lisbon and Madrid and smuggled into France by a French naval officer. By 1941 if not before, the Whaddon MkVII, suitable for parachute delivery, and thus often called the Paraset, became available. It was the simplest possible transmitter–receiver, a 6V6 crystal powered oscillator and two 6SK7 valves as regenerative detector and audio amplifier. With no 'fine tuning' control, the receiver was not easy to use by anyone lacking 'safe-breaker's fingers'. Nevertheless, the MkVII in wooden or metal containers, remained in use throughout the war and was used successfully by many agents.

It should be recognised that human problems can be more important than technology in determining the success or failure of a radio-agent infiltrated or parachuted into enemy territory. He or she may sometimes succeed in the mission, remaining at liberty for months or even years, and passing hundreds of cypher messages. That is the good news. In many instances, however, it is a case of the bad news. The agent may

MkVII agent's set in its early form in a wood frame. The MkVII/2 was called the 'cash box' version being in a metal frame, and also known as the 'Para Set'. The one illustrated is a MkVII and probably made at Little Horwood. It is shown housed in a small attache case – a true 'Suit-case agent's set'. The case would have been issued by father's wireless stores at Whaddon, and this particular one ended up with ISLD in Kunming with Tom Kennerley. It came into my hands back in Calcutta where I appropriated it as a camera case. I used it for over fifty years before I presented it to David White, and it is now on show in Hut 1 at Bletchley Park.

believe that he or she is succeeding when in fact he or she is operating under covert surveillance or transmitting messages fed by an enemy who may have already penetrated the network.

The agent may be quickly captured, possibly even met on arrival by an enemy-organised reception party. In a condition of shock, he or she may then volunteer or be persuaded to act as a controlled agent, possibly disclosing the security checks. The agent may already be a double or triple agent, using this means of returning to his or her true masters. He or she may reach his or her destination but then be unable to make wireless contact, possibly due to faulty or damaged equipment, inexperience, loss of nerve or loss of the crystals, or he or she may even have been given the wrong crystals for his signal plan.

It is not surprising that it was difficult to find experienced operators in the occupied countries willing to volunteer for this work. Pre-war amateurs in France, for example, cited the listing of their names and addresses in the Amateur Wireless call books, believing that this would make them an object of suspicion to the Germans. Some good operators were found, but it became increasingly necessary to provide training schools in the UK both for SIS and SOE agents.

Pictures taken inside R&D workshop at Whaddon showing the typical hut construction of our workshops. The benches were covered in thick linoleum, but the main problem was lack of heating in winter!

By the end of the war, good training in clandestine operation was being given, including exercises in which the trainees would have to make contact with training-stations over various distances. Unlike the Germans, the British did not run training links over their operational control stations. RSS had shown that if this was done radio-agents might be identified even before they became operational.

The lengthy business of training W/T operators encouraged Section VIII to develop R/T equipment that could be used by an agent to talk to an aircraft over much greater distances than was possible using the 350MHz S-phone. This equipment was developed in 1941 by SOE, primarily for use by reception parties. It came to have wider application for SOE and other units for shore-to-ship communications, etc.

Steve Dorman, a young engineer working at Whaddon, set about designing a low-power transmitter under the generic name of Ascension, to be used for this type of ground-to-air system. The first Ascension was thrown together in a few weeks and got as far as trials using Section VIII's own Avro Anson test aircraft, flown by Squadron Leader Maurice Whinney. The project was suddenly abandoned, however, possibly because it was realised how vulnerable it would be to interception. In any case, the intended range was only five to ten miles, and this requirement was thus largely met by the S-phone.

A second version of Ascension became the responsibility of Wilf Lilburn (see Chapter 36) and Alfie Willis, who were based in one of the R&D huts in the workshop complex built in the grounds of Whaddon Hall. This version. the MkIX, was designed for much longer ranges at frequencies of 30 to 35MHz and used on FM bands with American 10.7MHz wideband IF transformers. The transmitter power was quite high, about 25 watts. From 1942 onwards, this system was used quite widely in France and Belgium, the agent on the ground talking to a circling aircraft at distances up to 100–150 miles.

The early Ascension model was fitted into a few Douglas (A20) Havoc aircraft (also known as the Boston in the RAF). It was relatively successful, considering the restricted range of the wireless, although it was later withdrawn from service. The second version was installed in some of the Lockheed Venturas based at Tempsford. Later, a squadron (or possibly two squadrons) of North American (B25) Mitchell bombers were installed with Ascension and flew from Hartford Bridge in Surrey. The aerodrome is now called Blackbushe.

The aircraft flew off-centre relative to the agent. For the Normandy campaign of 1944, Ascension equipment, as well as MkVII RF W/T equipment, was used in the major 'Sussex' intelligence operation run jointly by SIS, the American OSS and the French BCRA. Some fifty two-man teams were dropped in a wide sweep from Brittany to the Franco-Belgian border.

Ascension was also useful for short missions where an agent was to spend just a few days or weeks in enemy territory without receiving Morse training. Ascension is believed to have been the only operational use of FM by the British in World War II.

Unfortunately, none of the MkIX equipment or even the full technical circuit details appear to have survived the wholesale destruction of Whaddon equipment after the war.

Electrical power for clandestine equipment was always a problem. Some depended on AC mains supplies although this imposed extra hazards. Some depended on 6V vehicle batteries in conjunction with electro-mechanical vibrator units or dynamotors for higher powers. A few Onan petrol-electric generators were deployed in the field, despite their bulk and noise. Many batteries were charged using stationary bicycles with generators clamped to the pedal-driven wheel. More exotic devices were also used including steam-driven generators and thermo-generators that could be fired by charcoal braziers.

Equipment for use by wireless agents was built in Great Britain by a number of organisations, including at Whaddon for the SIS and the French BCRA. From 1941, SOE used different resources. John Brown at The

Fryth, Welwyn, designed suitcase equipment for SOE much of which was manufactured by such firms as Marconi and Philco (GB). Some 7000 B-2 transmitter-receivers were made at an SOE factory at Stonebridge Park in north-west London.

Chief designer Tadusz Heftman of the Polish Radio Centre Workshops at Stanmore was responsible for a series of excellent transmitter–receivers, later augmented by Monitor Radio in Birmingham. Later, some of these Polish designs were produced by Monitor Radio at Birmingham who also made a modified Polish design, the MR3, for the SOE.

In 1941, the Czechs were supplied by SOE with Whaddon MkV transmitter–receivers and MkIII transmitters, some of which accompanied the assassination team which disposed of Heydrich in March 1942 on their long flight to Czechoslovakia (the assassination brought frightful reprisals on Czech civilians in its wake, including the destruction of Lidice). Later, the Czechs built a 60-watt transmitter in Britain for their clandestine operations.

The American OSS brought a range of clandestine equipment into Europe, including the SSTR-1 suitcase set and, in the final months of the war in Europe, the Joan–Eleanor hand-held sets. Like the S–phones these were UHF but were intended, like Ascension, for agents reporting to aircraft equipped with wire recorders.

Was all this activity worthwhile? In recent years, many doubts on the effectiveness of the French Resistance have been expressed, some of them justified, some less so. The German penetration of SOE networks delivered to the Germans the broadcast invasion warning messages of 1 June and the 'imminent landings' messages of 5 June, although fortunately the German High Command made only limited use of these warnings. 'King Kong', perhaps the most successful German double-agent, may not have betrayed the Arnhem landings, as if often believed, but he certainly warned the Germans that large-scale airborne landings in southern Holland were imminent.

Brigadier Gubbins once described wireless as 'the most valuable link in the whole of our chain of operations...without these links we would have been groping in the dark'. But doubts inevitably remain.

Wilhem F. Flicke claims that wireless agents, appearing for the first time in World War II, decisively influenced the entire course of the struggle. The USSR, Germany and England were the first to begin using radio agents, even before World War II. Germany was the first to employ wireless agents for purely military reconnaissance, first in the Polish campaign, and then on a much larger scale in the campaign in Western Europe. In the 1940 campaign, everything the intercept intelligence service could not supply, was reported at the time by wireless agents, dropped from German planes behind French lines.

But if Germany was the first to make use of agents they also laid the foundation for its military defeat. In the USSR alone, from June 1941 to June 1944, some 10,000 wireless operators were trained to work in German-occupied territory. About one-third of those committed to these operations succeeded in carrying on their work for varying lengths of time. The western powers did not use masses of operators in enemy territory, but in general the intelligence-gathering and standard of the agents were higher.

Flicke completed his typescript from the memory of his cryptographic work in the Funkabwehr (Radio Defence), covering in considerable detail his impressions of the Secret Wireless War. It was completed shortly after the end of World War II, but the manuscript was purchased by the American NSA, who suppressed it. It remained secret until a translation was released in the mid-1970s under the Freedom of Information Act.

Flicke shows considerable understanding and knowledge of the wartime clandestine wireless networks, although some details are difficult to reconcile with Western accounts and personal memories. He clearly shows the importance of the Polish contribution to British Intelligence in France and Belgium, and also the excellence of Russian cryptography. There is little evidence for his claim that the Germans routinely 'read' such British agent cyphers as the double transposition cyphers with keys based on poems (LMT cypher) or his belief that the entire SIS-controlled Jade network in France was broken up in January 1944. Despite losses, Jade remained an important intelligence network until the liberation of Paris.

Today, it is virtually impossible to compile a fully detailed account of the work of Section VIII or the Signals Directorate of SOE, and the situation is unlikely to change in future, even if more and more archival material from British Intelligence finds its way to the Public Records Office. Almost all the records of Section VIII and the Signals Directorate of SOE were pulped many years ago. Memories are fallible and can become distorted or exaggerated.

Malcolm Muggeridge, the broadcaster, journalist and wartime SIS officer, once wrote:

'Intelligence agents tend to be even bigger liars than journalists, and are given to exaggerating their achievements, as well as the importance of their opposite numbers in order to magnify the feat of getting the better of them. The manifold stratagems and knavish tricks they recount played little if any part in the war's final outcome, which might well have been pretty much the same if there had been no Abwehr, no MI5 or M16, and no OSS.'

Personally, I am not sure that this view can be supported, considering the wealth of information we have about the successes resulting from Ultra alone, quite apart from the known accomplishments of other wartime secret services, and the many brave agents involved in them. But then I seldom agreed with Malcolm Muggeridge anyway. However, the following statement by Pat Hawker, with which he chose to end his story of clandestine wireless, receives my wholehearted support:

'One cannot but admire without reservation the secret wireless agents who dared to tap out messages from enemy-occupied territory or, as in many occupied countries, secretly built equipment for others to use, and those who sheltered or assisted them. We remain in their debt.'

Chapter 15

Box 25 – The RSS and Hanslope

One of the most intriguing stories of the Secret Wireless War is the work of the Voluntary Interceptors (VIs) and the organisation that was built up around them that lead to the formation of the Radio Security Service. This later became part of Gambier-Parry's Section VIII and intercepted some of the most important of Germany's intelligence communications. This traffic was fed to the code-breakers and was of immense significance to those responsible for the conduct of the war.

BOX 25: A most secret address!

On 6 September, 1940 a lone German aircraft crossed the East Anglian coast and headed inland. Shortly afterwards, a parachute floated to earth near Aylesbury and a spy landed in a field, knocked temporarily unconscious by the wireless set which followed him down. He was soon caught, and agreeing to work on our behalf, was allocated the cover name Summer.

As Summer operated his transmitter, a British radio amateur stood close by to ensure that he was sending in Morse code only what we wanted him to send. It took a few days, and the expertise of the wireless amateur, before Summer managed to contact his German control and act as a double agent. How did a wireless amateur become involved? It was because he was a member of the highly secret Radio Security Service which employed at least 1,500 amateurs, making use of their special skills, on a voluntary basis, as experts in wireless communication.

Wormwood Scrubs prison

It all began even before the war, in the summer of 1939, when Lord Sandhurst (of MI5 and a director/owner of Hatch Manson wine merchants), approached Arthur Watts, then President of the Radio Society of Great Britain, to see if radio amateurs could provide a listening watch on the short wave bands. It was thought that enemy agents or spies might be detected by nearby short wave listeners because of the strong 'ground wave' and 'key clicks' produced. It was also considered that German aircraft might be guided to their targets by signals from this country. Wireless amateurs would be ideal for this task because of their wireless communication experience and also because they were widely distributed over the British Isles.

MI5 had set up a department to deal with spies in 'C' block of Wormwood Scrubs prison, having first removed the other prisoners, and Watts was invited to an interview which took place in one of the cells. Watts, an ex-naval officer who had come out of World War I minus a leg, was confident that amateurs would not only be suitable for the task but would also be keen to continue with the wireless activities which might be curtailed by the war.

Fortunately, although amateur transmitters were impounded on the outbreak of war the short wave receivers were not. Through personal recommendation, amateurs were contacted and formed into groups. They were given the name Voluntary Interceptors (VIs) and came under a section referred to as the Illicit Wireless Intercept Organisation (IWIO). As this name was rather too revealing, it was later changed to the Radio

Security Service, and along with Post Office operators, it was entirely responsible for seeking out clandestine wireless transmissions.

The officer-in-charge of the new service was Colonel Worlledge, and the postal address to which amateurs sent their log sheets, on which they had written down any suspicious Morse signals heard, was Box 485, Howick Place, London SW1. Mobile units, with direction-finding vans provided by the Post Office, were employed to round up the spies. The whole purpose for which the IWIO had been instituted was misconceived, however, as the amateurs were hunting a quarry which did not exist. No agents were found at this time and German bombers were being guided by beams from the continent.

What turned out to be of far greater significance was the discovery made by these listeners of some mysterious weak Morse signals which, although bearing some resemblance to amateur radio transmissions, were clearly not. The call signs were wrong, consisting often of three letters only, and the messages were being sent in groups of five-letter code. These messages were forwarded to Bletchley Park for the attention of the code-breakers but it was understood that they were of no interest. Whether this was a misunderstanding or an evasion is not clear.

The Abwehr exposed

Hugh Trevor-Roper (later to become Lord Dacre) relates how he and Major E.W.B. Gill occupied an insalubrious Wormwood Scrubs prison cell, which the evacuated prisoners resented leaving! Major Gill had served as a wireless intelligence officer in World War I and his disregard for convention was shown when he used the Great Pyramid in Egypt as an aerial support for his intercept work.

The two of them set to work to try to decypher this five-letter code. It turned out to be not too difficult as it was in a fairly well known hand cypher. Working in the evenings in the flat they shared in Ealing, they produced some German plain language. It transpired that this was traffic from the German Secret Service which, in this case, included the Abwehr, emanating from stations in Hamburg and Madrid. The former was conversing with agents on the Baltic and North Sea coasts, some of whom were preparing to land, by boat or parachute, in Britain. The Madrid control was communicating with agents in Morocco and other neighbouring countries.

Another station in Wiesbaden appeared to be training spies for infiltration, and valuable hints were gained from the laborious initiation of its pupils. The two officers were naturally excited by this coup, as was their commanding officer, Colonel Worlledge, who ordered Trevor-Roper to write a report about this most valuable discovery. Expecting to hear a purr of approbation, Worlledge had this document circulated to his normal contacts, thus illustrating that, although the Radio Security Service (RSS) had found no spies in Britain, it was certainly earning its keep.

The resulting unexpected explosion almost led to a court-martial for poor Hugh Trevor-Roper who, after all, had only been obeying orders. Worlledge was reprimanded and strict instructions followed that under no circumstances were RSS personnel to decypher intercepted messages as this was the province of the Government Code and Cypher School at Bletchley Park. The RSS had trespassed on to the territory of the Secret Intelligence Service (SIS or MI6) which was supposed to deal with all overseas matters and this fledgling RSS had only been set up to deal with internal matters!

In fact, the Abwehr, whose schedules were known, was continuously monitored by the RSS 24 hours a day so that Bletchley Park was able to achieve one of its greatest triumphs. In December, 1941, the Abwehr Enigma code was broken thus exposing the innermost workings and secrets of the German Secret Service for our people to see, and use to our own advantage.

Transfer to Arkley

As the RSS had been so brilliantly successful at Wormwood Scrubs, it was realised that the interception of this most important enemy intelligence simply could not be left to part-time listeners. The search began for a new headquarters, with its own analysis and monitoring facilities, no doubt hastened by the fact that bombs were falling on and around Wormwood Scrubs.

On 3 October, 1940, the RSS moved into its new headquarters at Arkley View, a large site two miles north of Barnet. This building was already being used by the Post Office as an intercept station. The 'View' housed the analysis, intelligence, direction-finding control and various administrative departments. Huts were erected in the grounds for intercept work, a teleprinter terminal and, later, the ever-expanding departments to identify, classify and collate the enormous secret intelligence enemy wireless networks. The secret cryptic address became well known to the select as **PO Box 25 Barnet,** this being the new postal address for correspondence from the VIs.

Arkley View was on the right-hand side of the road leading to Stirling Corner. Arkley Lane had the View on its left and Oaklands to its right. It is here that the orderly room stood, with a despatch riders' base for taking intercepted messages to Bletchley Park. Officers' and sergeants' messes were in Scotswood opposite the View. Other large houses such as Rowley Lodge, The Lawns and Meadowbank were used as billets, messing, transmitting and training schools. In Ravenscroft Park, High Barnet, a billet, operators' evaluation and a small intercept training station were established, run by CQMS Soames (later transferred to The Lawns).

The RSS side of communications was given the title of Special Communications Unit Number 3 (SCU3), SCU4 being used for the overseas section which was developed later. SCU3 was represented on various joint committees with MI5, MI6 and GC&CS from Bletchley Park. Worlledge was later replaced by Lt.Col. F.J.M.Stratton (formerly Professor of Astrophysics at Cambridge) until Gambier-Parry asked Major (later Lt. Col.) Kenneth Morton Evans to take over as Deputy Controller RSS in late 1941.

Morton Evans was a pre-war radio amateur with the call sign GW5KJ who supplied much of the information used in this account. Trevor-Roper worked with Gill, on the first floor of Arkley View, on traffic analysis until Gill moved on to do valuable work on radar and VHF communications for the Army. This left behind several Oxford dons who were active in various aspects of intelligence work at Arkley. They were Gilbert Ryle, Waynflete Professor of Philosophy; Stuart Hampshire (later knighted), who became Warden of Wadham College, Oxford, and Charles Stuart, a history don at Christ Church, Oxford.

At Bletchley Park (BP), Leslie Lambert, a radio amateur with the call sign G2ST, dealt with the hand cyphers used by some of the German Secret Service groups. When Morton Evans telephoned BP, he immediately recognised Lambert's voice as the well-known BBC storyteller who used the pseudonym A.J. Alan.

Arkley View in North Barnet, the first home of the RSS before the interception station was built at Hanslope. It continued to be responsible for receiving, examining and collating log sheets from all interceptors, RSS administration, and discrimination.

When Oliver Strachey took over this work, the material was always referred to as ISOS (Intelligence Summary Oliver Strachey) although he was later succeeded by Denys Page, Professor of Greek at Oxford and later Master of Jesus College, Cambridge. The machine (Enigma) cyphers were handled by Dilly Knox, hence ISK, and later Peter Twinn. Twinn, Denys Page, Morton Evans and others always attended the committee meetings with MI5, MI6, etc in London.

The Arkley Organisation

Arkley View was a large country house, which was well-known to all SCU3 recruits, as the induction to this branch of MI6 invariably took place there. The recruits, who were enrolled initially by Lord Sandhurst, soon to be followed by Captain (later Major) Bellringer (an undersized, black-moustached officer), took the oath and the King's shilling, which is where any resemblance to British army recruitment ended. The only other 'figleaf' of standard army enlistment was the issue of an Army Book 64 (AB64) parts I and II. Part I contained such personal details as date of birth, service number, rank with promotion dates (if any) and the date on which various chemicals were injected to protect the recipient from all the various bugs which arose from service to one's country. This concern for the soldier's welfare was followed by details of his next-of-kin and an invitation to write his will, (acceptable without a witness).

However, those of us joining as 'Special Enlistments' did not receive AB64 part II because this gave a record of pay – which in our case did not come from Army funds – but from SIS.

Under Morton Evans, the various departments were mainly located in one-storey buildings each about 100 feet long, at the rear of the Arkley View house. They were called huts and were frequently extended, as the need arose, in a similar manner to Bletchley Park. Under the general heading of Discrimination, were the departments General Search, Groups, Collation, Allocation, Mobile units and Administration. Reports from interceptors came to Arkley where, after processing as explained below, copies of the same messages from different intercepts were compared, to enable a good copy to be forwarded to Bletchley Park.

General Search

The principal work was the scrutiny of logs and the placing of the intercepted messages in the relevant groups. Box 25 (at Arkley) received up to a thousand log sheets daily from VIs and full-time interceptors. These had to be examined to identify new Abwehr services and to sort the familiar ones into their allotted groups. More than 14 different groups had been identified, each having a number of services from perhaps a dozen to a hundred or so. Thus, if General Search labelled a message 2/153 it would mean group 2, usually Berlin, and service 153 which could be a link to, say, Madrid, Oslo or Milan. The identification was by means of time, frequency, type of procedure (or preamble, if there was a message), or possibly the call sign. This latter was problematic, as call signs often changed daily. Bob King's work in General Search varied over four years between the various methods of identification.

One task was to examine logs for intercepts which had not been positively identified and to try to discover where to place them or even if they were wanted at all. If the operator sent in a previously identified station the details were sent to the relevant Group Officer, located in the next hut, who would then advise the operator whether it was 'already covered thanks' or 'still wanted'. He would have from two to a dozen staff, according to the group size.

A large wall map was kept in the 'Group' hut, with coloured wool stretched between points showing the location and working of the various stations. To prevent a casual visitor from seeing the extent of British discoveries, this map was covered with a curtain which was activated by an electric motor. No doubt the local wool shop did not ask questions. If it had, there would have been a misleading answer at the ready.

On other occasions, the 'suspect' or 'watch please' rubber stamp would be used on the log sheets to advise the

Name ....
Address ....
File No. ....

R.S.S. LOG SHEET

Sheet No. ....

Region ....
Group ....
Sub-Group ....

(1) DATE TIME G.M.T.	(2) CALL AS SIGNALLED	(3) MESSAGE AND REMARKS	(4) FREQUENCY KC/S	(5) WAVE LENGTH M.	(6) TYPE OF SIGNAL	(7) STRENGTH
		4				
		8				
		12				
		16				
		20				
		24				
		28				
		32				
		36				
		40				
		44				
		48				
		52				
		56				

A blank RSS Log Sheet for use by interceptors whether they be working at home as a VI – or in an intercept station – such as Hanslope Park.

interceptor. If the signal was not the enemy secret service the 'unwanted' stamp was used. Other stamps were: 'unwanted Hun', 'more please' and 'OK covered thanks'.

One method of determining which stations were 'working' each other, in order to establish a new service for our records, was painstakingly to enter all the unidentified but 'suspect' signals in large books divided up by time. This was because one thing in they had in common was the time of contact. The frequencies and call signs were always different. By studying these 'call books', it was often possible to pick out a new network. These were entered in card indexes where the data changed so frequently that each day long amendments were required in order to keep all our records up to date. Each of our sections worked on a given frequency band and had to keep its own card index of all the known groups for reference purposes. One section studied teleprintouts which contained reports from the full-time stations, called fixed stations, perhaps dating from the earlier days when mobile stations sent in reports. From these, suspect entries could be identified and added to the call books.

Another useful aid in identifying stations which changed their call signs regularly, and of particular value to the intercept operator, was by noting the peculiar and tuneless notes which some of the primitive transmitters produced. Cyril Fairchild, one of the first amateurs to work at Arkley alongside Lord Sandhurst, collected 196 different descriptions reported by interceptors. These included: a croaking frog, a fly in a bottle, a clucking hen, an Epsom salts note, and a painful and pathetic note!

The Groups

The various services were allocated to groups such as the following:

Group 1 (code-named Harry) included a number of separate Abwehr services controlled from a transmitter site in the vicinity of the port of Hamburg. These were usually long distance wireless links with clandestine agents operating from countries outside the German sphere of influence, and overseas continents, including the USA, South America, Mozambique and Angola. This group was supervised by the only naval member of this establishment, Chief Petty Officer Denis.

Group 2 (Bertie) near Berlin
Group 3 (Willie) Wiesbaden
Group 5 (Patrick) Prague
Group 6 a fairly small group associated with group 1
Group 7 (Violet) Vienna
Group 8 (Ivor) Italian
Group 12 Russian resistance groups
Group 13 Himmler's Sicherheitsdienst
Group 14 diplomatic (important group centred on Berlin)
These and other groups were not always static and changed in size and importance.

Group 2 services were mainly controlled by the RSHA (The German Intelligence Service) from a centre located near Berlin, code named "SCHLOSS", and this was probably the most important group. Sub-centres of the Group 2 network were located in Spain, Norway, France, Italy, Poland and the Balkans. All were directly linked to Berlin by daily and sometimes hourly schedules, depending on the tactical situation.

Each of these sub-centres in turn had its own network to various RSHA offices in the most important cities and towns throughout the particular country or territory. Thus, for example, a message originated by an agent in the field would be sent by that agent direct to his control centre in that country and thence relayed back to Berlin with little delay.

Group 3 was also RSHA with its centre located near to Wiesbaden and outstations in France and North Africa. Group 5 was basically a Central European SIS Czech network, with liaison links to Abwehr and Italian stations. The control for group 7 was located in Vienna with very busy links into the Balkans, especially to Ankara, and later Greece and the Ionian and Aegean islands. Group 8 covered the Italian Secret Intelligence activity, mostly in Spain and North Africa.

The Group 13 network was originally Abwehr, operating from Hamburg and resembling Group 1, but in 1944, following its takeover by the RSHA, the Centre moved to Berlin although the transmitter and receiving stations continued to operate from the Hamburg area. This was a very important group with a high standard of operation. Group 14 consisted of an undercover German diplomatic wireless network which was able to use the German Post Office point-to-point wireless facilities. The transmitter was at Nauen, and the German Foreign Office radio service was able to take over certain transmitters and to insert cyphered messages to its Embassy outstations at secretly scheduled times.

Spain and North Africa were particularly important to the Germans, because of their geographical position at the entrance to the Mediterranean and proximity to Gibraltar. They had reporting agents in many ports, most of them in hourly communication with Madrid and thence direct to Berlin. It was not unusual for information about ships entering or leaving Gibraltar to be passed to Berlin within the hour by the German intelligence post at La Linea. Similarly, convoys passing through the Straits of Gibraltar were reported by the German SIS station in Tangiers.

Collation

This small section carried out some very interesting work. Harold Brock relates how, by careful examination of the operators' Morse style, the use of initials and other clues, they were able to produce a wall-chart indicating the movement of these operators. This was of considerable value in understanding the entire structure, development and intentions of the German Secret Services. This section kept an account of anything known about the enemy operators and the type of transmission or, in fact, any detail which would assist in the mammoth task of classifying the networks.

It is interesting to note that many German radio amateurs and members of the Deutscher Amateur Sende and Empfangs Verein or DASV were employed in the service, much as British amateurs were recruited into the RSS organisation. However, licensed amateurs, although very well qualified by their hobby, were rarely employed as field agents by either country.

In addition, the location of the clandestine signals was of considerable significance and, for most of the war, this operation was directed from Arkley View.

Direction-Finding (D/F)

Knowledge of the location of enemy transmissions was vitally important. This not only assisted in building up our knowledge of the networks, but above all helped by providing knowledge of which cypher was being used. For instance, the Abwehr Enigma machine was a quite different model from that used by the services. Different networks used different Enigma settings and hand cyphers. At first, the RSS had relied mainly upon the GPO intercept and D/F stations at Gilnahirk near Belfast, Thurso in the north of Scotland, St Erth in Cornwall and Sandridge near St Albans. MI6 decided, however, that their performance had not been satisfactory and the contract with the GPO was terminated with SCU3 taking over from the autumn of 1941. With the addition of extra direction-finding stations at Wymondham in Norfolk and Bridgwater in Somerset, an extremely reliable system eventually emerged.

Major Dick Keen, who had been the D/F expert at Marconi in Great Baddow, visited Hanslope Park frequently to assist in the setting up of the D/F system. If the analysts at Arkley required to know a particular station's location, a request was passed to the intercept stations. When the station operator picked up the signal he could immediately transfer it to a supervisor who sat at a desk behind a type of switchboard. This device put him in touch with all the D/F stations by means of a dedicated land-line which was permanently connected to them. All the D/F station operators could hear whatever was being sent along the lines simultaneously.

Some D/F operators sat in a steel tank buried deep underground so that they would be screened from all interference. Above ground were four vertical aerials connected to rotating coils (a device called a goniometer) inside the tank. This formed the well-known Adcock Direction-Finding System. When a location was required, the intercepted signal (plus frequency) was transmitted, using Morse code, by land-line from Arkley control to the relevant D/F stations. The operator listened to this with one ear on his headphones, whilst he tuned for the same signal on his receiver, which he should then be hearing with his other ear. Then it was a quick operation (for a skilled operator), to rotate the goniometer to find a minimum signal strength and get the bearing. Under ideal conditions, this might take only a few seconds from start to finish, a necessity when the transmitter was only on the air briefly. The bearings from several D/F stations were passed to Arkley control where the 'fix' was plotted (by ATS women) on a large map-table in a room on the ground floor.

At Hanslope in 1943, an experimental D/F bullet-proof hut was built at the south end of the huge aerial field in the direction of the old Haversham village. This used a new system called the 'spaced loop' which consisted of two square loops placed at both ends of a metal boom on a short mast located on the centre of the hut roof. After a considerable period of testing, this was not deemed to be satisfactory and its use was discontinued. All the wireless direction-finding operators from the whole SCU3 D/F network were called together to pool their thoughts and ideas on this system but it was decided to continue with the existing manual, simple and reliable Adcock system. Captain Louis Varney realised that he had such highly skilled operators that they could produce a bearing on an enemy transmission accurately and more quickly than with the new automatically spinning goniometers which displayed the bearing of an enemy signal on the face of an oscilloscope.

Hanslope

Ted Maltby, a colleague of Gambier-Parry's in the wireless industry, was put in complete control of the RSS and instructed to set up a full-time intercept station at Hanslope Park. In August 1941, the Administration staff arrived, followed in September by two operators whose first task was to clean out the corn bins in the granary. These two radio amateurs, Wilfred Limb and William Chittleburgh, installed six American National HRO Communications receivers which were about the best that money could buy. They were eventually to be the mainstay for all intercept work in both the RSS and the military 'Y' Service.

The use of the Granary was discontinued in 1942 when the new intercept room in the Lodge was in full operation. Throughout the winter of 1941-42, many brick-built accommodation huts were provided for the operating staff; a large building at Bullington End, known as The Lodge, was used as an intercept station. Here, many HRO receivers were set up on trestle tables, the aerials being end-fed wires sloping upwards at an angle of 50° with one aerial used for two receivers. When the operating staff moved into the new station, they were confronted with a polished green cork linoleum floor; the same material was also fitted to the desk tops of all the bays. The sight of 66 brand-new HRO communication receivers was mouth-watering to all these radio amateurs who were keen to try out such sensitive sets.

The new, purpose-built intercept station was finally opened at the beginning of May, 1942. The first commanding officer of the station was Captain Prickett but he was soon replaced by Reg Wigg, who being a 'ham' himself, had a rapport with fellow radio operators who until only recently had been civilian amateurs. The staff were put into the uniform of the Royal Corps of Signals for security purposes, but were paid by MI6 and for this reason, no War Office record of their existence is available. There were three grades of enlistment paid at £7, £6 or £5 per week, according to Morse speed. 'A' grade required the ability to receive 23 words per minute in 5-letter code with an aptitude for intercept work. Pay was taxed, but in cases where army billets were not available, a tax-free allowance of £1.1s (£1.05p) was made for accommodation.

The aerial system consisted of 7 rhombic aerials 50 feet high and 5 Vee-beams 75 feet high as well as many end-fed general-purpose aerials sloping upwards at 45°, similar to the ones at The Lodge and Arkley View. An Australian named Ernie Buick was put in charge of the team of men who erected this huge aerial system using standard GPO telephone poles spliced end-to-end to bring them up to the required height.

An aerial specialist, Dud Charman, was called in to assist in setting up a new system of wide-band amplifiers. Along with many famous radio amateurs such as Lou Varney and Pat Hawker, a total of 126 radio amateurs were put to work at Hanslope on the full-time interception of the enemy wireless intelligence networks.

Besides being the Administrative HQ, Hanslope was engaged on other secret engineering projects which included the work of Alan Turing whose inventive mind went far beyond developing the world's first programmable computer. Hanslope was considered important enough for Lord Gort, General Alexander, Field Marshall Montgomery and General Eisenhower to pay it a visit, where they were conducted round by Brigadier Gambier-Parry. It was explained to the Hanslope Park visitors that when the VIs discovered new enemy 'wanted' signals, Hanslope was informed so that it could eavesdrop on these signals on a 24-hour basis. Other full-time intercept stations were constructed and evenly distributed throughout Great Britain, as will be detailed later. For instance, in early 1943, another large intercept station was commissioned at Forfar in Scotland, for the use of SCU3 and SCU4. There were 26 operating bays here with a D/F station nearby.

Never had there been such a large gathering of amateur wireless operators whose sole purpose was to eavesdrop on enemy wireless intelligence. No real recognition has ever been given to them or the VIs. Nevertheless, the authorities considered the RSS sufficiently important for a teleprinter link to operate between Wormwood Scrubs and Bletchley Park. After the move to Arkley, a teleprinter room, which eventually contained as many as eight machines, was manned by the ATS around the clock.

Wilf Limb the first RSS interceptor to move from Arkley to Hanslope

An aerial view of Hanslope in wartime with the mansion to the left of the picture

In a BBC broadcast entitled 'The Secret Listeners' (1979), Hugh Trevor-Roper described two instances of the usefulness of intelligence gathered by the RSS, as follows:

1. *'The material that we got was of great practical value. A lot of it, of course, was encyphered on the Enigma machine which the Germans thought was totally undecypherable. Therefore, they were pretty open in what they said in these messages and through them we obtained a very complete knowledge both of the structure and the daily working of the whole German Secret Service.*

   'This knowledge was valuable in itself and could be applied in many ways. For instance it enabled us to capture every spy who arrived in England as soon as he landed. It was of great value in deception which consisted of feeding false information into the German General Staff through the German Secret Service. 'This was, I think, one of the most important functions which this material played. I can give two spectacular instances. One was the famous operation Mincemeat when a corpse was floated ashore at Malaga with secret documents which deceived the Germans effectively into thinking that we were going to land in Greece and not in Sicily in 1943. That would not have been possible if it hadn't been for this material which first showed us where we could land the corpse so that the Spaniards would pick it up and hand the papers over to the Germans. Even after that, we were able to follow, through this material, the transmission of the documents and the extent to which they were believed through the whole general staff machinery. That was one operation which simply couldn't have been done without the added sensitivity which was given to us by a continuous knowledge of the operation of the German Secret Service'.

2. *'Cicero photographed secret documents in the Embassy and sold them to the Germans and we knew all about this and we saw our secret documents being sent to Germany from the German Secret Service in Ankara. But we were hamstrung because we couldn't communicate this fact to the Ambassador by the ordinary telegrams because it was precisely these telegrams which Cicero was photographing and sending to Germany. Therefore if we indicated that we knew about Cicero the Germans would know that we were reading the messages. So the whole Cicero affair had to be done by sending people out in order to convey personal messages because we simply couldn't afford to mention it in any radio communication we sent out. It wasn't that the Germans were*

Inside the interception room, Hanslope Park.

decyphering our traffic, we weren't frightened about that, we were pretty sure that they weren't. It was that when they were decyphered at the other end, the Ambassador's valet was taking them out of the safe, photographing and sending them to Germany.

'There were very few illicit or German Secret Service transmissions which escaped the notice of RSS, and even changes in procedure, employed by the Germans for security, were identified, in some cases before the enemy had become familiar with them'.

To quote from Hinsley and Simkins in the *Official History of British Intelligence in the Second World War*, *'... in all its activities, the RSS achieved a high and continuingly increasing degree of efficiency'.*

The Voluntary Interceptors

It would be presumptuous to suggest that but for VIs we would have remained ignorant of the German Secret Service radio networks. But a delay in this discovery, and the loss of early messages intercepted by the wireless amateurs, would have reduced the effectiveness of the RSS and its ability to assist in the prosecution of the war. Until the full-time operators began to produce intercepts in large quantity, the VIs were the principal source of intelligence. A detailed description of this unique operation may be appropriate.

Wireless Amateurs (known colloquially as 'hams') had an interest and skill in constructing wireless apparatus in order to communicate with other hams anywhere in the world. Essentially all of them could send and receive Morse code. This was the preferred method of working due to the simple equipment it required. Using Morse, it was also easier to communicate under poor signal conditions. The majority of pre-war amateur equipment was home designed and built, especially the transmitters. By communicating with other amateurs worldwide using similar equipment 'hams' became adept at reading weak Morse signals, where interference was also often present from background noise or other nearby signals.

The most common type of transmitter used a quartz crystal (similar to, but much larger than, present-day timepiece crystals), in an oscillator circuit. This not only simplified the transmitter circuit but also ensured an accurate knowledge of the frequency being used for both transmission and reception in the specially allocated frequency bands for amateurs. These crystals had to be surrendered in 1939 along with the transmitters.

As previously stated, when the help of wireless amateurs was sought, their task was initially to locate enemy agents or spies operating in the UK. For example, Jack Miller a member of the Society, living near the Clyde, was asked to look for strong ground wave signals and key clicks. Some Voluntary Interceptors were issued with an identity card, DR12, which carried a photograph and considerable authority. It was intended to enable a VI to enter premises from which he suspected unauthorised signals were being transmitted. It soon became apparent that there were no spies transmitting to Germany, or what few there were had either been rounded up and executed or were 'turned' and operated under our control. In some cases, a British operator (usually a wireless amateur), took over the transmissions and was accepted by the Germans as if he were one of their own agents.

Because the VIs had discovered the large numbers of distant-sounding signals, with unusual operating procedures which did not fit into the familiar classes of service and commercial traffic, they were each given a section of the short-wave spectrum in which to search for such signals. The problem arose of not knowing accurately the frequency on which they were listening. Those who were lucky enough to possess a commercial receiver, such as one of the American Hallicrafter range or the British Eddystone Superhet range or anyone rich enough to buy the National HRO (priced at $360 in 1942), had the calibration problem more or less solved. Later, the RSS issued lists of 'marker stations'. These were the more powerful transmitters such as Broadcast and Press stations of known frequency that were on the air most of the time; from these, the VIs could obtain spot frequencies and hence construct a graph to calibrate their receivers. Better still, a signal generator could be made (or bought) which gave greater accuracy. As the supply of HROs from the United States improved, many VIs were issued with these and received the bonus of being allowed to purchase them after the war for £5, which was not a fortune in those days, being a touch less than an average week's pay.

Many amateurs only owned home-built receivers or the popular Eddystone 'All World Two' which was also available in kit form. This was a simple two-valve receiver which required a degree of skill in use and one really needed both hands to operate it, one to tune and one to adjust the reaction control. In the case of some home-built receivers it was unwise for the operator to take his hand away from the dial as this usually meant loss of signal (due to the hand capacity effect). Clumsy as it was, many V.I.s took numerous messages using this type of receiver.

It became apparent that there was a vast network of these previously unnoticed call sign stations, usually consisting of three-letters. These call signs were frequently changed, hence the need for accurate observation of time and frequency. By doing this, a station could often be identified even though it had changed its call sign. Identification was assisted by the operators' 'fist' (the characteristic of an individual's Morse sending) and the procedure in use. There were several ways in which Bletchley Park had assistance in the extremely difficult task of breaking Enigma messages. One was the interception of hand cypher messages (in at least one case, being sent by our own double agent, hence the content was known to us), which were later retransmitted onward by the Germans to HQ, using Enigma. This meant we could compare the known message with the Enigma version.

Recruitment of Voluntary Interceptors

Throughout the early war years, hams (including at least one woman) were recruited on a regional basis. A Captain in the Royal Corps of Signals was put in charge of each region. Each VI was given a number, such as V/HN/358, for identification. HN stood for 'Home North', Home being London. Others were obvious such as SW and N. These VIs were recruited with great care but by different means.

Many radio amateurs holding pre-war licences, who also belonged to the Radio Society of Great Britain, received a letter from Lord Sandhurst. The amateur was then subjected to a security check by the police and was interviewed by the Regional Officer. If found satisfactory and able to devote time to the task, the VI was enrolled after signing a declaration under the Official Secrets Act. He was then given a number, some blank log sheets, postage stamps, envelopes, sticky labels addressed to Box 25, Barnet, Herts and a frequency band in which to search for signals using a certain type of procedure.

A volunteer may also have been given particular call signs to listen for and he was required to take down any messages which appeared in coded groups of five letters. As an indication of the precautions felt necessary by the authorities, it is worth noting that VIs had to place their completed logs inside an addressed envelope which was then inserted into another stamped envelope addressed to Box 25.

Some VIs were organised into groups, whose leader worked under the Regional Officer (RO). The group leader would arrange rotas and organise the covering of certain regularly required transmissions. Some VIs never met another VI and followed the directions sent to them by post. In some cases, the VI was even unaware of Box 25 as his logs were sent to the RO's office, which referred to the place dealing with the intercepts only as 'London'.

The VI worked mainly in the evenings because of daytime employment, but some who were retired or unable to work could listen during the day. Various 'covers' were employed, principally Royal Observer Corps (ROC) and, in at least one area, Special Constables, as reported by Stan Martin who made a valuable contribution as a VI for many years. ROC uniforms were issued, in some cases to people who could probably not distinguish between a Stuka dive bomber and a Blenheim! As one directive from headquarters said, 'Good relations must be preserved with the Royal Observer Corps – that is as far as possible – no relations at all!'

There appears to be no accurate record of how many hams were enrolled as VIs but it exceeded 1500 and may have reached 1700. The need for secrecy was impressed so strongly upon them that, even 34 years later when in 1979 the BBC broadcast 'The Secret Listeners' on television, they had never discussed their work with anyone, not even with each other. They did not know until then the nature of the enemy traffic (messages) and many felt acutely embarrassed that this 'taboo' subject should be made public in that way.

Eddystone 'Short Wave Two' wireless receiver usually built from a kit and widely used by pre-war Hams. Many VIs had sets of this type.

What was it like to be a VI?

The VI couldn't, of course, explain why he was unable to take part in duties such as fire-watching or the Home Guard because he didn't have any time to spare. Sometimes a small room in the house was used as the listening post, or in many cases, it was a shed in the garden, suitably blacked out of course. Many listened for long hours and prodigious numbers of messages, written on the RSS log sheets, began to arrive at Box 25 (Arkley View), peaking at several hundred per day.

The VI did not know the origin of the signals, although he could assume that they came from the enemy in spite of the absence of any 'clear' language except the usual wireless amateur abbreviations based on English. The German operators were strictly forbidden to use German 'clear' language or abbreviations, although lapses occurred on rare occasions.

The VI soon got to know the regular transmissions very well; he felt no animosity towards the operators, but just hoped that their signals would remain audible and contain messages. For instance, QTC2 (meaning, I have two messages for you) would immediately command the interceptor's attention, and probably increase his pulse rate as he prepared to copy down messages, perhaps lengthy ones, of 5-letter code. The intended recipient could ask for repeats of missed groups, a luxury naturally denied to the VI.

Sometimes a particular transmitter would not be heard at the expected time. Was the operator having a night off, or ill, was the signal just too faint to hear or had the transmission been moved to another time and/or frequency? An instruction to change time or frequency may have been contained in an earlier coded message of which, of course, the VI would have been unaware.

Sometimes 'Box 25' told the VI, through the Regional Officer, where to look for the missing link. This could be because the people at Bletchley Park had decoded the information or quite likely that it had been noticed by another interceptor during his routine search of an allotted band. Intercepts were made more difficult where the wanted stations used different call signs every day to an agreed pattern, although in certain cases it was possible to work out the changing call sign system. Often, the clandestine signal was identified because the VI recognised the operator's style of Morse code sending or perhaps the note of the sometimes primitive transmitters, as mentioned earlier.

Listening far into the night, especially when there were thunderstorms or other static interference, or perhaps weak signals partially swamped by more powerful ones, became very tiring. But the VIs were extremely tenacious and only gave up when conditions were absolutely hopeless. Les Proctor recalls how he was listening to a short transmission which soon closed down, but he was suspicious and stayed listening on the same frequency for a long time. Suddenly, this same station came up with no call sign or warning and sent a long message. Les received a commendation from Box 25 so he assumed that it had been something important.

Colonel Maltby (the controller of RSS) praised the Vis in a BBC broadcast, saying, 'I don't think anything but death or great unconsciousness would make them miss a schedule. There was one chap who was permanently on his back, a complete cripple, so he had his receiver rigged up over his bed and he was an absolutely first-class operator'.

The frequencies most used were between 3 MHz and 12 MHz and although some fell outside this range, the concentration was from 4 MHz to 9 MHz. Much of this band was occupied by broadcast stations and Morse code used by the services and press. With some 5 to 6 million cycles of band in which a Morse signal needed only a one thousand cycle space at most to be read separately from its neighbour, as many as 3000 stations (discounting the space occupied by broadcasting) could theoretically be operating simultaneously.

This is an over-simplification but it does indicate the value of having more than a thousand interceptors spread over Britain when a signal audible in Glasgow might well be inaudible in Dover owing to the features of propagation well within the experience of radio amateurs. Put simply and therefore incompletely, a wireless signal on short wave does not travel directly from transmitter to receiver. The receiving station depends on the radio wave being reflected from an ionised layer high above the earth's atmosphere and so back to earth much further from the transmitter than the direct or ground wave can travel. This reflection is unreliable and subject to change and fading, so not only may the signal grow weaker it can even disappear completely, possibly to return later in an unpredictable manner.

The VI was told to search a small selected part of the spectrum so that he would get to know which of its regular inhabitants were of no interest to 'Box 25'. Then he would take notice of a generally weak signal using, for example, a three-letter call-sign and a certain type of procedure. (For a more detailed description see Appendix 3).

The flavour of the VI's feelings may be obtained from the following verse which is attributable to Norman Spooner, a Bournemouth VI and Group Leader, and with apologies to John Masefield:

VI Fever

I must go back to the set again, to the Superhet and the phones
And switch off the broadcast music, the announcer's measured tones
And search again on the short waves, with loud calls blending
For the dim sounds of the Morse code that a far foe's sending

I must go back to the set again, for the time has come to seek
In the QRM and the QRN for my allocated squeak
And all I ask is a steady note, through the ether speeding
At a fair strength, in a quiet spot, at a nice speed for reading.

The Value

At its peak in 1943/4, RSS employed, in addition to the VIs, more than 1,500 people, most of whom were radio amateurs. More than half of these worked as interceptors at the 24- hour watch stations. A further few hundred were occupied in the investigation and establishment of the numerous wireless networks which were constantly being altered and extended by the enemy. A detailed knowledge of these revealed important information, even where messages were not decypherable.

So what was the value of the radio amateurs, and other RSS operators? There is no doubt that the work of the RSS was of assistance to Hut 6 at Bletchley Park in breaking the Enigma cyphers and also in revealing the innermost workings of the German Secret Service. Signals Intelligence was a vast undertaking, employing at least 50,000 people and RSS was a part of this. In addition to being of general assistance it played a vital part in deception.

The RSS kept a record of where and when agents were to be infiltrated into this and other countries so that, for instance, it was fairly certain that no spies went undiscovered in the UK. By taking over selected suitable agents (the double cross, XX, system) we could feed false information back to Germany and cause confusion. Thanks to our intercepts, we knew what the German secret service was accepting and believing and what its intentions were. Thus we were able to play a part in invasion deception plans, notably in Sicily and Normandy.

Others factors assisted Bletchley Park, such as the recovery of Enigma codes and details from equipment, retrieved from various ships, and the selfless work of the Polish cryptographers, not forgetting the sometimes lax procedures of the German Enigma users. Without the skill of interceptors, service and amateur, listening for long periods to weak Morse signals often disrupted by ear-splitting noise, BP would have been unable to develop its tremendous decryption expertise, in fact neither the RSS nor BP would have had reason to exist.

During 1941, decrypts rose from 30 to 70 per day to over 260 per day in December 1942 and to a peak of 282 in May 1944. A total of 268,000 RSS decrypts were issued by Bletchley Park during the war; 97,000 were in Abwehr hand cypher and 140,000 encyphered on the Enigma machine. The Sicherheitsdienst produced 13,000 decrypts. Seventy-eight secret German intelligence stations were transmitting by the end of 1941 and 147 by the end of 1942.

Establishment of the RSS

The backbone of the RSS recruitment consisted of intercept operators. The allocated enlistment numbers indicate a total of some 3,000 recruits, but this may include some who belonged to SCU1. No official record has so far been traced, but it is likely that at home and abroad, nearly 2,000 men were intercepting RSS material plus several hundred uniformed and civilian personnel, working mainly at Barnet, on recording and analysing the intercepts. The establishment included various support services and administration plus, of course, at least 1500 voluntary interceptors.

Intercept and D.F. Stations

It was customary for each full-time operator to use two receivers in an attempt to copy both ends of a transmission which would be on different frequencies. Three shifts were in operation and with allowances for sickness, supervisors, rest-days and leave, etc, 33 receiving banks (a bank usually comprised two receivers) would require some 150 operators.

Hanslope, Bucks	33 receiving banks	St Erth, Cornwall	17 receiving banks
Thurso, Caithness	20 receiving banks	Gilnakirk, Belfast	18 receiving banks
Forfar, nr. Dundee	26 receiving banks		

There were nine D/F stations operating in the UK under the RSS:

St Erth (two)	Bridgwater	Hanslope (experimental)
Wymondham, in Norfolk	Forfar	
Thurso (two)	Gilnahirk, N. Ireland	

Networks monitored by the RSS

These included:

Reichssicherheitshauptamt (RSHA): this covered all the non-military intelligence, including the Gestapo which was run by Heydrich, before he was assassinated, who was succeeded by Kaltenbrunner.

Sicherheitsdienst: Himmler's Security Service which also came under the RSHA heading.

Abwehr: (Literally 'defence'). This was Admiral Canaris's military intelligence, similar to our MI6. The Abwehr was absorbed into the RSHA in February, 1944.

I wish to express my thanks to Bob King for his tremendous help with the research into the RSS. (Bob's own story is Chapter 35).

Chapter 16

A Sea of Aerials

The various stations connected with the Secret Wireless War all required vast aerial arrays, and the following gives some idea of their extent. Alongside each one is a brief note on its use. The list was prepared by David White with additional information and notes from me.

Where I have shown an OIC (Officer in charge), he generally ran the station for a long period but not necessarily for the whole of the station's life.

1 Hanslope Park Radio Security Service
SCU3 – the RSS team intercepting Abwehr Enigma traffic (under Ted Maltby).

2 Calverton Transmitting Station
Operated in great secrecy by four engineers of the U.S. Civilian Technical Corps. OIC Lord Sandhurst (non-engineering section).

3 Upper Weald receiver site
For two-way working to MI6 agents in Norway etc (OIC Harry Tricker).

4 Nash receiving site
For two-way working to France, Belgium, Denmark etc. Calverton was the transmitting site for both the Weald and Nash Stations.

5 Whaddon Hall – Main Line station
'Prodrome' traffic (Foreign Office Diplomatic), 'Medal' traffic (MI6) and some Ultra.

6 Dower House, Whaddon
Transmitting site for Whaddon Hall – Main Line, on the road to Shenley Brook End.

7 Windy Ridge, Whaddon
Handling Ultra traffic from Hut 3 and run by the Special Operations Group of SCU1.

8 Tattenhoe Bare
The transmitting station for Windy Ridge, distributing Ultra traffic to British and American commanders.

9 Creslow Manor
Transmitter site for Whaddon Hall 1944–1945 and for Hanslope Park 1946–1993 (Diplomatic Wireless Service).

10 Stoke Hammond
Receiving station for Bletchley Park Typex (cypher machine) traffic.

11 Leighton Buzzard
The transmitting site for Bletchley Park Ultra and Typex traffic.

12 Drayton Parslow
A wartime station run by ATS, no further information. Was taken over by GPO who ran it until 1980.

13 Grendon Underwood
SOE station 53A, receiving station for agents operating from France.

14 Charndon / Edgecott
SOE Transmitter Site for 53A.

15 Poundon
SOE station 53B. The receiving station for SOE agents in Belgium, Holland, Denmark, Algeria, etc.

16 Goddington
The transmitting site for station 53B.

17 Poundon
OSS station 53C. The American OSS agents receiving site.

18 Twyford
Transmitting site for 53C.

19 Gawcott
Political Warfare Executive transmitting station. Two transmitters.

20 Potsgrove
The second PWE (Political Warfare Executive), transmitting site for so-called 'Black Propaganda'. It had two 7.5 kW transmitters, for short wave bands only.

In Other Areas:

1 Crowborough
First BBC then PWE, run by Harold Robin SCU3 for the engineering side, as was Potsgrove and Gawcott (medium wave only). Closed in 1982.

2 North Cerney
Near Cirencester in Gloucestershire. This was one of the first MI6 Section VIII transmitting stations, used for the earliest broadcasting of Black Propaganda. It was also a transmitter for Windy Ridge.

Note 1: The MI6 wireless station at Bletchley Park (known as Station X) started earlier on Gambier-Parry's orders, and was closed on 20 November, 1939. All the masts and aerials were removed on 27 November, 1939 and the station moved to Whaddon Hall.

Note 2: SOE did not have its own communications network and cyphers until June 1942. Prior to that, all communications were by SIS/MI6 through North Cerney and Whaddon. All SOE code work was performed on the first floor of Bletchley Park mansion under John Tiltman. All SIS code work was supervised by Miss 'Monty' Montgomery.

Note 3: Gambier-Parry had his HQ at Whaddon Hall where the first PWE discs were recorded and sent by car to North Cerney, and later to Potsgrove and Gawcott. Subsequently, he moved his own personal office to Wavendon Towers whilst retaining an office and bedroom at Whaddon Hall, which remained the HQ of MI6 (Section VIII). Black Propaganda broadcasts were by that time being broadcast from Milton Bryan in Bedfordshire.

Note 4: All the special enlistment staff at these secret sites were billeted out in private homes over the region but in mid-1944, they were asked to move into the new Tattenhoe hostel which had been opened specially for them. These billets had been known as the 'guinea billets' as the rent was 21 shillings (£1.05p) a week.

From 1944 onwards, SCU at Little Horwood supplied the transport from the Tattenhoe billet to all the secret wireless stations in the area, with the exception of Stoke Hammond and Leighton Buzzard stations, which were run by the RAF only. The Tattenhoe hostel closed in 1948 when the majority of Whaddon and Hanslope Staff were transferred to Bletchley Park, using huts, 3, 6, 8, 19 and the rear of block F.

Note 5: Many of these aerials remained in place for many years after the war, and those at Crowborough were particularly noticeable. Even when the masts were removed, the concrete bases for the masts and the stays were often left in place.

In August 2001, I gave a talk in the Whaddon village hall about Whaddon's role in the Secret Wireless War, and mentioned the transmitter in what we called the 'Dower House' on the road to Shenley Brook End. The station was in a field beside an area which is now a row of houses. One owner was pleased to learn from me the reason for the huge chunk of concrete he found buried in his back garden!

Chapter 17

The Packards of Whaddon

Before Section VIII took over Whaddon Hall towards the end of 1939, Whaddon was a quiet farming village that only came alive during village fetes and when the famous Whaddon Chase Hunt met. Traffic along the High Street was sparse, consisting of farm carts, tractors, bicycles and the occasional car.

As divisions of the unit started to move into the Hall from Barnes, Broadway and Woldingham, apart from removal vans there was still little military traffic as most of the senior incoming staff at the Hall were still civilians and used private cars. However, in the early days of 1940, there was a sudden increase in the number of army vehicles seen in the High Street. The closure of Section VIII's operations at Bletchley Park (including Station X), meant more traffic still in the village. The army vehicles came from the newly-formed military support unit, based at nearby Little Horwood, which started providing transport services for the expanding unit – motorcycles, coaches, lorries of all kinds and cars.

From May 1940, however, one would have been very aware of a constant stream of some of the world's most luxurious motor cars driving through the village and into the Hall grounds. These were the Packards of Whaddon and this is the story behind them.

Along with SIS's own intelligence gathering, Bletchley Park was becoming increasingly skilled at breaking the Enigma codes and listening in to the wireless traffic of the German forces. All this information was only of use if it could be delivered to commanders in the field at all speed after the messages had been decyphered. To do that, it was felt necessary to provide mobile units based at the various major command HQs, which were able to move with the unit if necessary. SIS had experimented with mobile wireless units in the late 1930s but, due to lack of funds (a constant problem for the head of SIS), the work was limited in scope.

1940 Packard Sedan as illustrated in their catalogue

In 'Spuggy' Newton and Bob Hornby, Section VIII had two men who had worked on designs of car radios at Philco. Alongside them was Wilf Lilburn who, I understand from his widow, Joyce, had been involved in providing the Glasgow Police force with early car radio equipment. This team designed wireless equipment that was not only small enough for use as agent's sets but that could be operated in cars or lorries. The first such vehicle was a Dodge car and it was taken to the BEF (British Expeditionary Force) in France in late 1939. However there was first a political angle to be played out. In his diary entry for 25 October, 1939 (See Chapter 21), John Darwin says:

'I think we must take the greatest care not to antagonise the Royal Corps of Signals. So long as our station is definitely for SIS work all should be well – but DMI Field Force (Mcfarlane) must not communicate with us over Paddy Nesbitt's head. Otherwise there will be hell to pay.'

Clearly, the army's own Director of Signals Communication would resent the presence of someone else at HQ passing information by wireless, hence John Darwin's concern. To operate properly, they needed the least disruption and the best possible co-operation from the military personnel based at the BEF HQ.

Bill Sharpe (later to be Lieut. Col. in charge of Section VIIIP at Whaddon and later still to command DWS), is known to have been an operator with an SIS mobile unit with Lord Gort's HQ in France before Dunkirk. Whether he used the Dodge is not recorded, but Section VIII also owned an Oldsmobile, a Humber and a Ford (amongst other vehicles), one of which was sent to Air Advance Strike Force.

The value of mobility in passing information to commands was made abundantly clear when it was realised that the Dodge was able to handle intelligence traffic whilst travelling back to Dunkirk.

As stated earlier, the Enigma information from Bletchley Park was by no means our sole source of intelligence. For many years, SIS had had a network of agents operating throughout Europe. In the early months of the war, established agents provided information but as Europe was gradually overrun anti-Nazi patriots were sent back to their own countries and bravely took their place. Initially agents worked back to the Section VIII's operators at Station X in the tower of Bletchley Park, but thereafter to the expanding Whaddon complex in such stations as nearby Nash and Weald.

The army returning to Britain after the debacle of Dunkirk had little or no command structure, so to facilitate its more rapid reorganisation, use was made of the one system that had not been affected by the withdrawal – the Regional Army Commands. These were in the regions and known as Southern Command at Wilton near Salisbury, Northern Command in York, Western Command in Chester, Scottish Command in Edinburgh, and Irish Command in Belfast. To each of these commands, Gambier-Parry sent a wireless unit built into a Packard motor car.

The main reason for these mobile units spread across the country has not previously been recorded. It must be realised that, in spite of Churchill's brave words to the public, there was great pessimism in the military high command. We had a disorganised and partially demoralised army, much of which had just been evacuated from Dunkirk without its equipment. The air force was then of unknown quality, and in spite of the overwhelming superiority of the Royal Navy, only a few miles of sea lay between German-occupied territory and our own shores.

The German invasion, which was expected at any time after Dunkirk, was expected to come from the Sussex and Kent coast, with the invaders making for London as soon as they could. It was essential, therefore, that SIS communications and Ultra traffic from Bletchley Park should be able to continue operating, even if the counties south of London fell, leading to the occupation of London and then up through the country. This is why the Packards were stationed at regional army commands, so SIS/Ultra communications would be available right the way up to Edinburgh, and even then they would be mobile and thus able to operate on the move and from any location.

It is also worthwhile recording that these mobile units from Whaddon took part in a major Army/SIS exercise in August 1940. It was based on the assumption that the Germans had indeed invaded and had rolled across the country in the way they had done in Holland, Belgium and Denmark, and as most military experts were predicting. The exercise proved that the units were still able to provide SIS and Ultra intelligence to Army, Navy and Air Force commanders, as successive parts of country were occupied.

To provide the mobile intelligence units for these Army regional commands, the Admiralty and various RAF commands, Gambier-Parry knew he would need a fleet of cars and a search was made to find the most suitable vehicles. The cars had to be large enough for the passenger compartment to be converted into a radio room, powerful enough to handle the weight of several men and heavy equipment, and with a capacious boot to house batteries and petrol-fuelled chargers and generators. These specifications all pointed to American cars. By the late 1930s, all the major American motor manufacturers had showrooms in the UK, and produced models with right-hand drive. A number of suitable makes were thus available.

The most renowned American car of the time – perhaps of all time – was the Packard whose UK distributor was Leonard Williams Limited, with showrooms in Brentford, Middlesex, on the Great West Road. Section VIII used SIS funds, apparently with a contribution from the British Army, to purchase Williams' entire stock of new Packard cars.

The Packards were later sent to Tickford's factory, now the Aston Martin factory, in Newport Pagnell, about ten miles from Whaddon. At Tickford's they were stripped down and repainted in camouflage colours. My

A Packard of our 'A' Detachment in Alexandria in 1941

1940 Packard Coupe as illustrated in their catalogue

father was at Whaddon Hall in 1940 and saw some of them before they were sent away for treatment. During my service at Whaddon, he described them to me as arriving in a wonderful selection of colours. The drop head coupes included one which he described as being 'lemon yellow'. My father remembers that the Packards were a 'drab lot' when they were returned to the unit from Tickfords. Most of the vehicles purchased were large saloon cars, known in the U.S. as a 'Sedan'. We had one or two drop head coupés, several 'business' (fixed head) coupés, and I believe, three Super Eight limousines.

As soon as the Packards came back from Tickfords, 'Spuggy' Newton (and anyone he could rope in who had radio knowledge), started to fit them out as mobile wireless units. Edgar Harrison was one of Spuggy's helpers (see Chapter 31). As each car was completed, it was sent off to a regional army command, the Admiralty and RAF Fighter Command. The Packards were manned by a crew of three, consisting of a member of Section VIII and two army personnel. (See Chapter 9 and the explanation accompanying Daily Orders Part 1 for 25 July, 1940).

The equipment in each vehicle was identical. The workshop teams installed a Mark III transmitter (of our own manufacture), and an American HRO radio receiver. The electrical power needed might come from several sources including local mains connections, batteries (using a Tiny Tim battery charger), or an Onan AC generator.

These were the mobile 'Special Liaison Units' (SLUs), a name given to them by Fred. W. Winterbotham, Head of MI6 (Section IV – Air), who was charged by Stewart Menzies ('C'), Head of SIS, to ensure the security of outgoing Ultra traffic.

Our unit had a large 'real army' support group based in a camp known to us all as 'Gees', close to the nearby village of Little Horwood. It included drivers, cooks, despatch riders, military police, etc. When using an army vehicle, we had official army drivers so I only drove a Packard saloon myself if a driver was not available, when our driver wanted a rest or sometimes if my immediate boss, Dennis Smith, felt I deserved a treat. I am sorry to admit that this was not a frequent occurrence.

On one occasion, I had the great good fortune to drive the drop-head coupé belonging to Major Freddie Pettifer who was in charge of our transport which included all the Packards. It was 'D Day' (6 June, 1944) and we had just finished installing new equipment on our test aircraft, an Avro Anson, based at Horwood aerodrome. Dennis Smith was called away urgently, leaving me to finish soldering the final connections to the latest version of our air-to-ground wireless, intended to contact agents from the air. I believe this updated model was designed by Dennis.

I finished the work earlier than expected. The perimeter and runway were clear (it being a secondary airfield mostly used for training and emergency landings), and since I was very bored I set off in the coupé down the runway, increasing the speed to something over 100 mph (according to the clock). I then swung round under the control tower and returned to where our driver was just waking up from a quiet snooze on the grass near where our plane was parked.

He saw me arrive and was about to explode but before he could say anything, the Wing Commander in charge of the field was driven over in a battered RAF Austin pick-up gesticulating wildly. He calmed down when I suggested he try the car himself – which he did – but not on the runway, just round part of the perimeter. Our driver very kindly did not report my prank to Dennis – probably as it would have reflected just as badly on him for going to sleep! However, if he reads this story and recognises himself in it, then let me say again how sorry I am to have given him such a fright. Even now, I realise how foolish and dangerous it was and cringe at the thought of what might have happened – even on a seemingly quiet aerodrome. *But I was a very young man and I really did enjoy the drive!*

Packards played a major role in disseminating Ultra traffic in the early war years and during my time at Whaddon, Packards were always going into and out of the Whaddon Hall grounds. It must be remembered we had about 70 cars of the 1939-40 ranges, all under Freddie Pettifer's care. Immediately after the war, the largest stock of Packard spare parts in the UK was to be found at The Pettifer Engineering Company, Paxton Place, Gypsy Hill, in South London!

Although there were no Packards at Bletchley Park itself, a daily run in a Packard to and from Whaddon to the 'Park' took one single passenger – Miss 'Monty' Montgomery – to Hut 10 at Bletchley Park and back. A section of this hut was entirely concerned with MI6 agents cypher work and 'Monty' was in charge of all the traffic that arose from our Section VIII stations such as Weald and Nash, all of which were within ten miles of Bletchley Park. So special was this lady that she was billeted in The Chase in Whaddon Village, within the security zone of our unit.

Our many Packards included three limousines for use by Brigadier Gambier-Parry, Lt. Col. Ted Maltby (Head of SCU3 at nearby Hanslope) and Lord Sandhurst. Sandhurst had earlier been Head of Radio Security Service (RSS) who intercepted the Abwehr and Gestapo Enigma wireless traffic – a task later carried out by SCU3 at Hanslope. He would be driven in his Packard limousine to SIS headquarters at 54 Broadway, in Westminster, once a week. Among the secret documents he carried were our agents' wireless plans. Joyce Lilburn, who was in the Planning Department (VIIIP) at Whaddon Hall, frequently travelled with him. This remarkable lady (see Chapter 24) attended a Packard rally with me in August 2001 which started at Bletchley Park. Afterwards, I led a convoy of seventeen of the various models of this beautiful car to Whaddon Hall where we parked in front of the Hall, the first Packards to be seen there for over fifty years.

Joyce is the widow of Wilf Lilburn, who was with our unit before the outbreak of war and she tells me that Sandhurst's Packard limousine had a military number 4304152. At the end of our rally, on the driveway at Whaddon Hall, I asked if I could take a picture of her in the rear seat of the car and opened the offside door to let her in. 'Oh no Geoffrey!' she said, 'Lord Sandhurst always sat on the right-hand side so he could use the speaker to instruct the driver.'

When the first Section VIII unit from Whaddon was sent abroad, it went to North Africa where it was called 'A' Detachment to handle Enigma traffic. Several fully equipped Packard SLUs were also sent over, but were found to be unsuitable for use in North Africa once they came off-road.

My unit at Whaddon was called 'Mobile Construction' and in 1943 we started fitting our wireless equipment into army vehicles such as the QL 4 x 4 as SLUs for North Africa and Italy. Later on, for the re-entry into Europe, SLUs were built into the British Guy 15 cwt stripped-out army wireless vans for Montgomery,

Dempsey and the British and Canadian commands, and into Dodge ambulances for Eisenhower, Patton, Bradley, Spaatz and the American commands. (See my own story Chapter 38).

It was always a pleasure to drive into an airfield to work on our aircraft and see the faces of the guards at the entrance gates who expected an Air Marshal or a General, at the very least, to be arriving in such a luxurious car. Certainly they did not expect it to contain just a sergeant and a couple of lowly signalmen – ostensibly from the Royal Corps of Signals.

Gradually, as greater use came to be made of military vehicles as our mobile SLUs, the Packard cars were returned to be used as passenger vehicles. It reached the point where we could call up transport at Horwood for a Packard to use for almost any purpose. Sadly, towards the end of the war, a number of these beautiful motor cars were permanently idle and parked up at Little Horwood. One of them was even ignominiously being used to tow the unit's auxiliary fire pump. How had the mighty fallen!

It is a matter of great regret that Packard stopped making cars in the 1950s. The important part played by these handsome, powerful cars in the Secret Wireless War will always be remembered. It is doubtful if any other make would have been as suitable for our use.

Chapter 18

Black Propaganda

Section VIII had many roles that were quite separate to its other, better-known duties. One of these was being the voice of Ultra via its station at Windy Ridge, Whaddon. Another is the most remarkable, however, that of broadcasting Black Propaganda. Responsibility for the transmission of the broadcasts of the Foreign Office unit, known as 'Department E. H.', came directly under the head of Section VIII, Brigadier Gambier-Parry and one of his senior wireless engineers, the quite brilliant Harold K. Robin.

Several colleagues worked on the Black Propaganda operation and I am fortunate in having Phil Luck, who worked in one of the wireless stations involved to advise me. He tells his own story in Chapter 29.

The most exacting research on the work of this unusual, but vitally important aspect of the war effort, however, was written by Mark Kenyon for the Magazine *After the Battle* and I am deeply indebted to its Editor, Winston Ramsey, for his kind permission to use it here.

Despite its excellence, there are two important points on which I disagree with the article. The first transmission was made from the SIS wireless transmitter at Rencombe Airfield, North Cerney near Cirencester which was operated by Section VIII. The station, which was later under the command of Brian Sall had been built by 'Spud' Murphy, and was up and running in the autumn of 1939. The initial decision to build Gawcott, which is much nearer to Whaddon than North Cerney, is alleged to be because Lisa Towse, Gambier-Parry's secretary (and later his wife), found the journey from Whaddon to broadcast the disks from her caravan there too tiring. That is highly unlikely though not impossible.

The second point concerns the person who made the first visit to the United States to view the transmitter which formed such a large part of the operation, and became known as 'Aspi 1'. I believe it was Bob Hornby who made that inspection rather than Harold K. Robin, and I refer to this again in the Bob Hornby story, Chapter 22.

Mark Kenyon first became aware that a secret operation had taken place in the little Bedfordshire village of Milton Bryan when his father was inducted as rector of the parish in the autumn of 1947. It was just over two years after the end of World War II and, in a fenced compound near the church, there was a camp for German prisoners-of-war, comprising former U-Boat crews whose homes were in the Eastern Zone of Germany and who could not be repatriated due to the Cold War. Occasionally, the camp warden and his family or one of the Germans went to the rectory for a meal and, in return, the hosts were sometimes invited to a film show in the camp canteen where they enjoyed Charlie Chaplin films surrounded by officers and men who still wore the uniform of the U-Boat arm of the Kriegsmarine.

The warden told them that the camp had been some sort of wireless station during the war and he assumed, from recordings of Hitler's speeches that were found there, that its purpose had been to monitor broadcasts from Germany. On one occasion, the warden took them into the main building and showed them the offices and studios that were still complete with their furniture and equipment.

Mark and his father were among the first ordinary civilians to see the Milton Bryan complex as it had looked during the war. In the years that followed, Mark often wondered about the clandestine wireless station at Milton Bryan but it never occurred to him that the wireless masts and buildings in the fields at nearby Potsgrove, also in his father's parish, had been anything more sinister than some sort of Post Office relay station.

It was not until 1987 that the name of Milton Bryan appeared in front of his astonished eyes in a page in Janusz Piekalkiewicz's book '*The Sea War*', and he first found a direct reference, albeit sketchy, to the real work that was done in the village during World War II.

An appraisal, prepared in connection with conservation by the Bedfordshire County Council, describes Milton Bryan as 'a place of exceptional character and charm, a virtually unspoiled example of a tiny English village'. It has some 50 houses and a population of around 140 and lies some 3 miles east of the A50 road to Northampton, isolated in the heart of the Bedfordshire countryside.

Milton Bryan, when I knew it, had but one small shop and a pub. It had no regular bus service and was a village that could only be 'discovered' by the inquisitive motorist. It is doubtful if even the villagers themselves know anything of the operation that was carried on in the former War Department buildings near St Peter's Church.

It would be logical to assume that the isolation of the village was the governing factor in the decision to use Milton Bryan, but the reason for its selection was, in fact, quite accidental. In the summer of 1939, it was assumed that when the war began, London would be subjected to intensive bombing by the Luftwaffe. Accordingly, plans were made to evacuate all government departments, together with a proportion of the population, particularly children. One after another, large country houses and stately homes were requisitioned by the authorities. The Duke of Bedford, horrified at the idea of his country seat being occupied by a horde of evacuee children from London's East End, offered Woburn Abbey to the government. The mansion was to be a rent-free loan but the Duke insisted that he did not wish to meet or even see any of the government employees.

Woburn Abbey was allocated to Department EH of the Foreign Office. This was a propaganda section set up not long before the war and whose cover name stemmed from its London base in Electra House. With only rare exceptions, cover names for buildings or places were their correct initial letters. Those people whose identity it was necessary to hide, were normally given cover names that began with the initial letters of their own names.

Department EH was evacuated two days before war was declared. On 1 September, 1939, the staff members were informed that they were to move to 'Country Headquarters' (CH or CHO), but the location was kept secret. They were to make their own way to Dunstable and report to a contact at the Sugarloaf Hotel, from which point transport was to be provided to CH. This, on arrival, was found to be Woburn Abbey.

The Duke of Bedford's palatial mansion was ideal. Woburn was far enough from London to be safe from aerial bombardment yet close enough for easy access to the Foreign Office either by car or motorcycle. The initial staff of 20 worked in offices in the riding school and lived in the flats above the stables that had formerly been occupied by the families of the ducal grooms. The fact that these buildings were separated from the Abbey by a spacious courtyard satisfied the Duke's requirement of no physical or visual contact. The Duke died in 1940 and the growing problem associated with a substantial increase in the number of personnel was resolved when the Abbey itself became available for use. Clifton Child, who worked in the Abbey, was surprised at the totally unsentimental attitude of the Foreign Office towards the requisitioned houses and their contents. He worked in a room surrounded by priceless Gainsboroughs and Canalettos and used Adam bureaux as filing cabinets!

The staff of Department EH settled into their new home at Woburn and were mainly engaged in the production of propaganda leaflets. The Department also operated a number of propaganda wireless stations that broadcast to Occupied Europe. Two of these broadcast in German and run by refugees, one right-wing the other a left-wing station operated by German communists.

Both the leaflets and the broadcasts were of the 'white' propaganda variety, that is to say their message was simply one of encouragement to obstruct the Nazi regime.

Initially, the broadcasts were recorded at Whaddon Hall but, as the department's activities began to expand, a headquarters with recording studios was established at Wavendon Tower in the same county. It was given the code-name 'Simpson' which was the name of a neighbouring village.

A site for the transmitter was chosen at Gawcott, south-west of Buckingham, and it was equipped with two 7.5 kW, low-powered, short-wave transmitter units. They were RCA 4750 models, built in the United States. Shortly after Gawcott became operational, it was found that more power was needed and an identical transmitting station, using the same equipment, was built at Potsgrove, a small village close to Woburn Abbey. Milton Bryan happened to be another village within the boundaries of the Woburn estate and that is how it came to be the home of Black Propaganda.

The organisation of the British Secret Service is complicated and confusing, sometimes deliberately so but, apart from the substantial involvement of Section VIII in handling the vital wireless aspect, SIS itself played little part in the story. It is sufficient to say briefly that in July 1940, Military Intelligence Research (MIR) and

The aerial array at Gawcott

Section 'D' of the Special (or Secret) Intelligence Service (SIS), which specialised in sabotage and subversion, were amalgamated to form the Special Operations Executive (SOE). Department EH was attached to SOE in 1940 but later became a Department in its own right known as the Political Warfare Executive (PWE).

Programmes were recorded at Wavendon Tower on American-made, 16-inch glass discs that were played at the then standard speed of 33⅓ r.p.m. To begin with, the recorded discs were played at Whaddon Hall and sent by Post Office landline to the Gawcott transmitter. This presented an obvious security risk, so when 'Simpson' came into being, the discs were taken from Wavendon Tower by road to the Potsgrove transmitter.

Many of the staff were foreign nationals with little or no English and this, coupled with the secret nature of their work, made it necessary to isolate them from the outside world. Enquiries from neighbours were answered with the simple statement that the houses were occupied by people engaged in research work for the government. The early broadcasting stations were in fact cover-named 'Research Units' (RUs). The staff were not permitted to make telephone calls, post letters locally or even visit the local pub. Outgoing mail was sent to London for posting and a London Post Office box was used for incoming letters.

In November 1941, when the Gawcott and Potsgrove transmitters were completed, there were 20 RUs in operation. Nine more went on air in 1942 and a further 16 in 1943, although some of these were replacement stations. In October 1943, there were 20 active Rus. One more, the final station, opened as late as January 1945, broadcasting under the name of 'Hagedorn' ('Hawthorn').

For the first 14 months of the war, Department EH continued with its work at Woburn. November 1941 saw the beginning of a large-scale development and a significant change of plan that was marked by the arrival of a man of whom it has been written that he was 'the nearest thing to a genius which PWE produced. In fact in his particular line he *was* a genius'.

Denis Sefton Delmer was known to his friends and close associates as 'Tom', a nick-name given him by his Jewish barber in Berlin, Moses Muhling, to whom all English-men were 'Tommies'. Delmer liked the name as it tended to distinguish him from his well-known father, Professor Frederick Sefton Delmer.

Sefton Delmer, as he was known professionally, was born in Berlin on 24 May, 1904. At the outbreak of war in August 1914, his father, who was a professor of English at Berlin University, was interned. The family was allowed to remain at home and Sefton Junior continued his education at a local school. In May 1917, Professor Delmer was released and together with his family was repatriated to England where his son was enrolled at St Paul's School, later winning a scholarship to Lincoln College, Oxford. In 1919, his father returned to Germany as a member of the Allied Control Commission and Delmer spent his holidays there.

After graduating from Oxford, Delmer decided to take up journalism as a career and, speaking German like the native that he almost was, became a European correspondent for the *Daily Express*. He soon made a name for himself, not only for his newspaper but also in Germany. He was personally acquainted with Hitler, Göring, Goebbels, Hess, Himmler and the other Nazi leaders, and was the only foreign newspaper correspondent permitted to travel in Hitler's entourage during the future Fuhrer's campaign for leadership in 1932.

Delmer left Germany when war was declared. When France fell in June 1940, he was evacuated from Bordeaux in the *Madura*, the last ship to sail for England. Delmer was keen to take an active part in the war but realised that his weight of seventeen and a half stone rendered him unsuitable as a front line soldier. Instead, he thought that his knowledge of Germany and the German language would make him useful as an interpreter or in some branch of intelligence.

Friends involved with secret work did their best to find Delmer a job but to no avail. The knowledge and experience that Delmer considered to be of value to his country were seen in a different light by the internal security authorities and the reason for their suspicion was, it seemed, the fact that in 1917, in the midst of war, the Delmer family had been allowed to leave Germany. After his escape from France in 1940, MI5 agents actually attempted to uncover Delmer's 'real' role as a Nazi activist.

In July, 1940, Duff Cooper, the Minister of Information in the new coalition government, asked Delmer to assist with improving the BBC's German broadcasts and, while continuing his work with the Daily Express, he began a series of programmes that were broadcast every Friday evening.

Sefton Delmer who was responsible for the Black Propaganda operation

On the day of the first programme, Friday, 19, July, 1940, Hitler made his famous 'final appeal to reason' speech. This was during the period of the 'phoney' war when those in positions of authority still went for long weekends in the country. Winston Churchill and the Cabinet had already left London and Delmer took it upon himself, without any authority whatsoever, to give Britain's reply.

Less than an hour after Hitler's last words had faded from the ether, Delmer broadcast live from the BBC. Speaking in fluent, colloquial German, he told Hitler: 'Let me tell you what we here in Britain think of this appeal to what you are pleased to call our reason and common sense. Herr Führer and Reichkanzler, we hurl it right back at you. Right back into your evil-smelling teeth'.

According to the American radio correspondent in Berlin, William Shirer, and to Mussolini's son-in-law, Count Galeazzo Ciano, the German reaction was a mixture of fury and disappointment. In Britain, Delmer's unauthorised broadcast raised a furore and Richard Stokes, the Labour MP for Ipswich, attacked him publicly in the House of Commons. It was only when the Foreign Secretary, Lord Halifax, replied officially two days later, and in much the same vein, that the heat was taken off.

Delmer continued with his Friday broadcasts and in September, 1940 he was asked by Leonard Ingrams if he would consider a full-time post with the BBC. Ingrams was a banker who had known Delmer in Berlin. He was now with the Ministry of Economic Warfare and also, secretly, with S02 (later to become SOE) and SO1 (later PWE). Delmer accepted on the grounds that this was better than nothing and perhaps a step in the right direction, but once again he was vetoed by internal security.

In November, 1940, the *Daily Express* sent Delmer to Lisbon. Before he left London he was asked by MI5 if he would be prepared to assist while he was in Portugal if required. Delmer gladly accepted.

In Lisbon, Delmer met a number of refugees from Germany and through them he was able to update himself on life in the wartime Reich. He also gained an insight into the tangled web that was the British Intelligence Service. When contacted, he found that the original approach from what he thought was MI5 was in fact from MI6 via MI5.

Delmer had barely met his MI6 contact when he received a telegram from Leonard Ingrams recalling him for an 'important job'. He flew back to London and was taken to Woburn Abbey where he signed the Official

Secrets Act and was interviewed by Richard Crossman, the future Labour MP, and at that time the Director of the German Section of SO1.

Delmer now had the sort of job that he wanted with a secret department but he had not been told what it was! He had an office in a secret building in Fitzmaurice Place off Berkeley Square but no work, and he was at the point of negotiating a transfer to Admiral Godfrey's Naval Intelligence Department (NID) when the new PWE, through Leonard Ingrams, offered him the job that he had been waiting for.

The right-wing RU had closed down when its operator was taken ill and PWE wanted Delmer to form and operate a new substitute station. He was to have complete control but was to work in close association with SOE, if and when required. In operating the station there were to be no limits and no holds barred; his only guideline was that similar German propaganda stations broadcasting to Britain used offensive language. 'Do something along those lines', Ingrams told him. The new RU was to be 'dirty' in more ways than one.

Delmer returned to his office and set to work. After studying the existing RUs, he came to the conclusion that they were preaching to the converted, broadcasting to those very few Germans who were already opposed to Hitler. Delmer felt it desirable, if not necessary, to reach the great majority who currently supported the Nazis but, as these people were highly unlikely to take any notice of appeals for rebellion from Britain, they would have to be tricked. In the same vein as such terms as 'black mass', 'black market' and 'black magic', Delmer christened his operation 'black propaganda'.

It was patently obvious that to simply go on the air and expect listeners to believe a station broadcasting from Britain was useless. Psychologically, it would be far better for Germans to find the wavelength by accident and, having done so, think that they were listening to a German military station whose bored announcer was whiling away the time by interpolating his own views, comments and pieces of information. For the sake of credibility, though, he had to give the impression that he was a loyal follower of the Führer though cynically critical of those in positions of authority.

In the knowledge that all German servicemen had taken a personal oath of loyalty to Hitler, the station would endeavour to drive a wedge between the men of the old Imperial Army and the new generation of fanatical Nazis. To sum up, the station would pretend to support Hitler but would expose a growing (and fictitious) rift between the traditionalists and the radicals. The objective was to make its listeners think and act in ways that would undermine the German war effort.

Delmer's chosen method of operation was a deliberate gamble, for audience reception was, at first, to be entirely dependent upon 'knob-twiddlers' – German military wireless operators and civilians who idly turned the tuning dials on the receivers and found the station by chance. When listeners did find the frequency it was important to grab their attention. The broadcasts would contain a considerable amount of truth to make them sound credible but the language used by the announcer would be coarse and often obscene. Delmer had a hunch that the use of barrack-room slang would encourage his hoped-for audience to keep listening and to pass the word on to others.

Delmer put his ideas down on paper and submitted them to PWE, after first obtaining and attaching written recommendations from both Richard Crossman and Reginald Leeper, his department head. The plan was accepted. Delmer called his station 'Gustav Siegfried Eins', the German Army signals phonetics for 'GS1', and what in the present day international system would be 'Golf Sierra One'. It meant nothing but was intended to give a ring of military authenticity. During his pre-war days in Germany when he moved in exalted circles, Delmer had noticed that the Nazi hierarchy referred to Hitler as 'Der Chef' – The Chief. His announcer, therefore, would use the same name.

Delmer left London and moved to Bedfordshire, taking up residence in a house named The Rookery in the village of Aspley Guise (RAG) where his initial team consisted of his wife, Isabel, Johannes Reinholz, Max Braun and 'Paul Sanders', whose real name was Peter Seckelmann.

Paul Sanders was a writer of detective stories from Berlin who had reached England in 1938. While a corporal in the Pioneer Corps, he volunteered for special duties and as a result of his interviews was passed on to Delmer for consideration. The tone of his voice, his accent, and the way he spoke were perfect for what Delmer had in mind, and Paul Sanders became 'Der Chef'.

On 23 May, 1941, Delmer and Sanders travelled by car to 'Simpson' to record the first broadcast for Gustav Siegfried Eins. The recording was then taken to Potsgrove where it was transmitted on two frequencies for ten-minute periods between the hours of 1253 and 2353. The same recording and transmitting procedure was followed every day. As the intention was to give the impression that GSI had been in existence for some time, there was no special introduction. 'Der Chef' simply launched into what was to become a daily routine, always using the same format. He would begin each transmission with a series of coded messages for various other, equally fictitious, GS stations. 'GS1 calling GS23' he would announce, read that station's message, and then call GS9 or GS12 or any other number that came into his head.

The messages relayed by 'Der Chef' were in a simple numeric code that could easily be broken. Delmer knew that the radio monitors of the Reich Central Security Office, the Reichssicherheitshauptamt (RSHA) would quickly track the location of the station to England and so every group of numbers translated into a simple message similar to those broadcast to the French Resistance after the BBC news at 9.00 p.m. Karl, for example, would be instructed to meet Rudolf outside the Union Cinema during the interval between performances. As the Union cinemas in Germany were as prolific as their Odeon counterparts in Britain, Delmer hoped that a large force of Gestapo agents would be out on the streets endeavouring to catch Karl and Rudolf red-handed at whatever it was they were supposed to be doing.

The 'Rookery' at Apsley Guise, Sefton Delmer's wartime base.

Everything that Delmer did was designed to be disruptive. 'Der Chef' was pro-Hitler and anti-British. He referred to Winston Churchill as a 'flat-footed bastard of a drunken old Jew' and it was hoped that by so doing, no German listener would ever believe that his broadcasts emanated from Britain. Delmer was himself anti-Semitic in a half-hearted sort of way and also disliked the French, the Russians and even the Italians.

Every night he attacked high Nazi Party and SS officers. Until September 1942, he left Hitler alone, but Himmler, Goebbels, von Ribbentrop and 'Sepp' Dietrich came in for their fair share of criticism. When he eventually included Hitler in his diatribe, 'Der Chef' was sympathetic and stated that the Führer was obviously being influenced by incompetent advisers. Generals of the old school such as von Rundstedt, Kesselring, Halder, von Kleist and von Bock were complimented but the popular Erwin Rommel was referred to contemptuously as a badly-trained, swash-buckling careerist.

The most popular part of 'Der Chef's' broadcasts was his revelation of sensational 'inside' information and his detailed descriptions of the sexual deviations practised by members of the Nazi hierarchy. 'Der Chef's' first broadcast went out 12 days after Deputy Führer Rudolf Hess had parachuted into Scotland and included in the programme was a list of suspected associates who were said to be under arrest. The names were real but the list was fictitious and Delmer was overjoyed when, several days later, German newspapers reported that some of the listed persons had actually been interrogated by the Gestapo.

Each broadcast ended with the words 'I shall repeat this broadcast, all being well, every hour at seven minutes to the hour', a deliberate suggestion that 'Der Chef' was taking a big risk and could well be arrested by the Gestapo himself.

Although the first recording was a little crude, it soon became very professional as 'Der Chef' gained confidence with experience and began to be involved with the scripts. Soon he began to attack local Party officials as well, telling of their black market dealings and other illicit activities.

The information used by 'Der Chef' was all true and was gathered from a number of sources. A file of names was compiled from German newspapers and magazines, conversations between PoWs were secretly recorded, and letters en route to the then-neutral United States were opened and read by censors. The best source was the Admiralty's Naval Intelligence Department. Admiral John Godfrey (whose PA was Ian Fleming who wrote the James Bond novels) had set up a special propaganda section (NID17) to work with Delmer, headed by Commander Donald McLachlan. Each of the names and addresses mentioned by 'Der Chef' was genuine. The accusations were invented, of course, but it was surprising to find out how many of them turned out to be true!

One such story concerned Robert Ley, the head of the Deutsche Arbeitsfront (German Labour Front) and the Kraft Durch Freude organisation (KdF, meaning 'Strength through Joy'). 'Der Chef' told his listeners that Ley was obtaining extra food rations under the counter for himself and his family. It was a complete fabrication but three days after the broadcast the story was told to his interrogators, by a captured Luftwaffe officer! Shortly afterwards, Ley and Goebbels attempted to scotch the rumour through the pages of the Party newspaper, *Der Angriff.* The seeds of doubt that Delmer was sowing in the minds of the German people were taking root.

After 18 months of operation, bigger and better plans for Delmer's talents were about to reach fruition, and Gustav Siegfried Eins came to an abrupt end. One night in October 1943 'Der Chef' had his dramatic swansong. Special sound effects were laid on for his last broadcast which ended when he was finally located by the Gestapo who burst in on him while seated at his microphone and gunned him to death. Play-acting, of course, but listeners were treated to guttural shouts of 'Schweinhund!', the German equivalent of 'Gotcha!' and bursts of sub-machine gun fire.

It must be remembered that 'Der Chef's' broadcasts were recorded and repeated at hourly intervals. The engineers at Potsgrove did not understand German and they had not been told of the significance of the final recording. Consequently, one hour after the 'death' of 'Der Chef', Delmer was horrified to hear a replay of the whole dramatic performance! What he said to those concerned is not recorded nor is there any evidence that listeners in Germany heard the same broadcast twice.

Unknown to Delmer at the time, exactly one week before 'Der Chef' recorded his first broadcast for Gustav Seigfried Eins in 1941, a plan for a far more complex 'black propaganda' radio operation was submitted to Winston Churchill and approved by him the following day. The plan originated when Hugh Dalton and Anthony Eden visited the studios at 'Simpson'. Eden was impressed by what he saw and asked if there was any way in which the output of the unit could be greatly increased. Harold Robin replied that all that was needed was a transmitter of much greater power.

The plan involved broadcasting on German wavelengths and was one that had interested the Prime Minister for some time, hence his speedy approval. (He simply said 'Pray proceed'.) It had originated in SO1 and was developed under the guidance of the three ministerial heads of PWE: Anthony Eden (Foreign Office), Hugh Dalton (Ministry of Economic Warfare) and Brendan Bracken (Ministry of Information).

Colonel Richard Gambier-Parry, the head of Section VIII, the SIS communications department, had heard of an eminently suitable transmitter in the United States. He sent Bob Hornby to see the manufacturer's United Kingdom sales manager and then submitted his purchase proposal. The 500 kilowatt medium wave monster was the highest powered transmitter in the world. It had been built for station WJZ in New Jersey but its intended use had been vetoed by the Federal Communications Commission because it exceeded the maximum limit of 50 kW laid down for commercial radio stations.

The equipment was inspected by Bob Hornby and an option for its purchase taken out on behalf of the British government. After approval by Churchill, the transmitter passed into the hands of PWE. The cost was estimated to be in the region of £165,000, including the cost of improvements and modifications. Being the most powerful transmitter in the world, it was promptly codenamed 'Aspidistra', in reference to 'The Biggest Aspidistra in the World', the song made popular by Gracie Fields.

The Chief Engineer for the Black Propaganda operation was Harold Robin from Section VIII. Our unit designed and built its wireless stations, as well as providing the technical staff to run them. This included the massive Aspidistra transmitter at Crowborough.

Harold Robin was then sent to the USA in the summer of 1941 and spent two months at the RCA factory at Camden, New Jersey, familiarising himself with 'Aspidistra'. Robin was another of the Philco team of radio 'wizards'. The son of an inventor, he did his early groundwork in his father's factory and soon developed a natural flair for radio. He cultivated this flair at school at Oundle and at university and then found employment in factories making wireless sets. While working at Philco, Robin encountered Richard Gambier-Parry who was responsible for Robin's entry into clandestine radio work in late 1939.

By the time Robin arrived at Camden, two of the three parts of the transmitter were complete, and Robin supervised the building of the third part which increased the power to 600 kw. He also modified the design so that the three parts operated in parallel. The three masts and the aerials were designed by G. H. Brown, an American antenna expert. The masts were 300 feet in height and were designed to stand on 25-foot pedestals. The completed transmitter and its masts were shipped to Britain in several consignments aboard ships of the Royal Navy, but the vessel carrying the aerial masts was torpedoed and a duplicate set had to be built and shipped at a later date.

Naturally enough, 'Aspidistra' was scheduled to become a part of the PWE broadcasting 'empire' in Bedfordshire. A large new studio complex was built in the tiny village of Milton Bryan, just inside the boundary fence of Woburn Park and a convenient mile or so from the Abbey. 'Aspidistra' was to be sited in Bedfordshire too, in a disused gravel pit, but Harold Robin insisted that the location must be as close to Europe as possible.

Several sites in Sussex were suggested but one after another they were turned down when objections were raised by either the Air Ministry (the height of the aerial masts would be a danger to low-flying aircraft) or the BBC (the transmitter would interfere with research work on secret RAF equipment). Finally, in exasperation, a meeting that was attended by representatives of the Home Office, the Post Office, the Air Ministry, the BBC and other interested parties was called in Whitehall. The delegates were confronted by a large map of Sussex on which all of the proposed sites were marked. Each was discussed and marked off when an objection was raised until a site to which no-one present objected was selected in Ashdown Forest. The site was at King's Standing, near Crowborough. At 620 feet above sea level it was the highest point in the Forest. It was public land and this time the conservators of the forest objected, but when threatened with a requisition order they gave way. 'Aspidistra' was to be housed underground. Some 70 acres of land were fenced off and a large hole 50 feet deep was excavated by a Canadian Army road-building unit which happened to be billeted nearby, waiting, so the commanding officer said, for the invasion of Europe! The hole was roofed with bomb-proof reinforced concrete four feet thick.

The concrete and steel were put in place by a civilian labour force 600 strong which worked 24 hours a day until the job was completed. Floodlights were used at night and only extinguished when there was warning of an air raid. The site took three weeks to excavate and the transmitter complex was completed in just nine months from the date of approval by Churchill. The construction and the installation of 'Aspidistra' was supervised by Robin assisted by Cecil Williamson whose pre-war employment in the British film industry was said to have influenced the interior decor.

While the site for 'Aspidistra' was under construction, a number of further objections to its use were raised by the BBC which now revealed that it was engaged in research on jamming apparatus that could be installed in RAF bombers. The BBC said that if transmissions started then the BBC would be bombed in retaliation. Heated discussions took place and eventually, as PWE had not actually worked out what it would do with its new transmitter, an agreement was reached that 'Aspidistra' would be used to supplement the BBC's overseas broadcasts. 'Aspidistra' was ready for broadcasting in early 1942 but was not used until 8 November, when it came on air to support the Torch Landings in North Africa with a broadcast by Roosevelt. This was recorded in the Oval Office by his nephew – a radio ham – on an 8-inch disc which was duplicated in great secrecy at Wavendon Tower.

Delmer must have known about this, but claimed that a use for the transmitter did not occur to him until one month later, just before Christmas 1942, when he was taken to lunch by Donald McLachlan. McLachlan's chief, Rear-Admiral John Godfrey, had had a shore posting in Germany in 1936. While there, he became aware of differences in German political opinions and of a rift between U-Boat ratings and their officers. In 1942, as Director of Naval Intelligence, he had the foresight and imagination to realise that one of the best ways in which the U-Boat menace could be overcome was to widen this rift which, on the evidence of the German Navy's mutiny at Kiel during the previous war, would quickly spread to the Army and the Luftwaffe.

The power house at Crowborough under construction.
Some idea of the size of the project can be judged by the thickness of the concrete walls.

It was with this in mind that Godfrey set up a special propaganda section, NID17z, which had been working with Delmer for some time. Now, through Donald McLachlan at the Christmas lunch, he put forward a plan for the development of his idea and suggested that Delmer open a new radio station that was to be aimed specifically at the U-Boat crews.

At the time, Delmer was unhappy with his work and its results to date. Godfrey's proposition appealed to him immensely but he realised that the present system of recording broadcasts for later transmission would not be good enough for the new station; programmes would have to go out live. In his book, *Black Boomerang*, Sefton Delmer states how at this point he 'suddenly remembered' that there was a vacant studio at Milton Bryan, a recollection that prompted him to tell McLachlan that he could put a new station on air, provided that he could obtain the use of 'Aspidistra'. This account is repeated by Ellic Howe, whose 'fakes and forgeries' propaganda printing unit is a story in itself (which he has recounted in his book *The Black Game*), but it is, in fact, incorrect. One can only assume that Delmer, as a top-flight journalist, thought that his book version made a better story.

Harold Robin was involved from the very beginning and knows the true story. In a recorded interview in 1982, and later in a letter written in February, 1990, Robin states that the Milton Bryan complex was built specially for Delmer's programme. After discussions with Delmer, the accommodation was laid out by Harold Robin and Squadron Leader Ted (later Sir Edward) Halliday. It was constructed by the Ministry of Works, using direct labour, and Robin with his staff installed the wiring and recording equipment for the studios. It follows that 'Aspidistra' was purchased for Delmer's use, despite the delay, and that he knew about it and the plan for the new station much earlier than he admits in his book.

According to Delmer, he put the revolutionary plan to his new chief, Sir Robert Bruce Lockhart, who had now succeeded Reginald Leeper. Lockhart was very interested but was afraid that a great deal of influence would be necessary to prise 'Aspidistra' from the hands of the BBC. How that influence was obtained is not recorded, but among those whose assistance was sought were Major-General Dallas Brooks of the Royal Marines and Delmer's departmental Deputy Director-General in charge of inter-Service relations, and Charles Lambe, the Deputy Director of Plans at the Admiralty. No doubt Churchill also had a hand in it. Whatever went on at top level, 'Aspidistra' was removed from BBC use and control and Delmer and his team took possession of Milton Bryan (MB) in January, 1943.

The MB studio complex covered five acres. The main building was a two storey, red-brick construction in the centre of a compound sited just off the road that runs through the village's Church End. The compound was surrounded by a 12-foot high, steel-mesh fence which was patrolled by Special Constables with Alsatian dogs. These security guards were retired members of the Bedfordshire Constabulary and were responsible to Colonel Chambers, the PWE Security Officer at Woburn Abbey. The guards were employed to keep unauthorised persons out rather than to keep the mainly German inmates in, although, because of the identity of some of the inmates, to say nothing of the effectiveness of the station's transmissions, there was always the threat of direct retaliation in that an attack might be staged by German commandos. In the event of this possibility, the guardroom contained an armoury of rifles and sub-machine guns. There was a firing range within the compound and the sound of staccato bursts of fire must have puzzled the villagers.

The MB studios were the most modern of their time and its telephone operators had years of previous experience with the GPO in London. Scrambler telephones stood on the desks of the senior staff, and there

The 'Aspidestra' aerial array at Crowborough

were direct telephone lines to the BBC and the Air Ministry. There were additional landlines to Reuters, the Press Association and to the PoW interrogation centres, as well as a Hellschreiber (see below). The latter were at Latimer and Wilton Park in Buckinghamshire (see *After the Battle No. 70*) and when Delmer learned of a prisoner who could be useful, he would drive down there immediately.

The delay in Delmer's use of 'Aspidistra' was largely due to the objections from the BBC who were afraid that the Germans might retaliate, either by jamming its broadcasts or bombing its headquarters at Broadcasting House in Portland Place. These objections had now been overruled and Delmer was ready.

The name chosen for the new station was 'Deutsche Kurzwellensender Atlantik', the English translation of which was 'German Short-wave Radio Atlantic'. The Germans soon abbreviated the name to 'Atlantiksender'.

Although Delmer had increased the size of his original team during the 18 months in which Gustav Siegfried Eins had been in existence, he needed a considerable number of extra staff for the Atlantiksender operation, and by the end of the war a total of around a hundred people were working at MB. Very few of them were British nationals, but two who were became Delmer's right-hand men.

Clifton Child was a 30-year-old education officer from Manchester who was fluent in the German language. He joined the army when war was declared and became a specialist corporal in the Royal Corps of Signals. Having just published a book on the subject of Germans in the USA during the World War I, he was posted to the Political Intelligence Department of the Foreign Office where he worked with Dr. John Hawgood on the preparation of both the German and American sections of the FO's Weekly Intelligence Summary.

In March 1943, when the PID was merged with a unit of Chatham House to become the Foreign Office Research Department, Hawgood recommended Child to Delmer and secured the necessary Army and Foreign Office approval for his transfer to MB, where he became Chief Intelligence Officer (Political). Clifton Child spent much of his time poring over German newspapers and intelligence reports and he had an uncanny flair for drawing the correct conclusion from the most vague and apparently unrelated news items. So good was he that Delmer was sometimes suspected by the SIS of having a 'mole' at Bletchley Park supplying secret 'Ultra' decrypts!

Child's opposite number at MB, the Chief Intelligence Officer (Economic) was C. E. Stevens, a Fellow of Magdalen College, Oxford, and an Ancient History don. He came from the Ministry of Economic Warfare and among his accomplishments were a photographic memory and the suggestion that the beginning of Beethoven's Fifth Symphony, which sounded like the Morse code letter 'V', should be used to introduce the BBC broadcasts to the Resistance in Occupied Europe that followed the news every night.

Delmer's deputy was Karl Robson, an Army major, and a fellow journalist from the *News Chronicle* who spoke German as well as he spoke English. Another Englishman who played a key role was Ellic Howe (see above). Howe was the only British national to use a cover name, probably for business reasons. Delmer called him 'Armin Hull' which was another unexplained departure from using correct initial letters.

The great majority of the personnel who worked behind the steel mesh fence at MB were German, some of whom had held prominent positions before the war. Max Braun, who blew his own cover on his first morning at RAG, had led the anti-Nazi campaign in the Saar before the plebiscite of January 1935. Philip Rosenthal was a well-known manufacturer of porcelain who became the Social Services Minister in Bonn in Willi Brandt's Social Democrat government after the war. Rene Halkett was a nephew of a former C-in-C of the German Army, the unfortunate General Werner von Fritsch, and Dr. Ernst Adam had been Chief of Staff to the Republican C-in-C during the Spanish Civil War. Fritz Heine had been secretary of the Social Democrat Party until it was disbanded by the Nazis in May 1933, and later on there was Dr. Otto John (cover name 'Oskar Jurgens'), a German Resistance leader and possibly the only one of the July, 1944 plotters who managed to escape. He was head of the post-war German equivalent of the British Secret Intelligence Service (SIS) until he fell foul of Reinhard Gehlen, West Germany's renowned spymaster.

The rest included some refugees and some deserters, but the rest of the staff were prisoners from the German armed forces – the Army, the Luftwaffe, and the U-Boat arm of the Kriegsmarine.

Exactly how these prisoners had ended up at MB is not quite clear. It has already been mentioned that there was a direct telephone line to the PoW interrogation centres in Buckinghamshire and at that time the Foreign Office had a temporary Prisoner-of-War Department headed by Colonel Henry Faulk, a German-speaking psychologist. It must be assumed that likely candidates were passed on to Delmer and it is known that PoW interrogation was also conducted at Woburn Abbey by, among others, the actor Marius Goring.

One of the most intriguing of the prisoners was a U-Boat Oberfunkmeister (Chief Petty Officer wireless operator) by the name of Eddie Mander. Although a fanatical Nazi when he joined the Navy, his treatment by his superiors turned him, and he willingly used his extensive talents to help destroy his former masters. Mander had been the Chief Radio Officer on the prison ship *Altmark* and, although badly wounded with a bullet in one lung, he evaded the boarding party from *HMS Cossack* by jumping overboard and escaping across the ice. After spending some time in hospital, he was called up for the Kriegsmarine and insult was added to injury when he was required to do his radio training again from scratch. He went on nine operational cruises before being arrested and court-martialled for 'black market' offences.

Finding him guilty, the court took the unusual step of giving him the choice of three sentences. Mander took what he thought was the lesser of three evils, returned to operational U-Boats, and was captured when his boat was sunk a few days out from St Nazaire. Mander became the leading scriptwriter for the Atlantiksender. To Delmer, Mander was invaluable due to his hatred for the Nazis, his knowledge of German radio and radar equipment and techniques, the U-Boat code-book that he brought with him, and his personal knowledge of the U-Boat men and their private lives. He was repatriated to Germany after the war but was murdered shortly afterwards, possibly by former U-Boat crewmen in an act of vengeance.

Even though Atlantiksender was ready to commence operations, negotiations for the transfer of 'Aspidistra' to Delmer had not yet been completed. The powerful transmitter was still in use by the BBC and when the first broadcast took place on 5 February, 1943, the programme went out live on the short wave via the Potsgrove transmitter. As with the first broadcasts of Gustav Siegfried Eins, evidence of feedback was eagerly awaited at MB and the first reports began to be received very soon.

Atlantiksender was based on Gustav Siegfried Eins but was very different in certain essentials. Unlike the earlier station, it purported to be a genuine German forces station. There were no obscenities or lurid stories and the broadcasts were basically a mixture of music and news, the music to make it contemporary and popular, the news to give credibility and to provide the vehicle for Black Propaganda.

On the very first evening, Atlantiksender was detected by the alert monitors of Reichsmarschall Göring's radio security service which realised that the station was British and attempted to jam it. Despite the fact that its cover had been blown within a few hours of its inaugural transmission, Germans continued to listen and a great many of them believed it.

Reference to Atlantiksender soon appeared in newspapers published in the few neutral countries there were left in Europe but, more significantly, captured U-Boat men and aircrew from Luftwaffe units operating in support of the U-Boat bases in France told their interrogators that there was little point in refusing to answer questions because Atlantiksender already knew the answers. Interrogation reports received from other theatres showed, for instance, that the station was extremely popular with the Afrika Korps.

Initially, the station broadcast for three hours daily from 8.00 p.m. to 11.00 p.m. Later, when it was linked with another Delmer station, the Atlantiksender increased its hours and later still the twin stations were broadcasting 24 hours a day.

The British Embassy in Stockholm acquired recordings of the latest German Top Ten hits, which were flown to England by RAF Mosquito aircraft, while the American Office of Strategic Services (OSS) obtained recordings of the latest hit tunes from the USA. The OSS also arranged a special recording session by Marlene Dietrich, leading her to believe that she was doing it for the Voice of America station. Some records for Delmer were made in the Royal Albert Hall by the Band of the Royal Marines, and the crowning musical touch was a genuine German band which recorded for Delmer at MB after its capture in North Africa where it had been entertaining the Afrika Korps.

It was essential that the news items were genuine and topical and a variety of sources were used to ensure that they were right up to date. The primary source was the interrogation of prisoners-of-war who provided up-to-the-minute details of military intelligence, gossip, troop movements and other items that could be used for subversive purposes.

Two members of the MB staff were employed in reading all the letters written by U-Boat PoWs as well as the births, marriages and deaths notices published in German newspapers. With these, they started what became an enormous reference file from which the Atlantiksender was able to offer congratulations to a U-boat commander on the birth of a baby or its sympathy to a Torpedomaat on the death of his father.

Reconnaissance photographs of German cities were taken by Mosquitos after bombing raids. The developed prints were taken by dispatch riders to MB where photo-interpreters in Clifton Child's team examined them. After comparison with a library of guide-books and large-scale city plans, the Atlantiksender was able to announce the names of the streets and the actual numbers of the houses in them that had been destroyed just hours after the event.

Agents in the U-Boat bases in France kept a watch on local events and radioed the results of football matches between flotillas, with the upshot that the Atlantiksender was able to broadcast the scores not long after the referee had blown the whistle for full time. Congratulations on the award of military decorations were often broadcast before they had been officially announced, a seemingly astonishing feat that was actually performed relatively easily as the staff at MB knew the tonnage required for the award of an *Eisernes Kreuz* (Iron Cross) and the scores of the individual U-Boats.

Despite these useful sources, the real key to the success of the Atlantiksender was Delmer's acquisition of a Hellschreiber. The Hellschreiber (literally 'bright or clear writer') was a teleprinter that printed by scanning across a paper tape. Dr. Josef Goebbels, as head of the official German News Bureau (the DNB), issued them to prominent newspapers and broadcasters, and through them the Reich Propaganda Ministry issued press releases and advanced texts of important speeches that were to be made by Hitler and other Party leaders, stating precisely which points should be emphasised and how editorial comment should treat them.

When the London correspondent of the DNB returned to Germany after the outbreak of war, he conveniently left behind his Hellschreiber in full working order. Delmer acquired it through the generosity of the head of Reuters, Christopher Chancellor, had it copied in the United States, and installed a machine on every desk in the newsrooms at MB. The information sent out by Goebbels' Ministry was thus received at MB at exactly the same time as it was received by German newspaper offices and radio stations and *before it was released to the German people.* As Clifton Child commented, it takes little imagination to realise the potential of the Hellschreiber machines in the hands of the Black Propagandists at MB.

Delmer broadcast information received from the Hellschreiber before the German radio stations did. Some items were repeated word-for-word, to give an authentic touch to what was, after all, supposed to be an official German station, while others were given a good measure of disinformation that was never intended by Herr Doktor Goebbels.

As the Atlantiksender was supposedly a German forces station, its news items were deliberately slanted so as to be of interest to Service personnel and military jargon was used. One of the objectives was to encourage surrender and desertion and this was done by suggestion.

Much was made, for example, of how prisoners in enemy (Allied) hands would have an advantage after the war because they were being taught new crafts and were already being paid good wages. A genuine International Red Cross report was repeated, stating that there was a noticeable increase in the number of deserters reaching neutral countries and obtaining good jobs there. Prisoners-of-war in Canada and the United States were also reported to be working for high wages.

Other objectives were to unnerve and to mislead. On one occasion, when information was received from the Admiralty that a number of blockade-runners were suspected of being about to sail for Japan, the Atlantiksender laid on a special programme of Japanese music for the ships, to let the crews know that their destination was no longer a secret. On another occasion, the Admiralty supplied Delmer with details of a new and secret anti-radar device. It was actually a completely useless piece of equipment but the Atlantiksender managed to give the impression that it was causing grave concern to the Allies and caused the U-Boats to have the utmost faith in it.

As the war progressed and 1943 turned into 1944, senior and experienced U-Boat commanders who had either been killed or captured were replaced by young men whose previous experience was limited to one or two cruises as officers of the watch. Aware of the gap that already existed between U-Boat officers and ratings, the Atlantiksender used the situation to drive the wedge deeper, suggesting that the drive for decorations and glory by the new, young and inexperienced commanders would result in the loss of boats and the deaths of their crews.

It is almost certain that Goebbels knew that his voice was being broadcast on the Atlantiksender. Using recordings of his speeches and the transcripts received through the Hellschreiber, the Atlantiksender retransmitted them, but the commentary that followed immediately afterwards consisted of a very different interpretation from the one decreed by the Propaganda Minister.

While Atlantiksender was aimed at the U-Boat arm of the Kriegsmarine, it was not long before Delmer was called upon to open a second major Black Propaganda station directed at the other two arms of the Wehrmacht, the Heeres (the Army) and the Luftwaffe. The new station was called Soldatensender Calais but, for obvious reasons, changed its name to Soldatensender West after the Allied advance following the Normandy landings overran the Channel ports.

Soldatensender was linked with Atlantiksender and also announced itself as such, but this time Delmer had the use of 'Aspidistra' and his new station blasted the air waves with 600 kw of power on medium wave. By this time, two 100 kw short wave transmitters had been added.

Initially, the Soldatensender broadcast for three hours daily from 8.00 p.m. until 11.00 p.m. but its hours were gradually increased until, when the planners of Operation Overlord required Delmer to undermine the morale of the German forces in France, it was on the air for 24 hours a day. Like its associate, the Soldatensender gave the impression that it was a German forces radio station based in the Pas de Calais area. Once again, its propaganda was subversive rather than by direct appeal.

Soldatensender's first transmission was on 24 October, 1943. There was plenty of popular music and much that was of interest to the ordinary German soldier in the way of sports results, promotions and decorations. A great deal of emphasis was laid upon how badly the war was going for the Germans, the devastation caused by RAF and USAAF bombing, the misery experienced by the families at home, the disruption of transport and other services. New and fictitious Allied weapons against which there was no protection were reported and the inexorable advance of the Red Army towards the heart of the Fatherland was announced repeatedly and in great detail.

The Soldatensender, as a 'German' radio station, could make no direct appeal for surrender. Its constant message, though, by covert suggestion, was that the war was already lost and that there was little point in the armed forces continuing the struggle. Plenty of cunning was employed too. In denouncing deserters, for example, Soldatensender told loyal soldiers what signs to watch for among their comrades and in doing so surreptitiously gave instructions on how to desert and get away with it. Other methods of encouraging desertion were to report on the numbers of men reaching neutral territory and to announce that certain well-known and high-ranking officers had not been captured but had crossed the Allied lines of their own free will.

Strange though it may seem, it was the Soldatensender that gave the first announcement of the Normandy landings! This was no breach of secrecy but a carefully planned attempt to break down the morale of the defending German troops. Delmer had been in on the plans for Operation Overlord for some time so that his operation could be used with the utmost effect.

As soon as the long-awaited news flash was received from the DNB over the Hellschreiber, the Soldatensender announcer broke into a music programme and broadcast his own prepared announcement. This took place at 4.50 a.m. on 6 June. It was a simple statement that the invasion had begun but, a short time later, the same announcer at MB was back on the air with a longer report, stating erroneously that the much-vaunted Atlantic Wall had already been breached by overwhelming Allied forces and that the coastal garrisons were surrounded.

The broadcasts of both Atlantiksender and Soldatensender were a continuous series of hoaxes but Delmer still had the ultimate hoax up his sleeve. What interested him most about 'Aspidistra' was the powerful transmitter's ability to change frequencies in seconds. It interested Churchill too, and the plan for the ultimate hoax, which Delmer referred to as his 'Big Bertha', had been approved in principle as early as December, 1942, but had been vetoed on the almost unbelievable grounds that it was considered undesirable for the Germans to be able to claim after the war that part of the cause of their defeat was a propaganda trick! The BBC also objected, fearing that the Germans might either jam or bomb its own transmitters.

It was not until shortly before the end of the war that Delmer's 'Big Bertha' was used and it came about quite by chance. Churchill, who was staying at General Eisenhower's headquarters, read in the American forces newspaper, *Stars and Stripes*, that Allied military radio stations were instructing the German people to stay calm and to remain in their homes. Churchill was furious. He wanted the German people out on the roads, obstructing the Army's movements, just as the French civilians had done to the British Army in 1940. He gave orders for immediate counter-instructions to be issued and sanctioned the use of 'Aspidistra' for this purpose. At last, Delmer could fire his 'Big Bertha' and how he did so was clever in the extreme – nothing less than the takeover of a German radio station!

In Germany, a single radio programme, the Reichsprogramm, was relayed from Berlin by transmitters located in twelve key cities. It had been noticed that when an RAF raid was 50 miles distant from the target city the local transmitter went off the air so that its signal could not be used by the approaching bombers as a navigational aid. It was this knowledge and the capability of 'Aspidistra' that had given Delmer the idea for his ultimate hoax.

The Air Ministry co-operated and the RAF would inform Delmer at lunch-time that a raid on Hamburg was planned for the coming night, giving him the time that the bomber stream was expected to reach the target.

On the night in question, half an hour or so before the stated time, the programme broadcasting from Hamburg would be picked up by the receivers at Milton Bryan and relayed by land-line to Crowborough. At the same time, MB would pick up the Frankfurt transmission and relay it on a second line to 'Aspidistra'. At the moment that Hamburg went off the air 'Aspidistra' relayed the Frankfurt transmission on the Hamburg frequency.

The change was made in 6 milliseconds and the listening audience was completely unaware that the music programme was not coming from the Hamburg transmitter. The team at MB continued to play the music for a short time before fading it out for Delmer's specially prepared announcements. These never took more than two or three minutes and when completed the music was brought back, played for several minutes, and then faded out again for the approach of the RAF raid. When the raid was over, the genuine programme from Hamburg continued.

Delmer's announcements were designed to cause confusion. Residents of Hamburg might, for instance, be instructed to report to the railway station where special trains had been laid on for evacuation to another part of Germany, and people in another city might be told that special ration vouchers for those bombed out of their homes were available at a certain Government office.

The first target for this clever intruder operation was actually Cologne. During the preceding two weeks, recordings of Radio Cologne were taken from the library at MB and carefully studied. Cologne used a man and a woman for its announcements and Delmer replaced them with Moritz Wetzold, who had once been a trainee announcer, and Margit Maass who was an actress capable of imitating any voice. The text of the announcements was prepared by Clifton Child, C. E. Stevens and Hans Behrmann, and rehearsed by Wetzold and Margit Maass until they could perfectly copy the voices of the Cologne announcers.

On 24 March, 1945, the RAF advised Delmer that Cologne was to be the target that night and that Radio Cologne could be expected to go off the air at 9.15 p.m. The operation was perfect. At 9.20 p.m. the programme being relayed from Berlin via 'Aspidistra' on the Cologne frequency was faded and Wetzold and Maass began their announcement. The Gauleiter, they said, had ordered people to leave their homes immediately, taking with them only essential possessions.

The men should report for duty to defend their neighbourhoods against the approaching Allied advance but the women and children must walk to a specified evacuation centre. Local Nazi party officials were ordered to marshal the evacuees. According to reports captured after the end of the war, the hoax was completely successful. Frankfurt and Leipzig were the next two cities to be targeted and by the end of the war 'Aspidistra' had been used for 'Big Bertha' intruder operations on ten occasions.

One further use was found for 'Aspidistra'. Although the war was nearing its inevitable end, the Luftwaffe's night fighter force was still capable of inflicting heavy losses, and the Air Ministry asked if Delmer's team at Milton Bryan could assist in any way with disrupting the German fighter controllers. The transmitters of the Deutschlandsender were very powerful and the German controllers used its network to broadcast running commentary instructions to their fighter pilots. Effective interference was obviously a job for 'Aspidistra'.

Delmer consulted the PWE psychological adviser, Dr. J. T. McCurdy, and learned that the greatest confusion could be caused if MB listened to a fighter control commentary, recorded it, and then re-recorded it several times over on the same disc with a small time difference between each. This was done and the resultant recording was broadcast by 'Aspidistra' on the Deutschlandsender frequency whenever Allied bombers were under attack.

Another, and very simple, variation of the same theme was to record the German fighter control commentary during an RAF raid on, say, Cologne and to broadcast it several nights later when Berlin was under attack. The

Luftwaffe pilots, believing the instructions to be genuine, used up their fuel circling in vain over a city that was not under attack. This very effective trick was code-named 'Dartboard'.

The Atlantiksender and Soldatensender West closed down for the last time on the night of 29/30 April, 1945, without any formal announcement. Harold Robin must have realised that Delmer's work would be of interest to historians in the future and he recorded the last two days' broadcasts on more than 20 discs.

Despite the lack of an official survey, there is a considerable amount of evidence that Atlantiksender and Soldatensender had a very wide and varied audience and caused great concern to the Nazi hierarchy, even though the broadcasts may not have actually accelerated the German surrender. Interrogation reports reveal an enormous amount of feedback and show that the programmes were listened to by members of all the German armed services in very large numbers, and in most of the war theatres.

In 1944, Hitler ordered Himmler and his deputy, Walter Schellenberg, to investigate the truth of Soldatensender news items and to search for a German origin. After the plot on Hitler's life in July of that year, a large team of military radio specialists was ordered into the Führerhauptquartier at Rastenburg in East Prussia to check the telephone system for line taps as it was thought that British propagandists might be obtaining information by this means!

Documents captured after the German surrender included several reports containing warnings against listening to Soldatensender, but perhaps the most telling evidence came from one of the men for whom Atlantiksender was intended.

Peter 'Ali' Cremer had joined the Kriegsmarine before the war and was one of only three senior U-Boat commanders to survive to the end. His badly-damaged boat, hit in a surface battle with a British corvette, *HMS Crocus*, was nursed back to La Pallice by an officer placed aboard from a U-tanker. After a period in hospital, he was appointed Second Staff Officer at U-Boat headquarters. He ended the war as a Fregattenkapitan and the commander of the bodyguard of Grand Admiral Karl Donitz.

Cremer relates that U-Boat headquarters had a retired commander, Gottfried Teiffer, who was responsible for analysing British propaganda. This officer recruited radio operators to transcribe Atlantiksender programmes. During Cremer's term as Second Staff Officer (February to April, 1943), U-Boat headquarters was in Berlin. At first, the HQ staff treated Atlantiksender with amusement but it was not long before the laughter was replaced by grave concern. Donitz referred to the station as 'Poison Kitchen Atlantic'.

Atlantiksender broadcasts gave the U-Boat men the impression that nothing was secret any more and they began to talk openly in the bars and cafés at the French bases. This careless talk caused the frequency of the regular monthly lectures on security to be increased to fortnightly and in some bases to once a week, for it was now apparent that the British interrogators worked closely with the staff at MB.

Despite the frequent lectures, the British interrogators continued to extract the information they required and they passed it on to Delmer. In a final attempt to counter the intelligence leak, the Oberkommando der Marine ordered a documentary film to be made. When completed, it was realised that its clear and unintended message was that the life of a prisoner-of-war in British hands was quite comfortable! The film was never screened. Aboard his own boat, Cremer pretended not to know that his radio operators listened to Atlantiksender sports results every day. On 1 June, 1942, he was unnerved to learn from the station that he would be taking *U-333* to sea on the following day!

After World War II had ended, a senior civil servant, responsible for former prisoners-of-war, possibly with a wry sense of humour, ended the story of Milton Bryan with an ironic twist. It may, of course, have been entirely coincidental but, when the period that was to become known as the Cold War began, many prisoners still remained in England because their homes were in East Germany and they could not be repatriated.

The compound at Milton Bryan became a camp for former U-Boat officers and ratings – the very men at whom the Atlantiksender had been aimed. The prefabricated huts where Clifton Child, the scriptwriters and others had worked, were converted into barrack blocks for the German inmates, and the canteen became their recreation hall. It is highly unlikely that the U-Bootbesatzungen ever knew that their temporary home in Bedfordshire was the place from which the programme they had so often listened to was broadcast.

'Aspidistra' continued in use on long after the war and broadcast World Service and BBC European Service transmissions. In September, 1982, Harold Robin attended a shutdown ceremony and pressed the 'off' button for the last time. The transmitter should have been preserved but Robin had retired from the Foreign Office in 1971 and had no influence in preventing its sale for scrap.

What an ignominious ending for a wireless station that had contributed so much to the successful prosecution of the propaganda war.

Chapter 19

Mobile Construction

This is seemingly a strange name for a section of SCU1 that was based in a hut in the grounds of Whaddon Hall, but it refers to the work carried out by the team that worked on mobile wireless units transported in aircraft, cars, wireless vans, MTBs and even the mysterious MFU. If it moved, it seemingly became the responsibility of 'Mobile Construction'.

Mobile Construction must have existed in embryo form as part of the Section VIII team at Bletchley Park, even before Whaddon was purchased. However, once Whaddon Hall was in our hands, its work started in earnest. F. W. Winterbotham claims in his book *'The Ultra Secret'* that it was his idea to provide tactical SIS intelligence (incoming and outgoing) near to the HQ of the British Expeditionary Force (BEF), in France

This one picture, taken at a landing strip in Libya near 8th Army HQ, illustrates the breadth of work carried out by Mobile Construction. A Packard of Whaddon's 'A' Detachment, taking Ultra traffic right into to 8th Army Command, is in front of a Lockheed Ventura flown in from Tempsford which has been fitted with our Ascension wireless gear. To complete the great importance of the picture, the main designer of Ascension – Wilf Lilburn – is standing centrally in the group.

during the so-called Phoney War (the first three months of World War II) by means of a mobile wireless unit. I am also told that Brigadier Gambier-Parry was involved. Whoever had the idea is hardly important at this stage, but the design, construction and operation were definitely left to Gambier-Parry's Section VIII.

The creators of the first wireless vehicle were Bob Hornby and Spuggy Newton but it was realised that the emergence of Ultra traffic made the role even more important. The first car was a Dodge limousine and it was sent to France early on (according to John Darwin, see Chapter 21) to join GHQ Field Force. The second car, sent shortly afterwards, was based with the Advanced Air Striking Force. Both Hornby and Newton had worked at Philco until 1938 and the company had been the leading designer of in-car wireless sets prior to World War II. These were receivers for civilian use, of course, but the firm was beginning to supply two-way wireless equipment to the police, including the Glasgow Police force. This was bulky apparatus at first, but new and smaller components were appearing that lead to the ability to 'miniaturise' wireless sets. An example of this can be seen today at the Bletchley Park wireless museum run by David White. It is a complete transmitter/receiver built into a Philco car loudspeaker of the period 1937/8, quite astonishing for the time. Its creator was Bob Hornby.

When the German Blitzkrieg struck in 1940, it rolled across France and Belgium at an incredible speed. Earlier, the German army had been able to use the ordinary telephone network but the sheer speed of movement ensured the need for greater use of wireless and therefore Enigma cypher machines. This resulted in a flurry of intercepted wireless messages at the Y stations and thus a rapid increase of work for the code-breakers at Bletchley Park. Increased German reliance on Enigma increased the role of these two mobile units even more. They were soon receiving sparse 'bits and pieces' of the early Ultra intelligence which, to be of greatest value, needed to be in the hands of military commanders at the earliest possible moment. Thus, the huge importance of mobile units was thus confirmed by these two early cars from Whaddon.

The Section VIII fleet of Packards were used as SLUs in the early days after Dunkirk, and was expected to provide mobile Ultra and SIS traffic to commanders on the move in the event of an invasion of England. The cars were also required to provide that service to such essential organisations as the Admiralty, the War Office and the Air Ministry. All of these ministries had one of Gambier-Parry's Packards on hand to provide a wireless connection with Ultra and SIS, in the event that they were forced to move out of their established offices. It must be borne in mind that an invasion by the all-conquering German army was considered a strong likelihood in the late summer of 1940.

Work on installing the wireless equipment into the Packards in 1940 was regarded as an absolute priority, even in a unit which by then was used to dealing with emergencies and crises. The installation work was performed as a matter of great urgency by 'Spuggy' Newton and the then infant 'Mobile' team. Anyone who could be found with the necessary wireless skills was called in to help in the rush to complete the work, including Edgar Harrison.

In 1941, the Section VIII 'A' Detachment was sent out from Whaddon to Cairo to handle the increasing amount of Ultra traffic emanating from Bletchley Park, which needed to be made more readily available to the commanders in the field in Libya. The vehicles sent were Packards but they proved unsatisfactory if they strayed off road, even if only into scrubland, as they became bogged down in the sand.

As a result, 'A' Detachment requisitioned several standard army 'Gin Palace' wireless trucks (I believe they were made by Morris Motors), and fitted them out as mobile offices. Our own wireless equipment, stripped out from the Packards, was fitted into three Humber Super Snipe estate cars that proved ideal for the terrain. These SCU/SLUs units provided Ultra directly to Eighth Army HQ, and at the nearby Tactical Air Force HQ.

The construction and use of the unit's fleet of Packard wireless cars has been dealt with in Chapter 17, but Mobile Construction soon became responsible for the early attempts at air-to-ground contact with agents from aircraft based at Tempsford aerodrome, and later at Hartford Bridge, near Camberley in Surrey. The design work for this apparatus was performed in the Whaddon workshops and a number of our R&D engineers were involved.

Aircraft fitted with Ascension communications equipment played an important role in contacting agents in Europe. Most of these agents probably had little training or aptitude for Morse, but they had other skills making them ideal as intelligence agents. Through the use of Ascension, they were able to convey their messages verbally, and rapidly, to an aircraft with less chance of being traced by direction-finding.

From the very earliest days, it was thought necessary to have shipboard wireless for interception, and in 1938 and 1939 experiments took place on several vessels, including the *SS Celica*, belonging to the film director Mansfield Markham. The later operational deployment of the *Celica* was vetoed personally by the First Sea Lord, and the idea abandoned for a time.

After a nine-month stint at the Little Horwood workshops making agent's sets, I was hugely delighted to be asked to join Mobile Construction back at Whaddon. By that time, it was run by Dennis Smith who was a friend of my father and I felt elated at the opportunity of working for him. When I arrived, they had just started fitting out three or four large QL 4 x 4 army wireless vans with our Whaddon-made equipment, after removing the remaining standard army wireless gear. I helped with the wiring on these and generally made

Three Guy 15-cwt SCU/SLUs just refitted with 'Whaddon' equipment for use in the British and Canadian sectors. The picture was taken near the 'Mobile Construction' unit hut close to the exit gate at Whaddon Hall, probably during April 1944 before the D Day invasion in June.
The ivy covered wall flanked the kitchen garden and our hut was just five yards to the right. I am leaning out of the door of the right hand vehicle and the four men in front of me (without braces!), are members of our team. Left to right Tony Wheeler, Wallace Harrison, Jock King and Jock Denham. Norman Stanton is in the left hand vehicle and the rest are members of SCU8 acquainting themselves with the layout.
The wireless gear has been taken out on its bench to be photographed and includes an HRO, a MkIII and a Marconi Morse key.

myself useful. There were a number of compartments inside the lorry, sufficient for six operators. These were used as SCU/SLUs in North Africa and later went on to Sicily and Italy. One at least went into Europe after D-Day with SCU9.

Early in 1944, we started to fit out a number (perhaps seven or eight) smaller Guy 15-cwt standard army wireless vans as SCU/SLUs, again after stripping out the army wireless gear. These were intended for SCU8 and to go with each Army Group, or Army, in the invasion of Europe. I believe one was sent to the HQ of the commander of the Second Tactical Air Force.

Our unit also made the SCU/SLUs for use in the American sector of the invasion. Here, we used standard US Army Dodge ambulances from which the stretchers had been removed. I think we made eight or nine Dodge SLUs. The same standard equipment was fitted in both the 15-cwt Guys intended for British and Canadian HQs, and the Dodge ambulances, for use by the Americans. Our carpenter built the benches for these across the front end of the cab and we wired them for aerial connections and alternative power supplies. They had transformers to deal with all the voltages they were likely to encounter, an Onan petrol-driven generator to give AC current and substantial batteries that could be recharged by a 'Tiny Tim' petrol-driven charger.

The crew for our wireless cars or vans usually consisted of a driver and two or three operators. The cypher staff (RAF personnel) were housed in other (matching) vehicles which, together with our wireless van/car, formed the complete SLU based at an Army HQ or Air Force Command.

It should be realised that these vehicles were only at the most important command levels where Ultra was to be channelled. For example, there was one at 21st Army Group, one with General Henry Crerar commanding the Canadian Forces on D-Day, and one at Second Tactical Air Force. On the American side, there was one with General Bradley's First Army and another with General Patton's Third Army. I am making the point that Ultra was only available at the very highest level of the Allied commanders, and the SCU/SLU vehicles we provided were always close by them in their HQ area.

As we finalised work on the vehicles, we had visits from the SCU8 wireless teams who were to operate the units in the invasion of Europe. They came over from their training at Little Horwood, one team at a time, to acquaint themselves with the layout of our vehicles. In the photograph, you will see three of the British Army Guy 15-cwt wireless trucks that had been fitted out with Whaddon equipment. The SCU8 chaps are easily identified, because they are wearing braces, which was probably the dress for the day!

A Dodge Ambulance, just as they appeared on delivery to us for fitting out with our wireless equipment, after the stretcher equipment had been removed. These were for use as SCU/SLUs in the American sector of the invasion of Europe.

The remainder of the men are members of Mobile Construction but sadly our boss Dennis Smith does not appear in the picture. Clearly, because of the secrecy abounding at the time, this was taken by someone in great authority. As the photograph came to me from Steve Newton, I suggest it was taken by his father Spuggy, who had overall responsibility for our unit. Whilst I have no recollection of the picture being taken, I am in it – looking out of the rear door of the right-hand vehicle. Other members of the mobile construction crew in the picture, are

Norman Stanton, Wallace Harrison, Tony Wheeler, Jock Denham and Jock King.

A Guy SCU/SLU at Little Horwood being used for training before leaving for France

The wireless equipment we had fitted had been taken out of the vans for the photograph to be taken and is visible on the benches. It consisted of the usual HRO receiver, a MkIII transmitter, a Marconi key, and a Terry lamp. Exactly the same layout was used in the Guy vans shown here for the British and Canadian sectors, and for the Dodge Ambulances intended for the American sectors.

The unit was responsible for fitting out ships from 'Slocum's Navy' used to ferry agents to and from the continent, as well as supplies of wireless sets and other equipment. At times they were used for interception. The equipment was always the same and John Darwin (Chapter 21), relates the story, recorded in his diary, of how he and 'Spuggy' fitted out a trawler, the *HM Trawler Hertfordshire* on 4 and 5 January, 1940.

Some of the Mobile Construction team of engineers.
Seated: Tony Wheeler, and standing left to right: Jock Denham, 'Jake', Norman Stanton and myself.

Before I joined the team, in 1943, they had installed wirelesses in a number of MFVs (motor fishing vessels) that plied the North Sea between the east coast ports of England and Scotland and southern Norway. A few were in Channel ports as well but mostly servicing the coast of France. Amongst their many duties would have been the task of weather forecasting and interception.

I recall there was a chart listing the names and code numbers of the various MFVs, detailing the equipment in them, on the wall of Hugh Castleman's office at Little Horwood. I would guess there might have been twenty plus names on it. I cannot now remember why it was there but I know that men from that workshops 'helped out' in the work of Mobile Construction, whilst I was still on its staff. That was before the team was enlarged, and chaps like Tony Wheeler, Jock Denham, and myself joined it.

We fitted our wirelesses into a number of MTBs that were based in ports along the South Coast including Brixham and Dartmouth. These, too, were utilised in the transportation of SIS agents and supplies into occupied Europe. I particularly enjoyed working on these fast and beautiful craft.

There was one more method of transporting wirelesses that needed the attention of the Mobile Construction Unit, and this was the strangest, and most secret, of them all. Its code name was MFU which stands for Mobile Flotation Unit.

I first encountered the MFU in 1944 at Tough Brothers boat yard, on the Thames at Teddington. My boss, Dennis Smith, and I were returning to Whaddon from a trip to Brixham, where we had been working on an MTB, and he then instructed the driver to divert to Teddington so that he could check on the progress being made. He had never mentioned the project to me and the shape of the vessel came as a complete surprise. Work was taking place in a very large shed on what looked like a big steel cigar, tapered at each end. Not like a torpedo, perhaps more like a submarine without a conning tower. I have described the MFU and the work done on it in detail in my own story, see Chapter 38.

I could not have had a more interesting job than the eighteen months I spent on the Mobile Construction team, and I felt myself to be very lucky indeed. The Unit's work virtually came to an end by April, 1945 as the war in Europe reached its final days and most of the staff were deployed on other tasks at Whaddon. I was sent to work with Jack Buckley, in the nearby wireless repair workshops. There I was to learn to service our wireless sets as part of the tasks I was due to perform during my forthcoming posting to the Far East.

Chapter 20

Stable Gossip

Stable Gossip was Whaddon Hall's house magazine, produced sporadically, that took its name from the stables of the Hall which before the war had housed horses for the famous Whaddon Hunt. The magazine was devised by Ewart Holden who was also its editor, reporter, publisher and print hand!

The stables were arranged round a courtyard which lay between the mansion and the road. During the war, they were used for storing wireless equipment of all kinds, including that used for the construction of agents' sets and those used overseas and for research. Ewart Holden had overall responsibility for the wireless stores and despatch but the daily running of the stores themselves and the issue of materials was in the hands of my father who was his deputy. The magazine had a restricted circulation and was avidly read by those were on the list or could borrow a copy.

Whilst it obviously helps enjoyment of the magazine's contents if you actually knew the people, the places and the various tasks being undertaken, there is still much of interest to the social historian, students of the Secret Wireless War and even just an inquisitive reader.

The cover of the 'Xmas Number' December, 1940

The issues are full of quite detailed comments about life at Whaddon, the people involved and glimpses of war-time Britain. Clearly, few of us are left to enjoy the 'in-house' jokes or savour the 'put-downs' so frequently contained in the articles but the humour was often of a high standard, much of it risqué in the gentle, almost innocent, way that existed in those days.

One is also struck by the understatements expressed by people engaged in a very serious war and often leading a dangerous life. Reading about dances, whist drives, 'boozy' parties, cinema shows, high jinks in the mess, snooker tournaments, darts matches and countless visits to local pubs, one might be tempted to ask whether there was there a war on at all! Yes there was, and this unit was very much part of it.

Many indications of the serious nature of the units' work can be picked up in *Stable Gossip*. In a 1944 edition, there is a series of cartoons showing our agents at work. One depicts an agent reaching heaven,

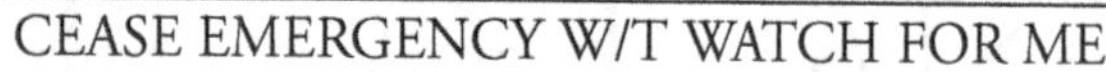

CEASE EMERGENCY W/T WATCH FOR ME

Slightly strange humour of the time – one of our agents caught. I think drawn by Ken Green.

PENETRATION

Another by Ken Green – an agent with a MkV wireless set reaching heaven.

carrying with a MkV agent's suitcase wireless with him. Another is sending 'Cease emergency W/T watch for me' as burly Gestapo officers stand over him. To bring that into focus, one of my colleagues was Duty Secretary one evening at the Hall, when she was told that an agent had come through *en clair*. His graphic message was to the effect: 'I am shutting down – the police are breaking down the door!'

Cartoons played a major part from the start. The first artist was Ken Bromley who joined the unit in 1940 and whose work features in several editions. His fine paintings (usually of pubs) could be found in Stony Stratford where he was billeted. The best known of these was his painting of The Cock Hotel which hung in the bar of that establishment for many years. Ken is now living in Bolton and tells me he usually worked for his beer. Later *Stable Gossip* artists were Bert Stacey and Ken Green.

Naturally, I like a cartoon by Bromley in a 1941 edition. It depicts my father as a pigeon, looking slightly the worse for wear and clutching some holly after 'Pop's' 1940 Workshop Christmas party, saying *'Wash – a – party!'*

Near the front of each edition, there was a message from 'Pop' – Brigadier Richard Gambier-Parry. His messages are a mixture of fun and serious comment but perhaps none more serious than when he refers in the Christmas 1944 edition, to the tragic loss of 'Jack' Saunders. Saunders was killed returning from a special operation behind enemy lines. He had given up his place in a Lysander for the return flight to an urgent courier for London, and then waited for the following night's flight. The plane is reported to have taken off safely but was never heard of again. Major 'Jack' Saunders had been in SIS radio communications since its earliest days before the war. His daughter Tessa married Ewart Holden in 1945 and the September, 1945 edition of Stable Gossip had a special feature headed: *'Editor marries Tessie – The Typical Technical Typist.'*

As an example of the sort of information about life at Whaddon that can be gleaned from *Stable Gossip*, I will give the following: the Mark III transmitter was one of the most successful wireless sets produced by the unit. Its range was astonishing and all sorts of true and fanciful stories about it exist. However, when I was working in the main workshops at Whaddon and at Little Horwood (1942–1944), the set consisted of the transmitter in an oak case with a separate power unit. It was originally allegedly on a single chassis and it must have been a cumbersome beast. However, the following poem from the December, 1943 edition of Stable Gossip explains quite a lot:-

The time has come,' the Brigadier said,
'To talk of certain things,
Such as, who made the first Mk. III
And whether its Morse has wings
And why does it shake the ether so
When you press the keys and it sings.

For its signals go a very long way
And bounce back many times from heaven,
Which makes one wonder what's going on
In the bulb of the poor 807.

Now Bob's design has lived very long
And Wilf has endeavoured to increase its song,
So into its innards he started to delve,
And found a corner for one more valve.

'Do you think' said the Brigadier,
With his infectious laugh,
'You could make it in two pieces
And cut its size in half? '

'I doubt it' said Wilf Lilburn
And let out a mighty roar
As Charlie West achieved this aim
By dropping it on the floor.

The reason for this attempt at rhyme
Is having no claim to its original design,
But good luck to both the old and the new,
Yours very sincerely, VIII/W.

So R.I.P. or V.V.V.
For the original model of the MkIII

It was signed 'P.C.' and under the initials it reads:-

(Editor's note: P.C. apologises to the ARRL Handbook but what about poor Lewis Carroll!)

The 'Bob' is Bob Hornby and 'Wilf' is Wilf Lilburn. The verses appear to be saying that Bob Hornby designed the original version, Wilf Lilburn improved its performance and Charlie West re-designed the chassis into two halves. The 'P. C.' who signed the poem must be Percy Cooper. Lieut. Cdr. Percy Cooper RNVR was in charge of the unit's workshops, including research and development, at Whaddon and Little Horwood, under the overall control of Bob Hornby. His 'Yours very sincerely, VIII/W' refers to Section VIII with the 'W' standing for workshops.

I must thank Irene Healey (neé West), for the loan of these treasured reminders of Whaddon Hall. I had them copied and bound as a record for several libraries including Milton Keynes, the County Records Office at Aylesbury and the Imperial War Museum. Irene's father was Charlie West who ran the various chassis and metal workshops, under Percy Cooper. Charlie was a kindly man and my first boss, personally supervising my early training when I joined Whaddon in 1942.

There are seven editions altogether in my compilation but the seventh is the 'odd-one-out' being dated 1947. It was actually produced when the whole unit, including the stores, had transferred from Whaddon to Hanslope. By that time, the unit had changed from its wartime role and was now handling all Foreign Office wireless services, under the title of Diplomatic Wireless Service (DWS). It remains as the base for confidential traffic, and still contains workshops and the wireless stores. I believe the Hanslope edition of *Stable Gossip* included in the compilation was the only one ever written there. Perhaps it did not travel well from its natural home – the stables of Whaddon Hall.

It is remarkable that these copies have survived all these years. Sadly, it is not known how many issues appeared,

My father after the workshop Christmas party 1941 – slightly the worse for wear but still clutching his holly – by Ken Bromley.

Entrance gate at Whaddon Hall – by Ken Bromley.

but I suspect time would have prevented Ewart from producing more than one or two a year, at the most. However, I do recall one other issue, because the cover had a cartoon by Bromley showing my father – again drawn as a pigeon – but wearing his beloved service cap.

The *Stable Gossip* in the collection referred to as No. 4 (dated December, 1943) has no cover so it is entirely possible that this was the one with my father's picture on it. Perhaps one will eventually surface from an attic and I shall find out. However, there is a clue to the numbers issued within the Christmas issue of 1944 where the editorial explains regretfully that the magazine was turning into a hardy annual instead of a periodical.

I should add that the numbering I gave to each edition is merely to keep the date sequence in order. The magazines were not numbered and that is another reason why one cannot be sure of how many issues were produced.

Stable Gossip was produced by various methods and the print quality varies dramatically from year to year, and even in the same edition! All this was before the age of photocopiers and the modern technology we have today. Some of the staples holding the sheets together have rusted away but as some are well over 60 years old that is not surprising.

I am fortunate to have found a *Stable Gossip* circulation list and the names of the lucky recipients. Clearly, the circulation was very limited but it reads like a roll-call of the leading Whaddon celebrities. Looking at the various ranks held at the time, I suggest it might be 1944. A photograph also exists of the stores team from The Stables which, I believe, was taken in 1944. It shows Ewart Holden sitting centrally, with my father on his left, complete with his famous peaked cap. It was taken in the stables courtyard, at the top corner of the drive before it turned down to the exit gate.

As far as I am aware, only Ewart, my father and Syd Wickens were in Section VIII. The remainder of the stores team being 'ordinary soldiers', they were mostly, though not all, from the Royal Corps of Signals.

The Stable Gossip circulation list

Looking at the circulation list (see over), one is now struck by its seeming formality in listing all the names with correct military titles when informality was generally the rule in the workplace. Certainly, undue ceremony was not in the make-up of the editor Ewart Holden. Reading the pages and recalling the man, one could say that an irreverent approach was more his style.

In those days, it was common practice to use surnames amongst colleagues in business, but there was widespread use of Christian names (or nicknames) in our unit, and that is reflected in the pages of *Stable Gossip*. I think this was encouraged by Gambier-Parry, who always signed himself 'Pop' and this informality is well illustrated in the list of nuptials, headed 'SCUppered', in the September, 1945 edition. 'SCUppered' was a list of Section VIII personnel who had married during their time at Whaddon, some such as Don Lee and Ann Trapman, having met there. Here are just a few from the list of over twenty names.

Maurice & Joane 19/2/43
Plug & Meg 18/9/43
Pop & Lisa 17/2/44
Em & Betty 20/6/44
Tess & Ewart 27/9/45

It could hardly be more informal than to list the head of MI6 (Section VIII) – Brigadier Richard Gambier-Parry, as 'Pop' and his bride as 'Lisa'.

Lt. Col. R. Pott.
Lt Col. K. MacFarlan
Major F.R. Hornby
Major A.A. Pollard
S/O. F. Seaward WAAF.
Miss. T. Saunders
Capt. G.E. Holden
Lt. C.P. Pugsley
Capt. W. Lilburn
Capt. J. Tricker
Capt. A. Gilles
Capt. D.A.T. Lee,
Major W.J. Sharpe
Major W.G. Duncombe Anderson
Capt. C. Crocker
Lt. Comdr. P. Cooper RNVR.
Capt. W.G. Furze
Lt. C.W. Lee
Major J.W. Saunders
Lt. R. Colborne
Wing Comdr. E.S. Adamson.
Lt. J. Angus
Capt B.J. Applin
Lt. H. Kempton
Major R.H. Tricker
Lt. J.H. Gerrish
Major. F. Cox
Mrs. F. Cox
Capt. J.W. Inman
Lt. Col. J. Russell
Lt Col. Lord Sandhurst
Major F. Ware
Sq. Ldr. M. Whinney

B. Walsh Esq.
Lt. Col. E. Maltby
Capt. A.G. Manson,
Lt. C. Herbert,
Capt. A.R. Miller
S/O. A. Lee WAAF
Capt. H. Fuller
Miss. J. Watson,
Lt. R. Seaward
Miss J.M. Kirk,
Lt. J.S. Darling,
Lt. G.A. Jury,
Capt. C.F. Bradford.
Mrs. V. Duncombe Anderson
Lt. T.N. Murray,
Lt. A.V. Durlin
Capt. C. West
Lt. E. Harrison
Lt. N. Walton
Lt. A. Willis
Comdr. J. Langley RN.
Miss. M. Mercer
Sq. Ldr. R. Mathews
S/O. P. Clive WAAF.
Capt. H. Castleman
W.O.II. J. Hill
CQMS L.J. Daniels
Capt. C. Emary
Capt. D. Lax,
C.S.M. H. Pidgeon
Lt. H. Wort.
Capt. C.H. Williamson
Mrs. C. Williamson
Capt. R. Watton
Capt. W.R. Chennells
Mr. H.K. Robin
C.W. Cox Esq.
Lt. L. Gregory
Mrs. L.M. Locke
K. Miller Esq.

PTO

Miss E.T. Towse.
Miss. B. Watson
Sgt H.E. White
Capt. T.N. Parish RAMC.

Capt. B. Sall,
Mrs. K. Hubbard,
Mr E. Daugharty.
Capt. W.G. Tutt.

Whaddon Technical Stores staff – circa 1944. Ewart Holden is sitting centre with my father (in peaked cap) on his left, to father's left is Syd Wickens, whose family ran a gents outfitters in Stony Stratford. To Ewart's right is 'Carter.' Whilst I recognise many of the others, I regret I cannot now recall their names. Only Ewart Holden, my father and Syd Wickens were actually in Section VIII – the remainder were 'real army!'

The *Stable Gossip* issue dated September, 1945, provides yet another glimpse of the informal nature of the place:

'Rumour has it that a certain C.Q.M.S. (Company Quartermaster Sergeant) who has been in the Army for nearly five years was obliged to draw his Army equipment from the Q. M. Stores before he could be demobbed. It seems no provision has been made for a man to be demobbed in civvies, it upsets everything and everybody – in fact, it can't happen!'

Perhaps the Stable Gossip circulation list, with its formal military titles, was drawn up by one of the army clerks attached to the Stores staff who was keen to make the right impression? Perhaps he also had the job of delivering them?

The list is a trip down memory lane and almost all of the names have a story to tell in their own right. Lt. Col. Lord Sandhurst was head of RSS and the VI's; Lt. Col. 'Jack' Russell was 'real' army and the unit's liaison officer with the military; Miss E. T. Towse was Elizabeth (Lisa) Towse who was Richard Gambier-Parry's secretary whom he later married;

Lt. Col. E. (Ted) Maltby head of SCU3 at Hanslope Park; Captain C. (Charlie) Crocker, the paymaster for Section VIII; Major W. (Bill) Sharpe then head of Planning at Whaddon and later as Lt. Col., the boss of my unit SCU 11/12 in Calcutta, at the end of the war against Japan; Major J. W. 'Jack' Saunders, who was killed returning from operations; Capt. D. A. T. (Don) Lee who had been an active SIS agent in Spain during the

Spanish Civil War – and so on. The list includes my father, H. E. C. Pidgeon, Edgar Harrison, Bob Hornby, Spuggy Newton, Harold K. Robin, Tessa Saunders, Percy Cooper, etc.

Even the editor of Stable Gossip himself is listed quite formally as Capt. G. E. Holden, but note the name immediately above his. It is that of Miss T. [Tessa] Saunders, who was later to became Mrs. Ewart Holden. Did the army clerk who wrote the circulation list know something?

Part III

Their Story: by some of those involved in the Secret Wireless War 1939-1945

Chapter 21

The Extraordinary Diaries of John Darwin

Charles John Wharton Darwin was born in Durham in 1894. He was a cousin of the naturalist, Charles Darwin. Following a conventional education (though with intervals in Germany), he went to Sandhurst, joined the Coldstream Guards and crossed to France with his regiment in August 1914, when he was nineteen years old. He had two years in the trenches, was wounded, and in 1916 transferred to the Royal Flying Corps, where he eventually commanded No. 87 Squadron and was awarded the DSO in 1918.

In 1919, Darwin was on Winston Churchill's personal staff in Paris for the Versailles Peace Conference and in 1920 he was appointed one of the first instructors at RAF Cranwell. In 1923, he was posted to the Air Ministry for Special Intelligence duties where he remained until 1928. There, he was in charge of liaison between Air Intelligence and the intelligence services of the other defence services, the Foreign Office and Scotland Yard (Special Branch). His service records show that his chief was a certain Admiral Hugh Sinclair!

In 1928, Darwin retired from the RAF and joined the Bristol Aeroplane Company, a job which involved worldwide travel. Later he became managing director of Saunders-Roe on the Isle of Wight. However, since his retirement from the Air Force in 1928, Darwin had been retained by Sinclair for undercover work. As his 1939 diary shows, he returned to work for 'C' early in 1939, and was involved in setting up radio posts in various European countries in a hectic race against time in the months leading up to, and immediately following, the outbreak of World War II.

John Darwin DSO

The diaries were found among the effects of his younger daughter, Susan Darwin, who died in July 1998. It was almost impossible to believe my good fortune when I learned of their existence just over a year ago. They are truly extraordinary documents listing the names and activities of the most senior executives of SIS and MI6 (Section VIII), in the run-up to World War II. Incredibly, we have here first-hand reports on meetings of the 'founders' of Section VIII and their travels in those hectic days. Extracts are reproduced here with the gracious permission of Darwin's other daughter, Vivian Kindersley. These extracts have been prepared by Griselda Brook (née Darwin). [She added the family details and the other comments are by me. All appear in square brackets.]

The first entry in the 1939 diary is dated Wednesday 15 February, 1939 and the last Sunday 31 December. There is also a 1940 diary which finishes on Thursday 4 April.

I have decided that, broadly speaking, I should leave each entry as it was made – even if it is difficult to understand. I do not feel I have the right to edit his diary, nor would it be easy, considering we do not know the full story behind every detail in it. Therefore, each extract is just as John Darwin wrote it at the time.

Tuesday, 21 February
Saw 'C' [Admiral Hugh Sinclair] at his flat 11.00 a.m. I suggested that being at a loose end, I should return and work with Fred [Winterbotham], Sect II. [This is strange mistake since Fred Winterbotham was head of Section IV (Air) not Section II (Military) which was then under Col. Stewart Menzies, who later became 'C'].

Wednesday 1 March
Meeting with 'C' at 11.00 a.m. Told him of the arrangement with Fred [Winterbotham] and said I'd see how it worked out.

Monday, 13 March
Lunch Sykes, Paymaster. [Cdr. Percy Sykes RN Head of MI6 Section VII – Finance]
RAC 1.15 Finance seems straightforward enough.

Monday, 17 April
C. S. S. 10. 30 He seemed reasonably amazed at my trip. I emphasised as well as I could the futility of our existing communications. Instancing the Albanian show and the fact that normally a letter from Gregory, for example, would take nearly a week to get home by safe means.

Met Gambier in passage. He [?] this as I do and is apparently in charge of W/T but would like to expand rather on the lines along which I am thinking.

Wednesday, 19 April
Met Gambier 10.30 and had a talk with him about joining his section. He will put it up to C.S.S. provisionally.

Thursday, 20 April
Left Dorchester Hotel with Gambier and Saunders en route for Geneva. Changed planes at Paris and Lyons. Arrived O.K. Spent night fitting up wireless equipment on the roof of the Passport Control office. Marshall operator (good) P.C.O. Brunell (rather wet – like the weather).

Sunday, 23 April
Discussed question of installing a set in Pontresina [Switzerland] for communicating possible Italian troop movements.

Tuesday 28 April
Left Pontresina by the early train. Changed at Zurich, arrived Geneva, dined with Burnett, took night train to Paris. Stayed Edouard VII. Connected up with Gambier at Avenue Flocquet. Lengthy arguments with Bill [Bill Dunderdale – SIS Head of Station in Paris] who thinks he knows rather more than he does regarding W/T.

Friday 28 April
Broadway. Just moved down to the offices on the 2nd floor. Such a difference.

Monday, 8 May
Gambier has returned from Paris. Had a talk about Pontresina, 11.30 Saunders in the afternoon re M.D.J. sets.

Wednesday, 10 May
Dick, 11.30. There is some scheme afoot about an Intelligence organisation in Denmark.

I must get Pontresina done first & incidentally get more experience with the M.D.J. sets in a really safe place, before I launch out.

Monday, 5 June
Dined with Gambier, Victoria Grill, at 7.30. He wants me to join the new communications section, to be known as Section VIII. Agreed.

Tuesday, 6 June
Went down to Florence House with Sibyl [his wife]. Collected Maltby and took him on to lunch at Sudbrooke. I think we will get on together but if I am to be Gambier's second-in-command, it is going to be a trifle difficult. [Those were prophetic words since early in 1940, John Darwin had returned to quite mundane duties in the RAF and Jourdain left shortly afterwards – leaving Maltby effectively as Gambier-Parry's deputy]

Thursday, 13 July
Office, 11.30. Collected my courier's passport and bag. One M.D.J. set was sent by normal King's Messenger & will be delivered to me when I arrive in Copenhagen. Left Liverpool Street 4.10 p.m. Harwich S. S. Esbjerg.

Friday, 14 July
Arrived Copenhagen 11 p.m. Emil Lassen met me.

Friday, 11 August
Parry to Berlin. Saw him off at Croydon by German Lufthansa. Hope he'll be all right.

Monday, 14 August
Parry returned from Berlin. Ware installed & everything O.K. Neville Henderson optimistic – but no great love shown to Parry who was 'trailed' all the time he was there. General feeling v. anti-British.

Wednesday, 16 August
Afternoon to War Station [Bletchley Park] with Gambier, returned with S. G. M. [Stewart Menzies, soon to become head of SIS].

Thursday, 17 August.
Newton back from 33 [The Hague]. Had hectic time but zero & VII Stations O.K.
Had first 'Directors' meeting G.P., Maltby, Self, Hornby, Jourdain. Concentrated on organisation. Meet Monday for all night (if necessary) sitting.

Saturday, 19 August
Hectic morning with Miss M. ['Monty' Montgomery head of Section VIII's own cypher department – entirely separate from GC&CS at Bletchley Park]. Partially solved difficulty. Called round to office. Various troubles re Berlin.

Monday, 21 August
5 p.m. 1st weekly meeting.
Parry, Self, Maltby, Jourdain, Hornby – Tutt. sec. Started off with my notes on the organisation of VIIIb. then VIII & VIIIc and then VIIIa. Most fruitful meeting – everybody very helpful. Sandwiches at 'Flowers', meeting finished 10.15 p.m.

11 p.m. Parry, Saunders & self listening to ZRO [The SIS wireless station in Poland] and zero & VI stations.

X's very bad [Barnes still operating at that time]. Signal from 3 received but atmospherics rendered 'try-out' v. disappointing. Home 2.30 a.m.

Friday, 1 September
Gambier rang up 7. German troops over the frontier 6.30. Our 77077 [Code – Germans invade Poland] worked v. well.

Monday, 4 September
01.28 XQ from the Hague E.A. Large numbers flying west. Informed Stevenson A.M. C. S. S. etc. checked on land line to Hague reported A.M. 01.40 – Air raid sirens started 02.35 descended to operation room 02.45. Teleprinters O.K. with Bletchley 02.50. All clear 03.25. Returned to office. Had a little sleep from 5.30. 7 a.m. news broadcast. S. S. Athenia torpedoed.

Sunday, 10 September
After lunch Miss W. drove me down to War Station [Bletchley Park]. From there to C. S. S. who gave the instructions re Whaddon Hall. Saw Selby-Lowndes [the owner of Whaddon Hall]. Everything v. pleasant.

Monday, 11 September
The great news today is that the boys have got ZRO on the air again after 36 hrs. located somewhere near Tornilow on the Polish-Rumanian frontier. A magnificent effort. Whaddon has been taken by C. S. S. and a number of circumstances have arisen to make it quite clear that we must move our whole outfit 'lock, stock & barrel'.

Tuesday, 12 September
ZRO still on the air. Poland seems definitely at its last gasp. Whaddon Hall now definitely ours.

Wednesday, 13 September
Details of our move to Whaddon taking shape.

Friday, 15 September
ZRO still on the air. Pathetic requests from the Poles for us to do something to help them from the air.

The unrestricted bombardment of open towns such as Warsaw, Lemburg, Tarnopol etc., is driving the population desperate. There is no doubt that the raids are being made with the deliberate object of terrorising the civilian population.

Tuesday, 19 September
Up early, took dogs for a run around the Park. Called at office & took G.P. home to breakfast. Back to office. Spoke to Hatton Hall – Ellis etc. Poland definitely down the drain. ZRO seems safe. Courageous sunk. Move to Whaddon arranged.

Wednesday, 20 September
Quiet night at the office. Home to breakfast. Spent morning preparing for our move. Lunch with Chris and Portal at Travellers. Packed a suitcase and motored down to Whaddon Hall. Parry, Jourdain & Maltby already installed.

Thursday, 28 September
Selby Lowndes [owners of Whaddon Hall] to lunch. They seem quite pleased to have us here and I think that perhaps they are lucky.

Friday, 29 September
Up to London with G.P. in the Ford 'Utility'– went home first – saw Sibyl & introduced her to G.P. then to office in small Ford. (ZRO) Gerrish and [?] Gavani back. C. S. S. received them. He was extremely nice and their magnificent show in Poland has helped Section VIII quite a lot.

Tuesday, 10 October
Office in the morning and saw Hatton Hall, Winterbotham, etc. Lunched Travellers and was driven down to War Station by Mrs. Barclay [Wife of Cecil Barclay, assistant to Henry Maine, the SIS liaision officer between the Foreign Office and Section VIII]. Stopped to visit M.I.5 on the way at Wormwood Scrubs and saw Harker and Alexander. What a ghastly place to choose for a War Station – they must be crazy.

Everything at Whaddon just going along nicely but there is a lot more to be done yet before we can switch over Bletchley here. None of the transmitting aerials are up yet, but the two receiving masts & triatic are erected.

Wednesday, 11 October
Whaddon – Spent most of the day going through my reports on the situation in 33 [Hague] and 13 [Brussels] lands. Gambier arrived back very tired from London and worked till late. Maltby & I got all the pegs driven in for the new radio masts.

Thursday, 12 October
Finished 33 & 13 reports and discussed them with Gambier – then, pegged out positions for the mooring stakes for the W/T. station. Ted took 'official' photographs of the operation.

Friday, 13 October
Office Broadway Buildings in morning. Saw Newton just back from Rome which is coming through R5. He is not very well and the doctor will see him. Discussed with him the equipment of a mobile station for 13000 in the light of his Prague experience. He will go over next week if fit. Back to Whaddon. The Barnes contingent have arrived. [Barnes Station X closed down]

Saturday, 14 October
Early news encouraging. 3 U-boats apparently sunk. Later news not so good – Scottish express crashed at Bletchley and one of our wireless masts down. Off with Ted to B.P. and view (a) damage to mast (b) damage to Bletchley station by Scotch express (a) rectified in a couple of hours. C. S. S. pleased when phoned to (b) a nasty mess. Met Miss Towse off the train.

Monday, 16 October
Lovely morning – out before breakfast – got on to the problem of R/T bilt [?] sets. Started off in the park at 700 yards – with Kempton, Whitemouth & self operating – Gambier receiving – G's semaphore not so good! Afternoon Whitty & Kempton & self off testing R/T and blackberrying – both most successful, Bletchley R5 & Buckingham ditto at 7 miles Tried again outside Aylesbury but got jammed by some station or other on our freq. after 1800 hrs.

Tuesday, 17 October
Pouring with rain. Repeated last test yesterday. They couldn't hear us at all not even carrier. Went back to 1½ miles and got R5. Then to Bletchley R5 to Buckingham R4 but bad background, probably due to wet aerial.

Gambier to London. Stewart M. rang up. Radio link to G.H.Q. seems to have fallen through. R Corps of S. jealousy?

The ceremonial digging of a hole for the first aerial at Whaddon bt Ted Maltby. He is accompanied by John Darwin holding the site map, and Richard Gambier-Parry checking with a British Army marching compass.

Thursday, 19 October
Broadway Buildings 9.30. Collected Newton to bring down & discuss Belgian problems. Collected Miss Towse. Met Gambier halfway down (He immediately borrowed £3 blast him!) and arrived at the War Station for lunch.

A certain amount of routine work during afternoon.

After dinner, a most important discussion between Jourdain, Maltby & myself on the organisation of Section VIII which is getting out of hand. Complete agreement reached between the 3 of us.[perhaps this is wishful thinking as the harmony between these three potential deputies to Gambier-Parry did not last]. A very useful document produced.

Friday, 20 October
Ted Maltby & I off at 10 to Birmingham to clear up the situation re wireless masts – Poles Ltd was difficult to find & we began to disbelieve in their existence. Ultimately they were run to ground and proved after suitable treatment quite amenable. Lunched with their sales manager. (Granby) Grand Hotel.

Home – just before Gambier arrived from London. Had a walk over to the site of the transmitting station. Evening – a long talk re Belgian war plans with Ted, Gambier, Newton & self.

Saturday, 21 October

Up early, breakfast 8 – off cub-hunting with Whaddon Chase who met at Thick and Bare Wood. What memories and what longings! Miss Towse [Secretary and P/A to Gambier-Parry] and self decided to disembowel Hitler with our bare hands!

Normal work – chiefly on organisation during the forenoon. After lunch off with G.P. to North Cerney to see out W/T station there. A very good show. Spud Murphy in charge. Various discussions in evening. Organisation approved by G.P.

Wednesday, 25 October

Up to London. Towse, Tutt & Gillies. Gillies to take wireless equipment out of 'Cecile' [Cecile was a yacht intended to be a floating intercept station but finally ruled out by the First Lord of the Admiralty].

Long talk with Fred Winterbotham and Hatton Hall re general situation and in particular the proposed station at GHQ Field Force [The mobile radio station or SLU that was being built into a Dodge car at Whaddon Hall, for service with the BEF to handle SIS intelligence].

For what my opinion is worth, I think that we must take the greatest care not to antagonise the R. Corps of Signals. So long as our station is definitely for SIS work all should be well, but DMI (McFarlane) [Director of Military Intelligence] must not communicate with us over Paddy Nesbitt's head. Otherwise there will be Hell to pay.

[This shows the jealousy already simmering between the established army signals route for intelligence via the Royal Corps of Signals and the 'new kid on the block'. Here was a signals unit to be placed at the centre of GHQ seemingly run by soldiers of the Corps, but entirely a law unto themselves.]

Thursday, 26 October

Gambier & I had lengthy discussion before he left for G.H.Q. I told him of my talk with Hatton Hall and warned him to be very careful.

Sunday, 29 October

Afternoon engaged in laying cable from 'Dower House' to Whaddon. I never realised how much a cable weighed until I had to carry one!

Everything O.K. by 6 p.m. Cable laid in trench – Ted & D. weary but triumphant.

Saturday, 4 November

Went to bed with a chill. Feeling like death. Shld. have gone to bed two days ago but tried to stick it out. Life has been made more bloody by the death of our beloved CSS. He died at 4.30 pm. He is quite definitely irreplaceable. There will never be anyone like him [This was Admiral Hugh Sinclair – 'C' – Head of SIS and obviously a friend for a number of years].

Sunday, 5 November

Had a bad night but pain slightly better. Everybody very gloomy over CSS death – But we must carry on as he would have wished.

Thursday, 9 November

Got the first di-pole aerial up, which marks the end of the second stage at the 'Dower House', as the W/T station has been named.

Humphrey has left us for France but returns Saturday. The Dodge car which we bought for converting into a 'mobile station' is fast assuming shape – Newton & Walt are working on it and the outfit should cross next week under Humphrey's protective wing.

Saturday, 11 November
Working on the 'GHQ' wireless car. It should be ready in time for Humphrey to take it over next Tuesday.

Monday, 13 November
Back at Whaddon 3.45 having dropped Humphrey at BP to do some code work with Miss Montgomery. Took Tutt and Saunders in later. Collected Humph. Inspected the Dodge (GHQ mobile unit), all OK. Connected R5 with Paris. 45000. G.P. sent signal that he had arrived at the Hague.

Tuesday, 14 November
Up betimes. Today is the great move of all the transmitters to the new 'Puissance'. A very great success. R-5 both ways from Athens to Gib., Stockholm etc. all on 120 watts instead of 500 watts. It shows what a properly designed aerial layout can do. Humphrey & the GHQ mobile unit pushed off for France after breakfast. They should be in touch with us tomorrow evening. Walt and 'Spuggy' Newton very smart in the new uniforms with sergeant's stripes!

Wednesday, 15 November
Jourdain had a successful trip to R. C.of Signals at Reading. We can apparently get people in *and* out of the Army as we please. A most satisfactory proceeding! [This prepared the ground for the later militarisation of the unit, seemingly as a real part of the Royal Corps of Signals].

Thursday, 16 November
Kirby blown up by mine 10 miles off Margate 4-XI-39. Quite OK.

Left Whaddon with Miss Craig. Pouring with rain. Made London in reasonable time. Stopped at No. 10. Phoned Whaddon, no change. Got some sandwiches from Travellers. Off to Folkestone & Dover. Weather awful 2 ft. of water on road in parts. Arrived at Steps Cottage, St. Margaret's Bay. Morris in charge. Seems an ideal place. Didn't stay long. Met Kirby at Dover filling up with petrol. He hopes to get sets working tonight.

Friday, 17 November
Arrived Whaddon 12.45. Station 'J' (Steps Cottage) came up R5 (+) but apparently couldn't hear us. I don't understand it as receiver was tested before dispatch.

Saturday, 18 November
Excellent results from directional di-poles. Rome & Athens R5 +.

Tuesday, 21 November
Slight trouble at the 'Dower House' owing to the storm twisting the 'lead-in' from the reflector di-pole. OK again now.

Wednesday, 22 November
Emary blown up by mine coming back from Gib to relieve Newton GHQ. Why do all our bloody operatives get blown up? (He is all right – three stitches in chin). His explanation is that when the explosion occurred, the table at which he was sitting – and himself went up together – the table beat him by 1/10 sec.

Caught 12.07 to London with Gambier. He lunched with me at Travellers. Walked over to office afterwards.

Friday, 24 November
Spent most of the morning getting out and keeping up-to-date our 'Plans' files.

Had a prowl round with Gambier & Ted just before lunch, inspecting 'Puissances' & workshops.

People overworked but cheery & willing as ever.

Sunday, 26 November
Hell's own gale last night. Arose crack of dawn & walked over to Dower House to see what damage. Everything OK bar one insulator. A very good mark for the technicians & Ted.

Monday, 27 November
Very heavy rain during night. Up early. Over to 'Puissance', found the unfortunate Kempton clad in a shirt, drying his trousers in front of the electric stove. Fixed up supply of oilskins for engineers on duty. Over to B.P. Spoke Taylor re Spanish problems. Lowered last of B.P. wireless masts to the ground. (What a wonderful job that old ramshackle affair has done). Present at the 'funeral' was Tricker, Ted and a few troops. RIP.

Friday, 1 December
Emil (B.X.O.) coming through (Loud cheers!) R3. 08.15.

Just heard the news that S.G.M. [Stewart Menzies] has been confirmed in his appointment as C. S. S. [Head of SIS] It is a tremendous relief to us all. Nothing can replace old Quex but SGM has never really had a chance with such an overwhelming character as his commanding officer, during the past fifteen years.

Saturday, 2 December
SGM to lunch. It's wonderful to think that, after the vicissitudes of the last few weeks, we have got somebody that one can trust at the head of things.

I really think that SGM [Stewart Menzies – the new 'C'] was genuinely impressed with our efforts. Gambier wonderful as ever, showing our stuff.

Tuesday, 12 December
North-east wind blowing last night. Doors banging. Hysterical women!! Gambier, Jourdain and I laughed ourselves sick this morning. Is there such a thing as 'justifiable 'feminicide?'

Thursday, 21 December
Agreed that I should write a book on our 'wares' in general to assist our representatives in communication problems with their agents.

Made a start – it's going to be a bit difficult to know how far to go.

Test with W/T and R/T to Gambier at B.P. on U.H.F. reasonably satisfactory.

Friday, 22 December
General routine in the afternoon. Walked over to the W/T station with Gambier. Bleeding women still fighting about something or other. Thank God I'm not involved. [The continued upheaval caused by squabbling between Mrs. Gambier-Parry and Mrs. Jourdain was rapidly developing into a serious matter. I was assured by Norman Walton, that it led to the eventual removal of Jourdain from his post at Whaddon].

Tuesday, 26 December
Caught 10.40 Lunch, Whaddon. Got on with preparations for Belgian trip. Tested the new small receivers O.K. Arranged for two to be packed ready for tomorrow.

Wednesday, 27 December
Up to London. Train very late. Lunch with Eddie Hastings at Travellers. [Later to become Director of NID – Naval Intelligence Division] Most interesting and very informal talk. I understand quite a lot of what was difficult vis-à-vis Section III. Pushing off tomorrow 9.25 a.m., Victoria for 13-land.

Thursday, 28 December
9.25 'Victoria' Arrived Shoreham Airport in God's good Time. No plane to Brussels! Took taxi to Palmyra Court (ma-in-law) She wasn't in so I went up to London. Snow everywhere – freezing hard – bloody!

Rang up Sibyl who feels much as I do.
My blasted son rings up. He has taken a Hurricane over & come back in a Rapide to Shoreham of all places. Twits Dad!!

Friday, 29 December
Rang up Shoreham 5.00 at crack of dawn. Weather seems propitious. Went down by 9.25. Lovely day. Snow on the ground. Left Airport on schedule. Arrived Brussels idem. Douglas DC3. Good trip.

Saturday, 30 December
Brussels. Very cold and snowy. Rang up Sabena. Weather on the English side apparently OK Left by Douglas DC3 10.30. Arrived Shoreham OK 12.15. Rang up Ma-in-law. Invited myself to lunch. Very pleasant, the old girl in great form. Mason motored me to the station.

Home in evening. Doctor. Slight pain. Injection.
Dined alone. Imlay got in late.

Rather weary but glad to have had a successful trip. Mackenzie Asst. Mil. Attaché wants help. Something has gone quite mad. Spoke Gambier telephone.

Sunday, 31 December
Arrived Bletchley with Pollard met by Mrs Jourdain. Picked J. up at B.P. & so home. A very hectic Parry to see the New Year in. All the engineers & operators about 45 altogether. I think they all enjoyed themselves. I retired to bed early, tired & some pain.

Monday, 1 January, 1940.
Up betimes. Rest of party suffering slightly from seeing the New Year in. A lot to do. Got out reasonable report on my trip to Brussels and the various consequences such as consideration of 13124 and 13134 plans as put forward by Calthrop and Barnes-Stott.

Trawler schemes etc. (Banwins and Blondé). 10.30 over to BP to discuss with Fred Winterbotham the RAF angle with GHQ RAF France. 'Ugly' Barnett [Bassett?] now in sole command. We seem to agree that (Mobile HQ unit desirable). (Army and Navy signal anti). Long talk in evening with Jourdain and Gambier. Wrote report on communications situation. I am not happy!

Tuesday, 2 January, 1940.
Hear definitely from Fred that 'Ugly' Barnett [Bassett?] is to be AOC in C in France, with rank of Air Marshall. I do hope it will work out all right, but he has had the habit of antagonising people who were really only too willing to help him. Up to London in evening, fog as per usual.

Wednesday, 3 January, 1940.
Lunch with Arnold Forster, Travellers. Then to Admiralty. Connected up with Archie Craig and one General Forster who is taking over ADNI from Archie who in some obscure way is coming to Broadway Buildings!

Dived into depths of Admiralty and met Commander Humphrey Sandwith. Found an number of mutual acquaintances. A very helpful and intelligent fellow [Sandwith was later to become Deputy Director of NID – Naval Intelligence Division]. Phoned Gambier office. Off to Portland to meet Spuggy Newton, HMT Topaz.

Thursday, 4 January, 1940.
Arrived Weymouth after slightly laborious journey (6 hours). Stayed Royal Hotel. Spuggy turned up having brought all the stuff, i.e. one MkIII transmitter, one HRO receiver and the usual aerial equipment. After dinner, Captain (anti-sub patrol) William Powlett RN turned up and the three of us had a very interesting talk.

HMT Topaz is **off** for some technical reason, so HMT Hertfordshire has been allotted to us for our experimental work. Powlett – a very nice fellow and very helpful.

Friday, 5 January, 1940.
Arrived Portland dockyard 9.30 am. Met by one Commander Farquhar. Boarded His Majesty's Trawler Hertfordshire. All lads from Hartlepool and South Shields so 'Spuggy' and I at home at once!

Given the chartroom to install our gear in. Lovely boat (750 tons) luxurious by our standards. (How nice it is to hear the good old Durham dialect again). Train to London.

Saturday, 6 January, 1940.
10.40 Euston to Bletchley. Mrs. Jourdain kindly acted as 'chauffeuse' picked up Pollard off the train and the 'VIIIA' at BP.

Skeds with 'Hertfordshire' OK R5 and + both ways. (V. interesting but annoying to think we could have done it last July!!).

Sunday, 7 January, 1940.
Slack day – thank God, so Gambier organised a walking 'paper chase', Miss Towse and Ted Maltby were the hares. Nice afternoon's exercise enjoyed by all. This ghastly party tomorrow night is beginning to loom up a bit. Feel I must support Gambier but would welcome any excuse to slide out. Played roulette all the evening with my usual luck (55 shillings in the red). (Gambier and I to BP in morning quite good work with Eddie Hastings).

Monday, 8 January, 1940.
Felt need for exercise, so walked half way to Shenley and back before breakfast. House upside down owing to the preparations for the party to which almost everybody we **don't** (repeat don't) want are coming so far as I can gather. The weather is incredibly good so I suppose they will all turn up. (Feeling thoroughly anti-social).

As I expected – being a pessimist – party was a huge success, but chiefly owing to sound staff work by Jourdain and Tutt without detracting from the gramophone technician, without whom we should have been sunk.

Tuesday, 9 January, 1940
Up betimes as no one else likely to do so! Helped the domestic staff to get the house shipshape again. They really do respond. Much though I loath parties, the spectacle of some of my ancient friends of the SIS (1924 vintage) really enjoying themselves was most satisfying.

A rather grim morning. The Lootenant Col. and Ted escorting the Whitcome to London 'En route' for Zurich. I did my best to comfort her and soften the blow but when I saw hysterics were about to arrive I incontinently fled.

Wednesday, 10 January, 1940.
Can't stick it out any longer. Pain very bad – dope makes it worse. Gambier is obviously worried. Well – I have stuck it out until W's departure.
Operation on teeth will be a good excuse for clearing everything up and starting again.

John Darwin never returned to Whaddon Hall or Bletchley Park. After his illness, he left the Service and returned to the RAF. He commanded RAF Kinloss until April 1941 when he again fell ill. He was on sick leave for the rest of the year and was still hoping to return to work when he collapsed and died on Boxing Day, 1941, a fortnight after his forty-seventh birthday. Sir Charles Portal, an old friend, wrote to his widow: 'Nothing mattered to him so much as his country. He was the best kind of patriot, unaggressive, quiet and tolerant but the bitter enemy of slackness and inefficiency.'

Chapter 22

Bob Hornby, Our Outstanding Wireless Engineer

When I started work at Whaddon Hall in 1942, Captain Bob Hornby was First Engineer and therefore head of the entire technical side of Section VIII, covering workshops, transmitters, power units, construction, research and development. However, it soon became obvious that he was regarded as a brilliant wireless engineer amongst his peers there, some of whom were themselves well-known figures in the wireless industry.

Hornby was 6'6" tall, broad but lanky, with dark-rimmed glasses, a full upper lip moustache and large nose (said to be a family trait), so he cut a very distinctive figure visiting the workshops or around the grounds. Bob Hornby answered only to Gambier-Parry and his authority was absolute but tempered. He had a number of able 'lieutenants', including Percy Cooper who ran the workshops and Spuggy Newton who was in charge of the installation and construction division.

Not long before he died, Bob Hornby dictated some notes to his daughter, Antoinette Messenger, and this chapter relies heavily on his statement to her. Unfortunately, it only takes us part-way through his career; perhaps he intended to continue the work after his illness.

I can piece together a fair summary of his great contribution to its work, with help from colleagues who worked there at the same time as me, and from his son Ben now living in California. It is not difficult to ascribe much of the achievements on the engineering side in Section VIII to his shrewd leadership.

Bob Hornby was born on 11 April, 1905 in Bolton in Lancashire. He went to Bolton Grammar school, and then on to Aylwin College, Arnside, Westmoreland. His 'splendid' Headmaster was known as 'Bandy' Monase. During his last year at school, he was Company Sergeant Major of the Cadet Corp 4th and 5th Border Regiment.

When Hornby left school, as Bolton was the centre of the cotton-spinning industry, his father found him a job as an apprentice lineset machinist with Dobson and Barlow. He spent about eighteen months there, and then moved to Great Lever Mills, as an assistant to the card room.

After about a year, the mill was bought out by the Pristone Card and Tyre Company who sacked many of the employees. By this time Bob had earned various certificates for engineering in cotton-spinning, weaving, bleaching and dyeing. He tried hard to get another job but did not realise that the bottom was falling out of the cotton industry.

In his statement, Bob says he then 'decided to go back to Manchester Tech' so perhaps that is where he studied for the engineering certificates. It was common practice, even then, for apprentices to be given time off to attend technical college during the day and perhaps he did that, or attended night school there. [Much later, this kind man was instrumental in arranging for me to have time off during the working day at the Whaddon

Bob as a school boy

workshops, to attend nearby Wolverton Technical College].

At Manchester Technical College, he met Dr. Palmer and told him he wanted to be an electrical engineer like his father. Palmer responded, 'No, everyone wants to be an electrical engineer these days,' and advised him to go into 'high frequency', of which Bob had never heard. He asked 'What's that?' To which Palmer replied, 'Wireless and that sort of thing.' So on Palmer's advice, Hornby started a two-year course which virtually covered every then-known aspect of the wireless world.

Upon completion of the course, Bob obtained a job with the agents for British Brunswick in the north-west, but they too went out of business (due to bad management, in his opinion). However, just before that happened, he met two people from Newcastle-on-Tyne and they offered him a job designing new products for them at Radio Recordian. Their main business was to design and manufacture quality electrical record-players. A large number of these machines went to firms such as Marks and Spencer to aid the sale of records in place of the old wind-up gramophones. They also made a ½ kW amplifier loudspeaker, especially for football games and for audio advertising.

The two directors were Pritchard and Simpson and the business was going well until Pritchard decided he wanted to move the whole manufacturing division to London and expand the market. The management team (including Bob, now described as a 'manager') disagreed with the move and told Pritchard it would not work, as there was already great competition for gramophone and radios down there. He was promptly fired for disagreeing with the Board. They then moved to London, and within twelve months they were bankrupt.

Before he left Radio Recordian, Bob wrote a reference, dated 16 July 1931, for one of his employees, a certain A. C. Newton – 'Spuggy' Newton, no less! The reference finishes with these words '… and it is with very great regret that we are compelled, owing to the state of the trade, to dispense with his services'. It goes on to say 'We are perfectly certain that whoever employs Newton will never regret it.' That was praise indeed and it was signed 'F. R. Hornby – Manager.'

Having left Radio Recordian himself, Hornby started to look around for a job in London, and happened to call on the advertising agents for Audio Advertising. Much to his surprise, they had just obtained an advertising account for Philco, an American firm, which had incidentally asked if they knew of any good wireless engineers. They promptly picked up the telephone to make an appointment for him to go and see them and to his obvious joy, he had the job within a month.

The appointment was no less than that of Service Manager to the UK division of the giant American company which rapidly became the second-largest manufacturer of wireless sets in the UK. It also undertook the manufacture of mobile police wireless sets for Newcastle-on-Tyne, and later the Scottish borders. Glasgow, Liverpool, the West Riding and others. Eventually, Bob became Philco's chief engineer.

During this time, Richard Gambier-Parry was General Sales Manager for Philco. In his dictated notes, Bob says that Gambier-Parry had been in the Royal Flying Corps (RFC), not the new Royal Air Force, and had been seconded to MI5 towards the end of the war. I have no confirmation of Gambier-Parry being in MI5 at that time, or in the RFC, so I repeat his remarks without comment.

Carl Dyer was Managing Director of Philco, and one day Bob was summoned to his office to learn that a government department wanted to interview him about a job. Dyer said that he did not know what it was about, but he had better go and find out. Bob Hornby duly went for an interview, and according to his notes 'found out it was MI5.' [I think this is confusion since it was plainly MI6 and perhaps the reference to Gambier-Parry being in MI5, is also a mistake for MI6?]. He was offered a job at £200 per annum, more than he was presently receiving at Philco. He reported back to Carl Dyer who said, 'You cannot refuse, but if it doesn't work out then you can return to Philco'.

Hornby joined the newly formed team at Section VIII in July, 1938. Clearly, this whole exercise was the result of lobbying by Richard Gambier-Parry who had been Sales Manager of Philco until March 1938, and was now the Head of SIS Communications. He had not wasted any time in picking the best engineers available for the embryonic unit, even if it meant poaching them from his old company.

The 'poaching' from Philco continued and on 29 July, 1938, R. Koller, Works Manager for Philco, wrote a reference for Spuggy Newton saying they were sorry to lose his services but understood he was leaving to improve his position, so wished him every success. This was extraordinarily generous of Philco, seeing they had lost their gifted Sales Manager, their Chief Engineer, and his most trusted assistant, in the course of a few short months!

At that time, things were hotting up in Europe, and some countries were putting artificial delays on telegram messages sent by the embassies in code to the Foreign Office (FO), so the FO decided to transmit more messages by wireless. SIS was asked to help them with wireless transmitters.

The new team's first task was to design and build transmitters that would pass unnoticed in diplomatic bags. There were no workshops at that time, so Bob Hornby designed and built the first transmitters in his garage at Hatch End. They were simple to work and operate, an essential facility. The first one was installed in Prague where he found an ex-Navy Chief Petty Officer who proved to be a first-class operator. Contact was made at the first attempt.

A new wireless station was being built for the SIS at Woldingham, near Caterham in Surrey. There, they had a small workshop where wireless sets could be designed and built, as well as in the newly acquired workshop at Barnes. The workshop at Barnes was only a short way from the Barnes station X, the established Foreign Office/SIS wireless station alongside the Thames. It later moved to Bletchley Park before finally settling into Whaddon Hall complex in late 1939.

The new, smaller wireless sets were installed in many capitals of Europe, including the Balkans, Eastern Europe and Scandinavia. Most of these were taken to their destination by Spuggy Newton but Hornby installed the one in Stockholm himself.

In the build-up to the war, Section VIII needed many more wireless sets for its growing commitments but no one in the UK, including Marconi, could deliver them without delay. So Hornby was despatched to the USA with a cheque-book to buy any sets that could be useful to HMG.

Having worked for Philco, Hornby knew the American market and he purchased quantities of wireless sets from Radio Corporation of America (RCA), General Electric (GE) and the National Corporation. His instructions were to get these back to the UK with all possible speed as it was considered that war was imminent. With the aid of the British Embassy, he was able to persuade Cunard to clear space in the luggage

Front of the Philco car loudspeaker – 1938

holds of the *RMS Queen Elizabeth* and the old *RMS Queen Mary*, and the equipment duly arrived before the outbreak of war.

One aspect of Hornby's work has never before been recorded. It appears in a reference in his notes to the fitting of three cameras to a 'pre-war reconnaissance Lockheed aircraft'. This can surely only mean the Lockheed 'spy plane' mentioned in Fred. E. Winterbotham's book *Ultra Spy*.

Bob Hornby lead the technical team of the SIS and was responsible for designing and supplying the latest communication equipment to them. However, like several others in the early months of the war, he was concerned that our military and government interception service was not really up to the job of intercepting all foreign and clandestine wireless signals.

He set out to prove their inadequacy by modifying a standard Philco car wireless set to receive short wave wireless transmissions. Over the rear seat of his own Rover motor car, he installed – inside the loudspeaker of the time – a wireless transmitter which could be used to send Morse code. As an officer of the SIS he visited various military establishments around the South of England and there would copy a completely unimportant document, off their administration notice boards.

Bob then drove the car to a quiet spot, threw an aerial over the branches of a nearby tree, and transmitted the contents of the document to his father's house in Bournemouth. He did this over a long period and asked the interception services if they had heard any of these secret messages, but found they had not.

Then he informed them of the actual day of his transmissions in advance, but still they did not intercept his messages. But many Hams had heard his signals, were suspicious about them, and informed the authorities. It was very embarrassing that wireless amateurs were picking up these clandestine signals, but the military and the government interception units were not.

Interior of the Philco speaker showing its conversion into a transmitter/receiver

This largely resulted in the recruitment of hundreds of wireless amateurs into the services and the governments intercept service. Many became part of the Radio Security Service (RSS) which was responsible for trying to locate enemy agents transmitting messages to the enemy. Later, they began intercepting enemy intelligence transmissions (See the full RSS story in chapter 15).

Because of the failures shown up by Bob's transmissions, it is believed that his work was mainly responsible for the government authorities banning all car wireless sets in the United Kingdom- from 1940 to 1944.

Considering the materials available at the time Bob's secret transmitter was a masterpiece of miniaturisation.

This remarkable set is still in existence today, and is on show in the wireless museum at Bletchley Park, in the remarkable Wireless Museum under its Curator – David White.

New transmission stations were being built in the UK, including a station at Potsgrove. At the same time, the Ministry of Information (MoI) wanted to build a propaganda station and asked the BBC to help but received a prompt rebuff. As everything Section VIII were doing was already working successfully around the world, it was asked to build stations for the MoI. It did so by providing a service at North Cerney, diverting the work at Potsgrove for the MoI and creating an entirely new station at Gawcott.

Bob Hornby's notes list the W/T stations he designed and had built, in order. The first was at North Cerney. In addition to communicating with China and South America, it was the first of the Black Propaganda stations.

Bob then worked on the SIS stations at Bletchley Park (in Hut 1 and in the tower of the mansion); the Dower House, Whaddon; The Bare, Whaddon; Windy Ridge, Whaddon; Nash; Upper Weald, Calverton; Potsgrove and Gawcott for Black Propaganda; Creslow and Forest Muir. He claims considerable input into the designing and building of Forfar, Bampton and Crowborough. Curiously he does not mention Calverton, or Main Line at Whaddon in the list but that is probably just an oversight. In his notes, Hornby goes on to mention other of his tasks so I have set these down, adding my own knowledge of his duties. It is a very impressive catalogue.

Hornby's wartime role covered overall responsibility for the workshops run by Percy Cooper with its research and design department and the manufacture of agents' sets. The design, construction of the W/T stations, and all the support services such as purchasing, stores, packing, and despatch. He also took responsibility for the supply of equipment for innumerable operations and he reports – 'some were abortive but many successful.'

The construction of the Mobile SCUs were included in his role, from the early ones sent to the BEF in France in 1939/40, those in North Africa, the fixed ones such as those in Fighter Command and the Admiralty, and the mobile SLUs constructed at Whaddon for the return to France.

The list is very extensive but not inclusive, it was an immense task for any one man but it was carried out efficiently, and seemingly to us mere mortals, without an apparent hitch.

Undoubtedly, he owed a great deal to the fine team he had assembled which I feel I must set down for the record: First and foremost, Spuggy Newton acted as his most able lieutenant, and was responsible for much of the construction work mentioned above.

The billiard room at Whaddon Hall. Bob Hornby standing left, next to him is Spuggy Newton. Far right is 'Colly' Colborne and sitting on the step Joe Inman.

Others in his team included, Alec Pollard, Ewart Holden, Percy Cooper, Alfie Willis, Steve Dorman, Wilf Lilburn, Bob

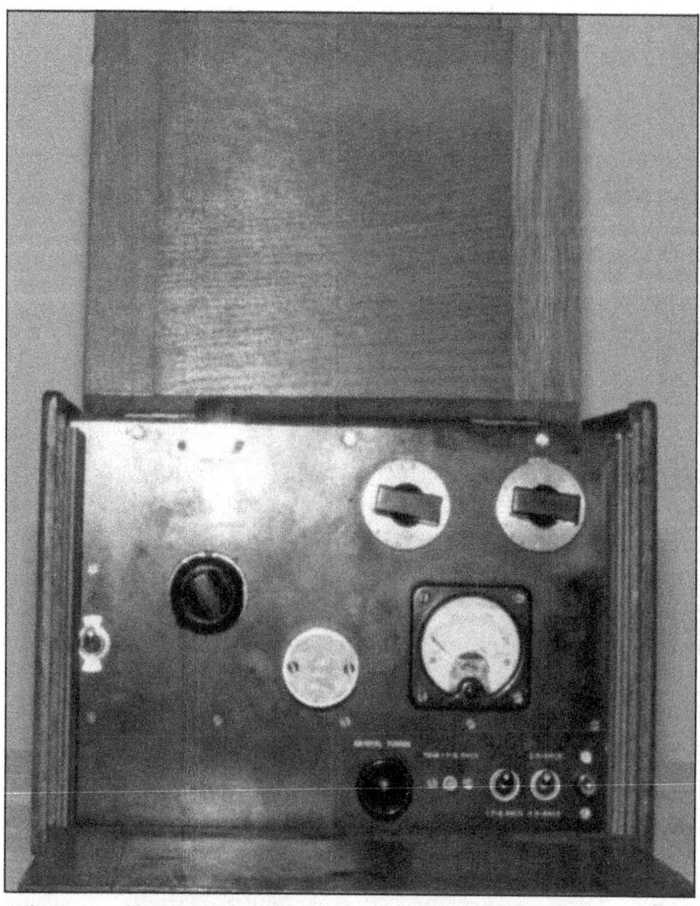
The marvellous MkIII

Chennells, Hugh Castleman, Alec Durban, Charlie West, Brian Sall, Tommy Ord, Douggie Lax, Dennis Smith, Charlie Pugsley, Hector Kempton, Harold K. Robin, amongst many others, and perhaps you will forgive me if I also include my father.

With Black Propaganda becoming even more important to the war effort, it was realised that a more powerful transmitter was needed. Hornby found that such a set was available in the States; he passed this information on and was told to purchase the transmitter complete. After purchasing the apparatus it was discovered that an aircraft carrier was leaving shortly from St. John's, Newfoundland for the UK and the Americans made a great effort and succeeded in getting it to the ship on time.

At an earlier stage, Hornby had been asked if he knew anyone who had the technical skill to supervise the operation of such large transmitting station and he says 'I recommended Harold Robin and he got the job against opposition, then transferred to the UK at the outbreak of war'.

I should add that there is some difference of opinion as to who actually went to America in the first place to examine and purchase the huge transmitter. According to some, it was Harold Robin but I do not think he would have been sent to make the initial decision as to its suitability. I believe Bob Hornby's account is the more likely. He claims that he was given the task by Gambier-Parry since, after all, by then he was the First Engineer of Section VIII with a formidable reputation.

More importantly, he was already in close touch with William Stephenson, the SIS representative in New York. Having made his appraisal, reporting back and obtaining permission to purchase the apparatus, Harold Robin was then sent over to take things further. The transmitter arrived in the UK in perfect condition. It was then a case of where to install it and the final choice was Crowborough, Sussex was finally chosen.

Bob Hornby joined DWS on its formation in 1946 and then moved over to Hanslope where he continued his remarkable career. This time, he had to deal with the technical challenges arising from the Cold War, but that is another story.

Lieut. Col. Frederick Robert Hornby OBE died on September 14, 1992, only six weeks after starting to dictate his notes. Unfortunately, they only take us so far but they include an insight into many parts of the unit's previously secret history. I feel sure that of his many successes, he would regard his early MkIII wireless transmitter as the one that gave him most pleasure, and it is fitting that there is now one in the Bletchley Park Museum, donated by his family.

Chapter 23

Arthur 'Spuggy' Newton

I returned to Whaddon Hall in late 1943, after a few months spent making parts for agent's sets at the Little Horwood factory, to join 'Mobile Construction' based just off the drive, near the exit lodge. My immediate boss was Dennis Smith. He, in turn, reported to 'Spuggy' Newton and this is the story of this remarkable man.

This chapter was prepared by his sons, Steve and Tony, and is based on a detailed report on his service with Section VIII that he wrote prior to his retirement from DWS at Hanslope Park. The chapter is built on that report, on other papers referring to his service in old files discovered in December 2001, and on my own additional comments since I knew him quite well. Arthur Newton was known universally as 'Spuggy' but nobody now seems to know why and certainly not his sons.

Arthur 'Spuggy' Newton was one of the original members of MI6 Section VIII founded by Brigadier Sir Richard Gambier-Parry in 1938, as the clouds of war gathered over Europe. Unfortunately, the veil of secrecy around his work meant that Spuggy never discussed his career, as the restraints of the Official Secrets Act were still being strictly enforced, even into the 1970s.

Spuggy Newton was born on 25 March, 1911 in Stanley, Co. Durham. His father was killed in 1916 whilst serving in World War I and Arthur was raised by his mother who managed to earn a living running a small haberdasher's. He was the middle of three children and really wanted to be an architect but family finances thwarted this ambition. Instead, he attended the North-Eastern School of Wireless Telegraphy from September 1927 to April 1928, obtaining a Government Certificate of Proficiency in Radio Telegraphy. He became a wireless operator with British Wireless Marine Service from 1928 to 1930 working as a 'sparks' on board merchant ships until the economic climate caused him to be made redundant.

Work in the north-east during the Depression was hard to come by but Spuggy got a job with Edison Electric in April, 1930 and moved in November, 1930 to Pritchard and Simpson, working as a wireless engineer, installing and servicing everything from two-valve sets to 800-watt power amplifiers. He was made redundant again in 1931, but his works manager at the time, a certain F. R. (Bob) Hornby gave him a glowing reference. Spuggy moved to Aitken Bros, Newcastle-on-Tyne, manufacturers of radio equipment, but was again made redundant in December 1932. He then moved south to work at Philco Radio & Television Corporation at Greenford, West London where he was again working under Bob Hornby.

There is a wonderful photograph of him that was taken at Durham Railway Station in 1933 with a party of friends. Spuggy is holding a briefcase and a parcel bound with string and is being seen off by his pals to his new job at Philco. Alongside him on his left is his old friend, Wilfred Lilburn. He and Spuggy had been at school together. Wilf was later to join Spuggy at Philco, and then followed him into Section VIII.

A reference for Spuggy, written by Bob Hornby and dated 18 January, 1938 (presumably prior to him leaving

MANAGING DIRECTORS { H. B. PRITCHARD
DR. R. W. SIMPSON

DIRECTORS { COLONEL O. W. NICHOLSON, M.P., CHAIRMAN.
W. LINDSAY EVERARD, M.P.

PRITCHARD & SIMPSON LIMITED,

MANUFACTURERS OF

EQUIPMENT.

13, 16 & 18, LISLE STREET,
NEWCASTLE-ON-TYNE.
TELEPHONE—28087/8.

REGISTERED OFFICE :
24, GROSVENOR GARDENS,
LONDON, S.W.1.
TELEPHONE—SLOANE 9218/9

141, LONDON ROAD,
LEICESTER.
TELEPHONE—59634.

Our Ref. FRH/JW.
Your Ref.

Please reply to

16th July, 1931.

To Whom It May Concern.

A.C. Newton has been in our employ since November 1930, during which time he has been employed on the construction, installation and servicing of our products, which range from 2 valve sets to 800 watt power amplifiers and has always shown himself to be a very resourceful, capable and conscientious engineer. He is absolutely honest and sober and it is with very great regret that we are compelled, owing to the state of trade, to dispense with his services.

We are perfectly certain that whoever employs Newton, will never regret it.

PRITCHARD & SIMPSON LIMITED.

F. R. Hornby.

MANAGER.

His reference on leaving Pritchard and Simpson (Radio Recordion) in 1931, note it is signed by Bob Hornby.

to join Section VIII) outlines Spuggy's career over the previous six years. At Philco, it appears he had been involved with installing police receivers and transmitters, as well as receiving apparatus in various types of commercial aircraft, and experimenting with fitting receivers to RAF planes at Northolt. This may have been in conjunction with work performed by MI6 Section VI (Air) and Fred. W. Winterbotham who wrote 'The Ultra Spy'.

Spuggy met his wife, Violet, the youngest daughter of a baker in Heston, West London and married her in 1938. Vi had worked for a German firm prior to the war and, as they were living at Stony Stratford from 1939, it was not surprising that she found work as a translator of decoded Enigma messages at Bletchley Park.

In June 1938, Spuggy was called in for an interview at Queens Anne's Gate in Westminster, part of the SIS headquarters at Broadway. This, presumably, was at the instigation of Bob Hornby. He was offered a job by

Spuggy – in trilby hat – being seen off by friends at Durham railway station in 1933 – en route to his new job at Philco. To his left is his school pal, Wilf Lilburn.

Richard Gambier-Parry who was forming a new special communications unit, subsequently called Section VIII of the Foreign Office. On 28 July, 1938, after a stringent W/T test at Denmark Hill and a thorough medical check, Spuggy was inducted into the service.

The new unit at that time was forming around Gambier-Parry and his trio of would-be deputies, Maltby, Jourdain and Darwin. Ted Maltby had a good technical background but the bulk of the wireless technology for the new Section VIII came from Bob Hornby. Technicians from a variety of sources were then rapidly recruited. Of course, due to their old association, Spuggy worked closely with Bob Hornby on the task of finding more engineers.

From the day he joined Section VIII until late 1941, Spuggy spent most of his time travelling all over Europe and the Middle East testing out wireless equipment, evaluating possible sites and installing two-way communications equipment. He visited most of the British embassies in Europe and had many narrow escapes as the Germans swept across Europe. He carried a very special hand-scripted Courier's Passport, personally signed by Lord Halifax, the Foreign Minister, and this probably helped him avoid imprisonment. 'Spuggy' certainly used one similar to the one illustrated. This was used during his visit to Istanbul in Turkey in January, 1941 when

PHILCO

PHILCO RADIO AND TELEVISION CORPORATION OF GREAT BRITAIN, LTD.

Perivale, Greenford, Middlesex
Telephone - - - Perivale 3344
Telegrams - Philcorad, Greenford

29th July, 1938.

To Whom It May Concern.

Mr. A.C. Newton has been in the employ of this company for the last five years. Since January, 1933, he has worked in the capacity of service engineer, production supervisor and radio technician in our department of Car Radio receivers.

We are able to state that we consider him to be a man of marked ability and integrity, and his theoretical and practical training qualifies him for any investigational or departmental work on radio installations. We are sorry to lose his services, but in view of the fact that he is leaving us to improve his position, we wish him every success.

Yours faithfully,
PHILCO RADIO & TELEVISION CORPORATION
OF GREAT BRITAIN LTD.

R.Koller.
Works Manager.

RK/RH.

R Koller

Directors: Carleton L. Dyer, Chairman & Managing Director, E. C. Baillie, Viscount Hinchingbrooke, H. W. McAteer, U.S.A., C. U. Neison, J. M. Skinner, U.S.A., T. F. Williams, U.S.A.

A reference from Philco dated July 1938 presumably to support his impending enlistment into Section VIII he installed an up-to-date SIS wireless to connect with Whaddon, at the British embassy. The film called *Five Fingers*, starring James Mason is based on the true story of Elyesa Bazna – known to German intelligence as 'Cicero' – who was a German spy and valet to the British Ambassador, Sir Hughe Knatchbull-Hugessen. Cicero was a source of invaluable intelligence to the Abwehr. The film actually mentions the embassy's wireless room.

Spuggy's talents were obviously in great demand, as not only was he a very competent wireless engineer, he was also a qualified wireless telegraphist. He rapidly obtained a reputation as a fixer and procurer of equipment to ensure the job was completed. Over a three-year period, Spuggy was away from the UK almost continuously. This included three trips to Brussels during which he installed a two-way communications link with the UK, purchased a caravan and converted it into a mobile wireless station capable of communicating with UK, trained a number of Belgian nationals in the use of the equipment and fitted out a saloon car with radio equipment capable of communicating with UK.

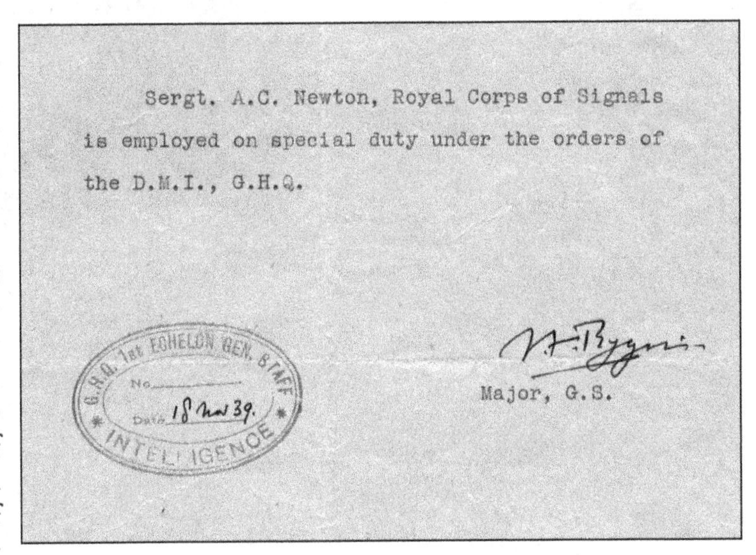
Sergt. A.C. Newton, Royal Corps of Signals is employed on special duty under the orders of the D.M.I., G.H.Q.

G.H.Q. 1st ECHELON GEN. STAFF
No.
Date 18 Nov 39.
INTELLIGENCE

Major, G.S.

A pass signed by Director of Military Intelligence for GHQ (BEF in France before Dunkirk)

Four round trips were made to The Hague, again installing two-way radio links to Britain, touring Holland establishing two-way communication installations in three separate villages, including Delfzi on the North Holland/German frontier. Spuggy was finally escorted out of Holland, the day after Commander Best and Mr. Steven were captured by the Gestapo on the Dutch frontier.

Spuggy's next venture outside England was to Nuremberg and then on to Prague and Warsaw, installing two-way communications and performing tests with Station X at Barnes, Prague and Brussels. He journeyed on to Talinn in Estonia, then to Helsinki and Stockholm.

Spuggy made another trip to Prague was to install radio equipment in the British Embassy basement, but this time he was caught up in the German invasion of the Sudetenland and ushered out of Prague with Sir Walter Runciman's abortive mission. There were two trips to Rome, then on to Lisbon and Madrid, and then Istanbul installing two-way communications and testing wireless transmissions with the UK, Rome, Lisbon, etc. He made two trips to Athens on an Imperial Airways flying boat and installed two-way communications at the Embassy and on a motor-yacht at Patras.

On the rare occasions when Spuggy was in England he worked with Bob Hornby and assisted in the design and construction of new radio equipment for overseas and in the UK at a little bungalow radio station at Woldingham, Surrey, known as 'Funny Neuk'. He had helped expand the small factory at Barnes but shortly after the declaration of war in September 1939 all this was moved to Whaddon Hall, where he assisted in setting up workshops and R&D laboratories. He then became involved in installing wireless equipment in boats, cars, caravans, helping the police track down illicit transmitters and designing and constructing fixed transmitting and receiving stations, all within a 20-mile radius of Whaddon Hall.

These included Nash, Dower House, Main Line at Whaddon Hall, Windy Ridge, Calverton, Gawcott and Upper Weald. Spuggy had two huts at Whaddon – 4 and 5 – and to this day few know what went on in there.

Courier's Passport.

SIX PENCE 12 6 40

No 592

This passport is valid only for the journey specified and must be given up with the despatches

DIEU ET MON DROIT

We Edward Frederick Lindley, Viscount Halifax, Baron Irwin, a Baronet of Great Britain, Knight of the Most Noble Order of the Garter, a Member of His Majesty's Most Honourable Privy Council, Knight Grand Commander of the Most Exalted Order of the Star of India, Knight Grand Commander of the Most Eminent Order of the Indian Empire, etc. etc. etc. His Majesty's Principal Secretary of State for Foreign Affairs.

Request and Require, in the name of His Majesty, all those whom it may concern to allow *Mr A. C. Newton, proceeding to Istanbul, via Takoradi, — Cairo and Athens. Belgrade and return* charged with Despatches to pass freely without let or hindrance and to afford him every assistance and protection of which he may stand in need.

Nous Edward Frederick Lindley, Vicomte Halifax, Baron Irwin, Baronnet de Grande-Bretagne, Chevalier du Très Noble Ordre de la Jarretière, Membre du Très Honorable Conseil Privé de Sa Majesté Chevalier Grand Commandeur de l'Ordre Très Exalté de l'Etoile des Indes, Chevalier Grand Commandeur de l'Ordre Très Eminent de l'Empire des Indes etc. etc. etc. Principal Secrétaire d'Etat de Sa Majesté au Département des Affaires Etrangères

Prions et Requérons, au nom de Sa Majesté, tous ceux à qui il appartiendra de laisser passer librement *Monsieur A.C. Newton, se rendant à Istanbul via Takoradi, — Cairo, et Athènes, Belgrad et retour* chargé de Dépêches, et de lui accorder en toute occasion l'aide et la protection dont il pourra avoir besoin

amended at Athens 17/4/41

Given at the Foreign Office, London the *29th* day of *January 1941*

Halifax

FOREIGN OFFICE 29 JAN 1941

A remarkable document. A Courier's Passport to allow Spuggy to travel to Instanbul. It was personally signed by Lord Halifax and is dated 29th January 1941.

In their notes, Steve and Tony suggested that one was for the repair, maintenance and tuning of his various personal cars. As author of this book, I can confirm that this was the case and his Rover (along with that of his boss, Bob Hornby), was the best equipped, maintained and polished Rover in creation!

Shortly after the war broke out, Spuggy was given the job of designing and building a mobile two-way station in a Dodge car which he drove to GHQ in Arras in France. He tested and commissioned this vehicle and handed it over to Colonel Plowden. Later, in 1940, he was sent to the Auxiliary Intelligence Organisation which, under the command of Captain Peter Fleming, had set up its H.Q. in an old farmhouse named The Garth, in Ashdown Forest. He installed wireless equipment in a number of hideouts and dugouts in the forest, to be used by resistance forces in the event of invasion. This is connected with the 'Stay Behind Army' referred to in the glossary.

Spuggy Newton's most circuitous, eventful and dangerous trip was to Yugoslavia in early 1941. The Germans and Italians controlled the Mediterranean but Spuggy was directed to take some urgently needed radio equipment to Belgrade. The route was, to say the least, complicated. The trip was scheduled to last six weeks but in fact took him almost right around the world and lasted five and a half months! He started off from London on 30 January, 1941 and took the train to Greenock in Scotland. With him were one small suitcase of personal effects, three large wooden crates and nine very large Foreign Office bags, all containing equipment for various jobs in Yugoslavia and Greece.

The following is a direct extract from letters written to his wife:

'Left Heston, London, January 30th and proceeded to Glasgow. Went on board and sailed in a hell of a snow storm.

Spuggy in his element – fitting a wireless set into a car

Sailed for days in one hell of a gale. There was ice all over the ship. We were very near the North Pole before turning South. For days it was B cold! We just missed an armed raider by 2 hours. Had a few false alarms re: submarines.

'Eventually the engines broke down and we were sculling around in the middle of the Atlantic for a day, drifting in a hell of a northeasterly gale. We eventually limped into Freetown which was a lousy dump with one rundown hotel that was, in any case, full.

'It took the chief a week to repair the engines and we set sail for Takoradi. The expected plane was awaiting and we took off with the crates and diplomatic bags and two other passengers on a long haul across the Belgian Congo and up the Nile via Lagos, Doubla, Massaka, Stanleyville, Juba, Malakal, Kossti to Khartoum and Cairo.

'The aircraft was a German three-engined Junker, impounded in Africa by Sabena after war was declared. The aircraft met with a good deal of suspicion at a number of airfields and in two cases was fired upon by over-enthusiastic local native troops before landing safely'.

Somewhere between Stanleyville and Kosti the plane ran into a tornado.

'After a few minutes of being tossed around like a cork, things looked pretty hopeless. We were all strapped down tight to our seats. The radio operator got a bit panicky but the Belgian pilot managed to pull the plane out of this ruddy awful black tornado and we had to run like hell away from it. We eventually left it behind us but we were miles off course and it was nearly dark and impossible to make any airport before dark. The skipper had ten minutes to find a spot to land before twilight turned to pitch black'.

The pilot found a small village and managed to make a miraculous landing. He dropped the plane in a narrow street, missing trees, cesspools, etc. by inches with the wingtip shaving bits off the native huts.

'He pulled up with six inches to spare between the nose and the chieftain's hut. First off was the radio operator, not because he was brave but because he needed to leave the room'.

The natives were scared stiff as this spectacular giant bird came out of the sky and came to rest with its nose lodged in the chieftain's hut, one wheel broken and the other stuck in the sewage pit. Native women wearing G-strings were running around everywhere, peering from the jungle goggle-eyed, was a frightening sight.

Eventually, a Belgian missionary appeared and Spuggy's fellow passenger, Captain MacAdams, walked forward and in a loud English voice said, 'Dr Livingstone, I presume' and gravely shook his hand. The missionary only spoke French so this witty riposte went over his head!

The Belgian crew and passengers were escorted to the missionary headquarters where having been ushered to sit on home-made chairs, a bottle of Scotch was produced and rapidly emptied. Jungle drums had been beating continually since their arrival and it transpired that this was to summon the other white men.

Shortly afterwards, two white men arrived in a very tired Model T Ford. Spuggy managed to transport all the wireless equipment from the plane to the mission. He recounts having a shower produced by two local boys pouring water down a tube with a syringe on the end! He was well looked after until several days later a squad of Belgian troops arrived to transport the pilot and Spuggy to their unit headquarters at Lisala. Spares were eventually flown out and the Junker was able to take off after a ten-day delay.

They had sent a telegram to the next airfield warning of their arrival in a hastily repaired plane but it had not arrived and to their horror, the airstrip was blocked with oil drums and old cars. The plane had woken the locals and who were soon speeding to the air strip and jumping into trenches and machine gun nests.

It transpired that it was a close shave as to whether they shot the plane down, thinking it was the enemy, but a quick-thinking local sergeant who was in charge had his doubts and ordered the troops not to fire. It transpired he had shot up two visiting Hurricanes the week before and had been severely reprimanded, so he was a little less trigger-happy!

The plane pushed on again, arriving at Khartoum for a quick refuel before pushing on to Cairo. On take-off at Khartoum, there was a loud bang and a tyre blew. The plane slewed to a halt. Passengers and crew had to wait for a car to take them to a hotel for the night. Arriving in Cairo four days late, Spuggy had a long bath, some cold beers, a haircut and slept for twelve hours solid, before embarking on a boat with a contingent of New Zealand troops bound for Athens. He was the only civilian on board and he obviously had difficulty fielding questions about his reason for making the journey.

In a letter dated 7 April, 1941 from Athens Spuggy wrote:

'The days after Yugoslavia told Germany to go to hell, imprisoned the Government, turned out Prince Paul and started mobilising I had to go to Belgrade. Had stacks of baggage too! Caught a train Monday 4.30am and arrived in Salonika at 2 am the following day. Went to hotel – had two hours sleep – then up again to catch another train for the Greek-Yugoslav frontier. Got across border and changed for Belgrade. Arrived Belgrade 4.30 p.m. Wednesday after a nightmare journey. Train was packed with troops and women and children crying in the corridors. There was no food.

'I delivered all the bags but was told to get out fast and to take the midnight train out from Belgrade to Greece. It turned out to be the last train. Crossed the Greek frontier at Salonika only a few hours before Jerry invaded Greece. Arrived Athens 3 a.m. Sunday'.

The Athens Embassy was in a panic and Spuggy was shipped out on a refugee boat of women and children bound for Alexandria.

'I had half an hour's notice so had to leave all my laundry behind. Scrambled down to docks in an air raid, hoping to jump aboard a destroyer. Well destroyer did not show up, so had to scramble aboard a ship full of evacuees. Had loads of bags and equipment and had to load this myself, which meant heaving out a derrick, tying all my boxes on to the end of the guy and then running up the ship and working the winch. After sweating and struggling for half an hour in the boiling sun I was a mess.

'Scrounged a wash from the engineer. There was accommodation for only six passengers yet we had 200 evacuees, no food and limited water and toilet facilities. Managed to scrounge two bottles of brandy and one of scotch. Had a couple of quick tipples and went off to scrounge some food off the second engineer, before settling under a lifeboat cover. Existed for three days and nights like this. Jerry tried to sink us by dive-bombing the ship, then spraying the deck with machine gun fire. No one hurt. Two mines went off close by but we escaped again. Arrived Alexandria and then headed rapidly by train to Cairo, looking like a tramp. Hope to God I can get out of Cairo soon, as I am fed up to the teeth with this place'.

On arrival in Alexandria, Spuggy pushed on to Cairo and delivered all the diplomatic bags from Belgrade and Athens. Going down with a bad dose of dysentery he spent a week in bed in a hotel.

When he recovered, he reported to the embassy and found the two operators in some trouble. The number one, Commander Barby, had ordered the main transmitter to be shipped to Rangoon, as that was where it was proposed to evacuate all the staff, it having been decided to leave Cairo to Rommel and his advancing Panzer Divisions. The low-powered emergency transmitter was unable to make good contact with UK so Spuggy persuaded Commander Barby to let him track down the transmitter and re-establish it.

Spuggy, as I remember him in his office at Whaddon, circa 1944/45.

With Navy assistance, it was located on a ship in the Bitter Lakes of the Suez Canal. Eventually, the equipment was returned to Cairo and the link to London was made, enabling a backlog of traffic to be cleared.

Rommel's advance along the desert road had been halted, the communications network in Cairo was stable again and Spuggy was sent on a boat trip up the Nile to Wadi Halfa, Atbara, and Port Sudan, surveying for possible wireless station sites. Returning by the same route to Cairo, he received orders to return to England via the same route, across the Belgian Congo to Lagos in West Africa. This time it was an uneventful journey although he had to wait two weeks for a flying-boat to take him to Lisbon and then to London, where he touched down on 30 June 1941, after twenty two weeks and thousands of miles travelling.

It is not surprising that Spuggy had a reputation for fixing and scrounging things as the story of his trip to Belgrade and back to Cairo demonstrates. He clearly showed determination to get kit and equipment through against many odds and at a time of great disruption. He must have had a considerable negotiating skills or some very special papers to ensure doors were opened for him. Perhaps his collection of diplomatic passports were a help in these escapades?

There is a story about Spuggy's return to England that may be apocryphal but allegedly, he staggered back to Whaddon Hall, having lost most of his baggage and covered in dust and grime where he found that the hierarchy were all in a high-level meeting. Unable to restrain himself from announcing his triumphant return, he flung open the doors and marched in, only to be greeted with something like 'Oh you're still alive, for God's sake go and get yourself cleaned up'.

From July 1941 to 1945, Spuggy was located in England working, from Whaddon Hall and resuming his previous duties of wireless and construction engineer responsible for the design and building of two-way radio stations. These were installed in cars, lorries, vans, scout cars, minesweepers, motor launches, midget submarines, cruisers and one battleship, the *HMS Nelson.* He also maintained and expanded the various workshops and radio stations previously built at Calverton, The Bare, Windy Ridge, the Dower House, Upper Weald, Nash, etc.

In 1944, immediately after the invasion of Europe by the allied armies, a number of trucks and wireless equipment vehicles were sent from the UK in the wake of the advancing army and airforce commands. These were SLUs, engaged in the dissemination of Ultra traffic via Windy Ridge to SCU8. All were built by Mobile Construction, under Spuggy's direction, into Guy vans and Dodge ambulances. The same team were also responsible for the work in minesweepers, motor torpedo boats, midget submarines and aircraft.

After the German surrender, Gambier Parry asked Spuggy to take over all the Special Communication Units at Little Horwood, reduce the fleets and supervise the demobilisation of the MT staff. The disbanding of the 300 men and 120 vehicles was successfully completed in early 1947. Spuggy then moved to Hanslope Park where he formed a new Services Department charged with running the Diplomatic Wireless Service site and radio stations at Hanslope Park, Creslow, Gawcott, Poundon, Poundon House, Lawencekirk and Montreathmont Moor.

On a personal note, I would like to record the last time I saw Spuggy Newton. In early March 1947, I returned to England from our station in Singapore. After a short leave, I went to Whaddon Hall in April, determined to leave the unit and join the family business.

At that time, I held current army driving licences for each of the army cantonments of Delhi, Calcutta and Singapore. However, to obtain a civilian driving licence without taking a test, it was necessary to hold a 'green certificate' from the army to present to the local licensing authority. Having been forced to take three or four tests already in my 'military' career, I wanted to avoid yet another, if I could.

About the only senior person left at Whaddon was Spuggy. In spite of seeing all my collection of licences, he still would not grant me the valued document without his personally approving my driving ability. He sent for a car from the diminishing collection at the transport pool and a senior driver, whom I had fortunately known years earlier, turned up in a Humber Super Snipe estate car. Clearly, my test, in today's parlance, was going to be 'a doddle'!

I drove through the villages around Whaddon for a while, then I took advantage of having such a nice car to visit Hanslope to see Major Charlie Crocker (previously head of SCUs finance – now head of finance for DWS), to pick up my last salary cheque.

We returned to Whaddon and there Spuggy presented me with my certificate with a smile since he already knew from Charlie Crocker that I had used the time on my 'driving test' well. Spuggy wished me good luck in the future and sent his regards to my father. I never saw him again.

Arthur Coates Newton was awarded an MBE in recognition of the work he had undertaken during the war. This was presented to him by King George VI at Buckingham Palace in 1950. His sons say it is sad that much of what he did will never be known, unless records remaining in the bowels of MI6 are ever released. I am not sure, however, any records could be as fulsome as his very own history of the unit. Incidentally, he only set down the story of his career with Section VIII because the 'powers-that-be' were so parsimonious about his level of pension; otherwise we can be fairly sure this tale would not have been written.

It is likely that Section VIII's staff records, and those pertaining to the war years and earlier, were destroyed in

1946, at the time of the move over to nearby Hanslope Park, when the unit took on its new role as the DWS. Indeed, I am assured by several colleagues who are apparently 'in-the-know' that this was the case.

Spuggy retired from MI6 in July 1971, buying a bungalow in Devon, but sadly not being blessed with good health, he died in 1978 from serious bronchial asthma, a legacy from years of smoking Players Uncut cigarettes.

Chapter 24

Joyce Lilburn and the Remarkable Hill Family

This is the story of a remarkable family with very strong links to our intelligence services, from World War I right through to the 1950s. Her mother, Edna, her father Charles, her sister Anne and Joyce herself all worked at some time in our secret services. Her husband was Wilf Lilburn of Section VIII. Joyce's family story was related to me by her; she is now over eighty. I met her husband, Wilf Lilburn, in 1939, whilst he was working at the SIS 'Funny Neuk' wireless station at Woldingham, near Caterham, Surrey. I later served with Wilf in 1946, at the units station in Singapore.

Joyce Hill was born at Bridge of Allen in Scotland in January, 1918, to the skirl of the pipes played by Pipe Major Daniel Laidlaw, VC who was from the Western Front in France, where he had won the Victoria Cross playing his pipes under fire and encouraging the troops forward. He was stationed at the army camp near the end of Fountain Road in the town and his task was to wake up the soldiers billeted outside the camp in houses nearby. Joyce was born during his march around the streets; her birth certificate records her birthplace as 'Hafton', Fountain Road, and the time as 7.00 am.

Their real home was in Moscow but mother and Anne had moved to London in 1916, to temporary accommodation. The Zeppelin raids had been heavy towards the end of 1917, and so the family moved up to Scotland to live with her mother's sister at Bridge of Allen. Joyce's family had strong connections with Russia. An ancestor of her maternal grandmother had gone to St. Petersburg in the 1760s to lay out the Winter Palace garden for the Tsar. From then on, each generation of the family sent one member to live in Russia. Joyce's grandmother had married a Scot by the name of David Maxwell; her mother was born in Russia in 1884. Her father, Charles Battersby Hill, went to Russia when he was in his twenties, to refurbish Mrs. Maxwell's home. There, he met and married Joyce's mother, who was the youngest Maxwell daughter. They lived in Moscow and there Anne, who later worked at Whaddon Hall, was born in 1909.

When World War I broke out, Charles Hill left Russia to enlist in the army, passed through OCTU and was commissioned. His wife and the young Anne followed him to London in 1916, when the troubles started in Moscow. The family home and everything else, including their bank accounts, were sequestrated by the Bolsheviks, and her aunt was imprisoned by them for a while.

Joyce reports her mother had been 'brought up like royalty'. There was no question of attending school, instead she had tutors, who didn't even teach her to add and subtract. To her dying day, she could not handle money properly and apparently would claim that 'only common people do that!'

As an aside, in the 1980s Joyce received less than 1 per cent of the millions due to her from the Russian government. Her grandmother Maxwell was 'exceedingly rich' but her fortune was totally lost.

Anne was always away from home at boarding school somewhere in England and was thirteen before Joyce met her for the first time.

A photograph of the three Hill children taken in Germany in early 1920s.
It shows Alex who was posted missing over Germany whilst serving with Bomber Command,
Joyce in the middle, with Anne to the right.

During the war, Hill was in the Royal Engineers (Signals Section) where he worked on Russian liaison. He was offered and accepted a seven-year short service commission. In 1918, he was sent to Salonika with his batman, Corporal Bradley, who was also a wireless operator. Their task was to monitor Bulgarian wireless traffic as it was thought that Charles' knowledge of Russian would be of help. He came home at the end of the war, and then worked as aide-de-camp to General Haking with the Inter-Allied Disarmament Commission, based in Hanover. It is certain that he was working for various intelligence services for most of the war and continued to do so until he was demobilised in 1924.

Incidentally, whilst waiting to proceed to Germany, Charles Hill was based for a time at a Signals Training Unit at Fenny Stratford, only a mile or two from Bletchley Park, that was to become so famous in World War II. Joyce and her mother were billeted in the vicarage, in the village of Simpson, where so many BP girls were to be billeted twenty years later. Even more astonishing is the fact that Anne, when she started working for Section VIII at Whaddon Hall, was first billeted in the same village.

The family followed Charles to Germany in 1920, and lived in the Kastens Hotel in Hanover where they stayed until 1924, when they returned to London and several difficult years. Anne's real name was Gladys, which she hated, but apparently there was another Gladys in the department at Selfridges in which she worked, and as no two staff could use the same Christian name, she chose 'Anne' from Anson Road where they were living at the time. I have referred to her only as Anne, as that is how she was known at Whaddon.

Anne worked at Selfridges from 1927, until she met and married Adrian Trapman in 1937. He had been British Consul General in Addis Ababa during the Italian invasion of Abyssinia. He was awarded the Empire Gallantry Medal – a very high award of the time – for his work during the Italo-Abyssinian war. In Addis Ababa, Trapman worked with Don Lee who was listed at the British Embassy as an 'archivist' but was fairly obviously working for SIS. He later went to Spain as an agent during the Civil War. One can reasonably assume from the closeness of his friendship with Don Lee, that Adrian was the SIS 'Head of Station' for that region.

Charles Battersby Hill in World War I – as an officer in the Royal Engineers (Signals corps) – the forerunner of the Royal Corps of Signals.

After the wedding, Anne and Adrian spent their honeymoon driving through Europe en route to his new post in Ankara. He was tragically killed, however, in a motor accident while they were driving with the British Minister in Athens, who also died. Anne was injured but made a full recovery.

After finishing school, Joyce went to St George's College, then part of London University, to take a course in secretarial and commercial studies. After a short period working at Rowley's Art Gallery, she moved to Hatfield, Dixon & Co., Chartered Accountants, which she found most interesting. They were accountants to the Duke and Duchess of Kent and other members of the Royal Household. On the outbreak of war, the firm was evacuated to Sanderstead but Joyce's mother would not allow her to go with it, so she found a job with the Procurator General's office in Whitehall, in the Admiralty Prize Division.

Anne Trapman – as she was known at Whaddon.

During this time, her father, Charles Hill, worked at a German travel bureau from where it is believed he gave information to MI5 about possible German undercover agents. It is also known that he acted for Commander Blackburn of the Metropolitan Police, Special Branch. With the outbreak of war in 1939, he was employed as a civilian at the Home Office, most probably for MI5. Joyce's mother, Edna, was also multi-lingual; with her knowledge of languages and especially of Russian, she was recruited into MI5, although Joyce is not sure what her work entailed.

Charles Battersby Hill was sent to Scotland when Hitler's deputy, Rudolf Hess, landed by aircraft, but being extremely security conscious, did not explain why to his family. The family followed him to Scotland, and found a home in Glasgow, as it was thought Charles would be there for some time. Again, Joyce had to find a job, and this time it was as a shorthand secretary in a naval unit known as *HMS Startiate* but was actually the St. Enoch Hotel, Glasgow, not a ship at all. Its role was as Admiralty Shipbuilding and Repair Department and apparently quite fascinating. Besides the main work, they arranged for ships to be degaussed to protect them from magnetic mines.

Joyce's father's stay in Scotland having been cut short they all returned to London. Charles later appeared in RAF uniform as a Flight Lieutenant, and worked at the Air Ministry on what he described as 'liaison duties' in the Belgian Section. He was fluent in French and his boss was Wing-Commander Kenneth Horne, who went on to become a popular radio comic.

Charles was then posted to an American intelligence unit which followed front line troops into France. His role in this operation was to ensure that German intelligence documents were not destroyed. Being attached to the American Army, his command of German, French and Russian later led to his being engaged as an interpreter at the War Crimes Tribunal at Nuremburg. His interpreter's pass for that, and for the International Military Tribunal, now as Squadron Leader Hill, are illustrated.

Knowing she would need a job on her return to London, Joyce approached the office in Glasgow before leaving. They apparently contacted a special Labour Exchange department dealing with senior civil service postings, who knew of a vacancy for secretarial staff at the Admiralty in London. After interviews at the Admiralty, and no doubt good references from Glasgow, Joyce was invited to join Naval Intelligence as a secretary. She quickly found she was working in the Operational Intelligence Centre (OIC), which included the Signal (Y) Department, and NID9, based in the underground complex called 'The Citadel'. The overall commanding officer of OIC was Admiral Sir William James who, as a child, had been immortalised as 'Bubbles' in the painting by Millais that became famous throughout the country because it was used in an advertisement by Pears Soap.

Other officers in the unit were Captain Eddie Hastings RN, who was director of NID and the Signal unit, Captain Sandwith RN, the deputy director. Lieutenant Commander Clive 'Joe' Loehnis was the unit's liaison with MI6, Bletchley Park and Section VIII at Whaddon.

Joyce worked as a secretary, in the department covering naval operations worldwide, including submarine tracking and reconnaissance by RAF Coastal Command. For the latter duties, they had Wing-Commander Cohen who was attached to the Admiral's staff. Admiral James and his small staff were in charge of the 'A' watch and others staffed watches 'B' and 'C', each watch covering 24 hours. They also had a wireless room staffed by a Captain RN with a midshipman and petty officers.

There was a photographic unit with a civilian photographic officer, whose role was to interpret the reconnaissance pictures in 3D. Other duties included handling Ultra traffic sent by teleprinter directly from Hut 3 at Bletchley Park. The operators in the Ultra teleprinter room were nicknamed the 'Secret Ladies.' They had to lock the door whenever they went in and out of the room.

The units main task was to produce, at 03.00 a.m. each morning, a detailed intelligence report, including the estimated position of all convoys at 08.00 a.m. that day, mine-laying activities, position of naval craft and status. This went daily to the Joint Intelligence Committee marked for The Prime Minister Winston Churchill, the First Sea Lord, all other Sea Lords, Army and Air Force Commanders in Chief. Joyce discovered recently that a copy went to His Majesty George VI.

Although Joyce was working in the heart of the Admiralty, she was not put into Wren uniform but was issued with Merchant Navy identification papers. Apparently, Dame Vera Loughton-Mathews, the commanding officer of the Wrens, refused to allow 'her girls' to work in the Admiralty itself – and certainly not in this holy of holies. This paints an entirely different picture to the war films we see depicting Wrens climbing ladders and putting pins on convoy positions on the wall charts, or sat at a desk furiously typing! There simply were no Wrens in the Operations Intelligence Centre.

INTERNATIONAL MILITARY TRIBUNAL

0563
Pass No.

Date Issued

HILL, CHARLES B.

Name

Other Data

is authorized to enter the Area of the

PALACE OF JUSTICE

SECURITY OFFICER

signature of bearer

OFFICE OF CHIEF OF COUNSEL FOR WAR CRIMES

PASS NO. 0557

DATE ISSUED

SIGNATURE OF BEARER

Squadron Leader, R.A.F.

NAME

OTHER DATA

ISSUING AUTHORITY

Passes issued to Squadron Leader Hill for his role as interpreter in The War Crimes Tribunal at Nuremburg and the International Military Tribunal

After a time, working long hours underground, and without heating of any kind in winter, Joyce's doctors advised that she be moved out from the Citadel into a unit working in a more normal environment. Here, Commander 'Joe' Loehnis came to her rescue, and she was introduced to Brigadier Richard Gambier-Parry, head of Section VIII, during one his visits to Broadway. He asked her to join his headquarters which he said, were based in a country house in Buckinghamshire.

On arrival at Whaddon Hall, she was shown around and then came to the office in which she was to work, that had earlier been the library. Imagine her complete astonishment to see her sister Anne behind one of the desks! Joyce had absolutely no idea where Anne worked or what she did, other than it was war work, 'in the country somewhere'.

At first, she was disappointed to find that, instead of being in the centre of a busy 'front line' operation, here she was merely expected to file reports, and burn sensitive papers in the boiler in the cellar of the Hall. Soon, however, she started to work in Section VIIIP headed by Lt. Col. Lord Sandhurst, with Major Bill Sharpe ('Sharpie') as his deputy.

The team consisted of Bill Sharpe, her sister Anne, who by now had married Don Lee, Evelyn Watts from the ATS, and Joyce.

VIIIP at Whaddon was 'Planning' and this linked with 'P' or Planning Officers based at Broadway. Each of these 'P' officers covered a sector of the theatre of war and was responsible for the installation of agents into their area, as well as their control. For example, P1 was French Planning, P3 Belgium, and so on.

The task of the VIIIP, at Whaddon was to provide the wireless co-ordinates, suggested frequencies, call signs and times for contact for the agents sent out by the 'P' officers. An SIS agent's W/T traffic would only come back to one of the Section VIII stations at Nash, Weald or Forfar, depending upon his location.

VIIIP was based in the library of Whaddon Hall and the picture on page 78 shows Lord Sandhurst and some of the staff on its steps, in early 1944. They kept in 24 hour communication with the MI6 Planning Sections at Broadway and the W/T stations at Weald, Nash and Forfar.

Weald, dealt with France, Holland, Belgium and one Danish network, Nash, with Holland, Belgium and some agents in France. Forfar dealt with Norway at a later stage. There was obviously an overlap but it depended upon the agents and/or network being contacted. Joyce told me there was also a couple of women acting as agents in Germany.

Lord Sandhurst (whom Joyce always refers to as 'the boss') and Bill Sharpe visited SIS HQ at Broadway twice a week by car to discuss the communications needs of the various operations being planned in enemy territory. W/T plans were then prepared together with the necessary agents' wireless equipment. Lord Sandhurst had a Packard limousine and Joyce and Anne were sometimes allowed to travel in the car to London, if they had a day off.

Training of these agents in the use of wireless was normally carried out at an SIS 'safe house' in Hans Place at the back of Harrods in Knightsbridge, under the care of Bert Gillies and others (see chapter 36 – Dave Bremner). Before the agent left, a record was made of the agent's Morse style or 'fist' which is always distinctive to the trained ear. This record was then taken to Whaddon so that it could be compared with the first signals as they came through from the SIS agent. Many operators were captured as they parachuted in, and a few were betrayed by the locals. The fear was that their codes and the W/T plan might be seized and used by the enemy, pretending to be the agent. If the transmission was suspect, MI6 would then either cease communication or continue to send messages but giving false information. VIIIP also received en *clair* copies of all traffic between MI6 and its agents so they could check for anything affecting W/T operations or security. These messages were then burnt in the boiler in the basement of Whaddon Hall.

Anne in WAAF uniform with Don Lee

All the agent's traffic received at our W/T stations was decoded or encoded exclusively by Miss 'Monty' Montgomery and her team, who had a secure SIS office area in Hut 10, at Bletchley Park. They were not in anyway concerned with the Ultra traffic, nor with the 'codebreaking' operations of BP. Transport of these messages to and from the stations to 'Monty' in Hut 10 was by our large team of despatch riders based at Gees.

A 'crack' signal would be sent to controllers by the BBC which, for France, was called 'Reseau.' This later became the name for controllers in all countries. After the war, Joyce worked for Charlie Bradford. She transferred to Hanslope on the closure of Whaddon, and later to Broadway, retiring from the service in 1961.

Following his work at Nuremburg, her father, Charles Hill, was employed in the Russian section of GCHQ at Bletchley Park, then at Eastcote, and finally, before his retirement, at GCHQ at Cheltenham. He was awarded the MBE for his services.

It is surely remarkable that four close members of the same family, mother, father, and the two sisters Anne and Joyce, should have been employed by MI5 or MI6, at the same time. While at Whaddon, the widowed Anne was in the WAAF and was known as Flight Officer Trapman. Later she married Don Lee, who had served with her late husband in Abyssinia (Ethiopia).

Joyce married Wilf Lilburn, a long time member of Section VIII, in 1951 (see chapter 36).

Chapter 25

Wilf Neal in SCU8 with General Patton's Third Army

Wilf Neal was born in the Ward End district of Birmingham in March, 1925, and still lived at the time of his call-up for military service. He had had a Grammar School education but had to leave when the school ceased offering full time education on the outbreak of war, when many pupils were evacuated (though not Wilf).

In February, 1940 he started work in the buying department of the Valor Company in Birmingham, a long-established company that had made oil stoves before the war but which had now switched to making ammunition boxes as well. At the age of 15, Wilf started fire-watching duties and at 17 he joined the factory's Home Guard company. In January, 1943 he registered for National Service and was called up on 21 October, 1943.

Wilf Neal's association with SCUs, the United States Third Army and beyond, commenced on Tuesday, 12 October, 1943, when he received his Enlistment Notice with an invitation to present himself at the Royal Corps of Signals, No. 1 SCU, Bletchley, between the hours of 0900 and 1200 on Thursday, 21 October. This notice arrived while he was at work, and he recalls that his mother kept it hidden until after he finished his tea, presumably thinking the shock would put him off his food! This was not the case, however. Wilf had registered for National Service in January of that year, and in the interim turned down a request to join the Navy or work in the coal mines. It was at last a relief to know what the immediate future held in store.

During is initial interview after registering, he was asked if he had any preference as to which service trade he wished to pursue. He did not understand the logic of this question since, as far as he knew, you were 'put where you were put' and that was the end of it. He asked not to be put into signals as, although he had developed a schoolboy interest in radio, it had by then waned and he held ambitions of becoming a driver, and so it was a bit of a let-down to learn where he was going.

Memories of Little Horwood

On arrival at Bletchley station, well within the appointed time, Wilf waited around with a number of other sad-looking characters until a three-ton truck arrived to take them on to the SCU1 establishment at Little Horwood, affectionately known to all who have served there as 'Gees'. He was in the second recruited intake of some 40– 45 men.

Until recently, Wilf had been under the impression that this date in September was the start of operator training at Little Horwood, but in conversation with old colleagues at reunions in recent years, he has discovered that training started as far back as 1940/41. These earlier trainees were men who had achieved certain operational speeds at other establishments but had come down to Gees for advanced training before being posted to duties within the SCU 1 orbit, such as Whaddon, Weald or Nash.

NATIONAL SERVICE ACTS.

ENLISTMENT NOTICE

MINISTRY OF LABOUR AND NATIONAL SERVICE REGIONAL OFFICE,
(HALFORDS BUILDINGS)
239, CORPORATION ST.,
BIRMINGHAM 4
11th Oct 1943 (Date)

MR. W. D. Neal
217. Sladefield Road,
Ward End
Birmingham 8.

Registration No. WFZ 19672

DEAR SIR,

In accordance with the National Service Acts, you are called upon for service in the Territorial Army and are required to present yourself on Thurs day 21st October 1943 (date), between 9 a.m. and 12 noon, or as soon as possible thereafter on that day, to :—

~~Primary Training *Centre/*Wing~~
Royal Corps of Signals
No1. S.C.U.
Bletchley, Bucks
Bletchley (nearest railway station).

Every endeavour should be made to report between the hours stated above.

* A Travelling Warrant for your journey is enclosed. Before starting your journey you must exchange the warrant for a ticket at the booking office named on the warrant. If possible, this should be done a day or two before you are due to travel. If your warrant is made out to travel from London you may obtain a railway ticket at, and travel from the most convenient station to your address.

If you have been transferred on or after 1st June, 1940, beyond daily travelling distance from your home by or with the approval of the Ministry of Labour and National Service to work of national importance, and you desire to travel home before you are required to report for service, you may apply for a free travelling warrant for this purpose. If you wish to apply you should go immediately to the nearest Local Office of the Ministry of Labour and National Service, and take with you this enlistment notice [and the enclosed travelling Warrant (A/cs. 617)].

A Postal Order for 4s. in respect of advance of service pay, is also enclosed.

Immediately on receipt of this notice, you should inform your employer of the date upon which you are required to report for service.

Yours faithfully,

W. D. HILL

for Regional Controller.

YOU SHOULD READ CAREFULLY THE NOTES OVERLEAF.

* *Delete if not applicable.*

N.S 12A M28767 200M 3/43 C.N.&Co.Ltd. 749 (9591) [P.T.O.

Wilf's enlistment papers

At this stage, recruitment was being carried out surreptitiously in army signals units, and not necessarily in the Royal Corps of Signals. Most army units had a signals attachment and anyone showing exceptional talent was quietly tested, without being made aware of the fact. This was done by a number of officers who had been briefed only to the extent that a number of above-average operators were required, without any idea of what their duties were to be. It can therefore be assumed that September 1943 saw the first of many enlisted intakes, mainly for involvement in the forthcoming invasion of Europe. Not all were trained as wireless operators, some became drivers, teleprinter operators and other ancillary trades.

The earliest trainees spent the first six weeks living in the stables at the far end of the depot. These were brick-built and had bare concrete floors, so the inmates were bitterly cold for the whole of their time there. In addition, the sanitary facilities were rudimentary and there was no hot water. The time was spent approximately on three weeks learning the rudiments of sending and receiving Morse code up to a speed of 10–12 words per minute (one word being the equivalent of one group of five random letters), and the remainder in trying to make these raw recruits into reasonably presentable soldiers.

In the early part of 1944, the Gees stables were closed and new recruits carried out their first six weeks training at Ashby's in Stony Stratford, then going over to nearby Little Horwood to join SCU 7 for completion of their course. The Morse training was given in one of two wooden huts in the stables area. There were about six instructors, of whom two were Sergeant Garland, at whose Morse training table Wilf sat, and Sergeant McShane (?), a red-faced older man from Ireland. McShane was a most accomplished operator who, the trainees were led to understand, had spent time working at the GPO station at Rugby.

The drill instructors were Corporals Martin and Varney. Both were lance-corporals at the time of the first intake, and were promoted to sergeant at the completion of the training of the intake. Wilf also came under a Sergeant Clayton (mainly for weapons training), and a Lance-corporal Jordan, who seemed to be much involved with gas warfare instruction. He has recollections of them standing many times under the archway at the stables, dressed only in thin denim uniforms, in bitterly cold weather, waiting for Lance-corporal Jordan to emerge and march them down to breakfast. Such was the authority wielded by the lowly rank of lance-corporal.

The Other Ranks mess at that time was in a long, low brick building, presumably an old cow-shed, on the far side of the Training Wing parade ground, running parallel with the main drive. The replacement building on the guardroom side of the drive would not have been erected until mid- or late 1944.

The C.O. of SCU 7 was Squadron- leader R. Matthews, RAF. Of the many other members of the depot staff whom Wilf remembers, a certain Captain Marsh of the Royal Artillery stands out in particular. Marsh was a very tall man, who mainly appeared on the Saturday morning drill parades, often in Wellington boots. He would stand near the sergeant's mess and have the recruits march up and down the

Signalman Wilf Neal, Royal Corps of Signals.

main drive for about two hours. What his real function was in the unit during the rest of the week, they never did discover!

After Wilf's first spell of leave, he returned in December, 1943 to the comparative luxury of the SCU 7 Nissen huts and a top bunk-bed alongside the heating stove. Tony Whiteside occupied the bottom bunk on his arrival, and when Tony left Wilf moved into this coveted position. He soon found there was a real disadvantage to this, however, as in the evening when it was cold, his bed would become the centre of attraction for all who wanted to sit on it and try to get warm!

Part of the advanced training in SCU 7 involved participating in schemes whereby a driver and two men would travel out in one or more of the Packards and work back to other cars located on the parade ground or to a control station set up in the training room. Wilf finally attained his B3 rating on 15 March, 1944.

On return from leave on 12 April, Wilf was posted to SCU 8, which had just been formed at Little Horwood, and on 9 May he was posted again, this time to XIX USTAC (19th U.S. Tactical Air Corps), at Bushey Park. He went down with Sergeant Povey and a fellow named Gerrard, and on arrival they teamed up with several others. At this station, they took letter-code from the Windy Ridge station at Whaddon, but what happened to it he doesn't know, as he cannot remember any evidence of an SLU cypher team being present.

The station was in a wooden hut next to the main gate, against the inside of the perimeter wall skirting Sandy Lane. The equipment used was an HRO receiver and a Whaddon MkIII transmitter. This was regarded as a good posting as it enabled the team to get into London during their time off.

On the way to Bushey Park, Wilf and his fellows thought that they were about to have the ultimate in good postings when they were told to unload their kit at a station known as The Dugout, which happened to be in St. James's Park, in the centre of London, directly outside the War Office entrance. However, it was not to be. They were soon told to get back on the vehicle and resume the journey to Bushey Park. The Dugout was manned by SCU 1 personnel as a backup to the teleprinter lines from Bletchley Park and many years later Wilf met Jimmy Gee, who told him what took place there. On 4 June, 1944, he left Bushey Park with Gerrard and Sgt. Povey and they made their way back to SCU8 at Little Horwood.

'Zeta' and the U.S. Third Army

On 8 June, a small party left Little Horwood in a Dodge 6x6 truck, and a Dodge American ambulance which had been adapted at Whaddon (by the Mobile Construction team at Whaddon Hall) to accommodate the radio gear. Their destination was Peover Hall at Knutsford, Cheshire, where they were to be attached to the G2 Section (Intelligence) of the Forward HQ of the U.S. 3rd Army. On the way, they had to meet up with the RAF cypher personnel at Stony Stratford. In charge of the whole unit was Captain C. W. Hutchinson, Royal Corps of Signals, who had previously seen service with an SLU in Italy. The other members of the signals team at that time were Sgt. Povey, three operators – Neal, Gerrard and Britton– and two drivers, Bill Bailey and Jack Croucher. The code name for the SCU wireless unit was 'Zeta'.

Almost all of Captain Hutchinson's time was taken up with the cypher section operations, leaving Sgt. Povey to handle signals matters. After they had been in France for a few weeks, there was one change in the signals team with the addition of another operator, but there were more changes to the RAF personnel. At the outset, their team consisted of Captain Hutchinson, Captain Whitfield (both Royal Corps of Signals), Sergeants Douglas Jackson and Jack White, and two corporals, John Mawer and Ron Chatfield. Apart from Capt. Whitfield, the others remained with the SLU throughout. There were also two American lieutenants on their section, Lieutenants Hull and Brown.

During the latter part of 1944, Captain Whitfield left for Rheims to take command of another SLU, and over a period of time, other minor changes were made in the compliment of the RAF personnel. At the onset, and

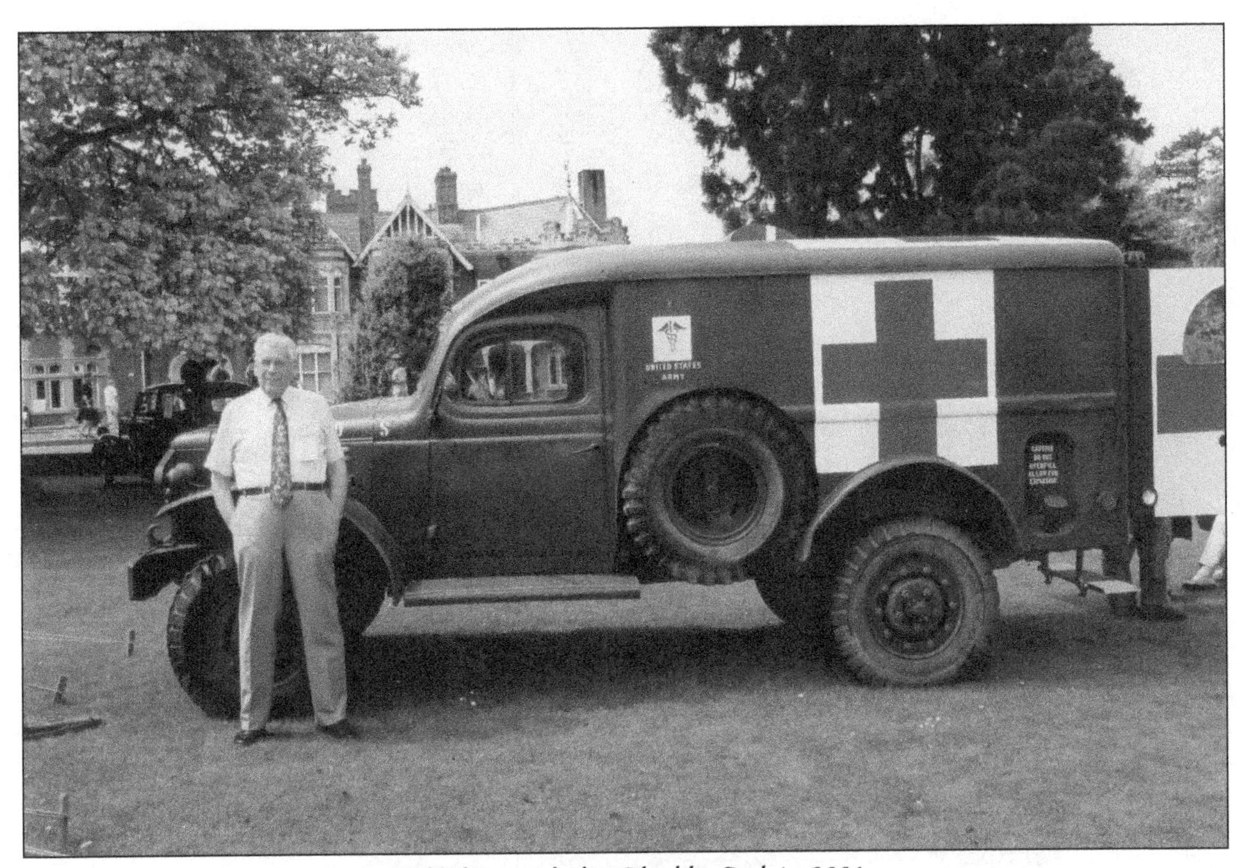

Wilf photographed at Bletchley Park in 2001
alongside a Dodge Ambulance of the US Army, similar to those used in France.

for about the first two weeks at Knutsford, the SLU cypher vehicle was a Humber Station Wagon. However, this was exchanged for a Dodge vehicle with a chassis that resembled a horsebox in which a person could stand upright. It was taken from the unit during August and passed to another SLU who used the more bulky TYPEX cypher machine, as opposed to their compact one-time pad system. As a replacement, the SLU had a Dodge ambulance of the same type as the one used by the signals team but adapted for their needs, and this remained with them until the end of the war in Europe.

Whilst most of the Forward H.Q., including Gen. Patton, operated from within Peover Hall, (the Rear H.Q. was a mile or two down the road, nearer to Knutsford, at Toft Hall), the unit was located in the surrounding park area just outside the boundary wall of the Hall. They were at Knutsford for three weeks and apart from regular contact schedules with Whaddon, they were non-operational until 25 June when the traffic started. This was the commencement of a 24-hour-a-day, seven-days-a-week shift pattern that Wilf would follow right up to within about the last week of the war.

Wilf describes the SLU radio equipment in the Dodge ambulance as follows:

'Inside our converted ambulance a 'desk' was fitted crossways just behind the cab seats and a dark-coloured curtain strung between. The operator's seat was a chest with a hinged top used as a bench. Inside it, the various radio spares and WT Red Message Pads etc. were stored. On the desk, on the left-hand side, there was an HRO receiver with its attendant plug-in-coil units, then a Whaddon MkIII transmitter and power pack. Also installed were a clock, anglepoise lamp, a large grey Marconi Naval-type Morse key and the inevitable ashtray.

'The earphones used were American made 'Trim' lightweight models. We also carried a Syko portable cypher machine, used daily for the purpose of changing call signs, and for encoding or decoding the occasional "internal" message that

HEADQUARTERS THIRD U.S. ARMY
APO 403
WILFRED NEAL, 2392216, Sigmn, British Army
Name
Is hereby / est / ist hiermit
Permitted to proceed to / Autorisé à aller à / Erlaubnis gegeben weiterzufahren — In army area
via / par / über — At discretion
within the area occupied by the American army / dans la zône occupée par l'armée américaine / Innerhalb des von Amerikanischen Truppen besetztem Gebiet
and return to / et retourner à / und zurück zu kehren — At discretion
via / par / über — At discretion
This pass is good from / Ce permis est valable seulement du / Dieser Pass ist erhältlich von — 4 February 1945
To / au / zu — 5 March 1945
By command of Lieutenant General PATTON
No. 812
Date issued 3 February 1945
R. E. CUMMINGS,
Colonel, A.G.D.
Adjutant General.

Pass issued to Wilf by US 3rd Army

may have passed between Whaddon and ourselves. Also on board was a 'spy suitcase set' for use in the event of equipment breakdown, or if we had to make contact during a long move. As far as I can recall it was only used once, and then for the latter purpose.

'Inside the back of the vehicle, there were two storage compartments over the wheel arches. The offside one (right-hand side) contained a canvas bag holding all the aerial gear – sectional mast, aerial wire, guy ropes and mast base. The nearside compartment contained the heavy-duty batteries, as I recall, two banks each of 6 volts.

'When working in the field, in the vehicle, as we did for the first three months, we used an Onan generator during the day and evening, and then changed to battery operation from 2200 to 0600hrs. Usually, at around 0300, we would change the leads feeding the receiver, which used a 6-volt supply to the other battery, to avoid excessive drain. The transmitter leads were not changed, of course. The batteries were then recharged during the day using a "Tiny Tim" petrol charger. Both the Onan and the Tiny Tim were carried inside the vehicle when moving'.

On 29 July, Wilf was moved down to the village of Breamore, some seven miles south of Salisbury, in preparation for the journey to Normandy. The HQ functions were contained in Breamore House but the signals vehicle and cypher vehicle were parked on the verge of a lane close to the House. They were billeted in a barn in the village. The mess was situated in the village hall, which fronted onto the A338, opposite a large pub.

The unit was fully operational for the whole of the time at this location and on 4 July, it left Breamore at about 1800 and made its way to Southampton docks. The journey took Wilf and his fellow-soldiers through parts of the New Forest and Wilf vividly recalls the vast amount of boxes of ammunition stored there, not only along the roadside verges, but also well into the wooded areas.

On arrival at one of the dockside warehouses, they had to set about waterproofing the vehicles in readiness for going ashore on Utah Beach. The following day, they boarded the Liberty ship *J.F. Steffen*, which sailed at around 2100, then anchored off the Isle of Wight for the night. The ship resumed its journey at 0900, in a convoy of about thirty ships, many flying barrage balloons, and along a swept channel marked with buoys at each edge.

The convoy arrived off the Normandy coast in late evening and Wilf recalls:

'Conditions on the boat were very primitive; we slept in one of the holds, lived on K Rations and all shared washing and toilet facilities intended for use solely by the crew. During the night, bombs were dropped on the assembled ships.

On the evening of the 7 July, we disembarked down a rope-ladder over the side, and onto a flat Rhino landing barge, which deposited us in our vehicles, in the water, a little way from the beach. All made it safely ashore except for our Dodge 6x6, which had to be towed in by a tractor. We moved a short distance inland and spent the night sleeping out in the open, after going through the task of removing the waterproofing material from the vehicles'.

On the 8th, they moved inland to an orchard on the outskirts of the village of Nehou, this being the area where the divisions and components of the U.S. Third Army were assembling in readiness for their forthcoming role in the liberation of Europe. At this time, the unit was five or six miles from the front. It remained at Nehou for around three weeks whilst the build up took place, and as strict secrecy covered the whole operation the soldiers were not allowed out of the site except on official duty.

With only three operators to cover the 24-hour stretch this did not allow much spare time, but there were occasions when they could go out in one of the trucks to collect the daily water ration. This was more often than not drawn from wayside pools, sometimes covered in thick green algae, and then purified. It worked out at probably a little less than two gallons a day per man, and Wilf remembers it 'smelt and tasted horrible'. Later, they received a fourth operator and one of the existing operators was replaced. The two men were :ance-corporal Ron Parsons and Signalman Bert West, both from the first intake at Gees.

Life at Nehou very much set a pattern for their existence over the period up to the end of September, as for the whole of this time, they were operating as a mobile unit and living in two-man tents. They slept for some time on ground sheets and palliasses, but later managed to find themselves German army beds or American army folding cots. Meals were taken with the Americans, usually in a marquee. though for the first few days at Nehou they had to take them out in the open air where it rained and rained incessantly. Food was mainly the American equivalent of C Rations but K Rations were very much in use and they were always issued when they were on the move. Horsemeat was also 'on the menu' from time to time.

The SCU/SLU with Patton's 3rd Army. Note the Dodge ambulances converted to wireless vans by Mobile Construction at Whaddon.

The unit members were entitled to the same free weekly PX rations as the American troops, the most important item for some of them being the 140 cigarettes. For the whole of the time with Third Army, the British soldiers did their own laundry and for this they used the petrol-fuelled pressure stove that 'came with the job', and a large tin that came from the cookhouse. Wilf recollects 'The results were not spectacular, but it was better than nothing!'

Discipline within the unit was good but there were no parades, guard duties or the like, and Captain Hutchinson took the view that as long as they did their work satisfactorily and kept themselves in good order, then that was all right by him. Wilf remembers that during the first three months whilst they were mobile, there was nowhere for them to relax during the little spare time they had. The two alternatives, as he recalls, were to either lie down on your bed or sit in the cab of one of the vehicles.

Another difficulty was personal hygiene. 'There was the occasional enterprising G.I. to be found who had rigged up a form of shower, probably in the shape of a bucket filled with warm water, pivoted on the top of a pole so that on pulling a length of string, the contents of the bucket poured down on you. An alternative was a wash-down from a bucket in a field!'

Being a highly mobile unit and operating more at the 'front end', the cypher section used one-time pads. as opposed to the Typex machine associated with more static stations. All the traffic was in long figure code, which made the job much easier. The transmitter had an output of around 35 watts and was controlled from a selection of about five fixed frequency crystals. During the day, the operators used frequencies up in the 7,000 or 8,000 M/cs range, and as night fell they went down to around 3,000 M/cs. In those days, the short-wave spectrum teamed with stations of all nationalities transmitting Morse, and skills were required to separate the station to which one was listening from others operating on, or very close to, one's own chosen frequency.

Windy Ridge at Whaddon transmitted on precisely the same frequency as the mobile unit. The unit's crystals were identified by letters of the alphabet and to change frequency it was only necessary to send the appropriate Q Code signal (QSY), followed by the crystal identification letter. The station's three-letter call sign was changed daily, possibly at midnight. It was absolutely forbidden to send any signal in plain language, or make any other mark which might be of use to enemy intercept operators. It is surprising what information can be obtained from unofficial operator chat! Transmissions were liable to be intercepted by our own security people. Periodically, transmissions would be made using taped Morse messages, still in figures, but these were not always successful. During manual transmissions, if the receiving operator lost a few groups, due to bad interference for instance, it was possible to stop the other operator by depressing the key and keeping one's hand there until he realised what was happening. With tape transmissions, this was not possible, the tape would have to run its full length before it could be repeated.

Messages were all given a certain degree of priority. The very lowest or routine were classed as either 'O' or 'Z' in the message preamble, others being two, three, four or five 'Z's', according to urgency. It was the duty of the operator on watch to take messages to the cypher people. Wherever they were, either working in the vans or indoors, after knocking on their door, the door would be opened to a crack, the messages taken from them and the door immediately closed.

It is worth noting that security was so tight on the communications side that Wilf was not even aware of the nature of his work, until the story became public in the mid-1970s. The operators maintained three-watch patterns that worked out as 2200 – 0800, 0800 – 1400 and 1400 – 2200. Naturally, only one operator was on duty on each watch. The cypher section worked similar watches, usually with one man on duty during the night and two during the day. Although they 'virtually lived in one another's pockets' for the whole service with Third Army, there was little direct contact between the wireless operators and the cypher staff. Apart from the vital security aspect, it was accepted, in those far-off days, that one did not bandy words with anyone with the rank of sergeant and above!

When decoded, the messages would be passed, possibly by one of the SLU officers, to a Major Helfers (U.S. Army) in the HQ G2 Section, who would evaluate them and use the information in his situation briefings to General Patton, or the other three or four officers in G2 who were authorised to receive decoded ULTRA traffic. At intervals during the day, any paperwork used in decoding, including that passed to Major Helfers, would be collected, burnt and the ashes well raked over. This was the sole responsibility of the cypher people. High priority messages, such as those in the 4 'Z' and 5 'Z' category would be passed, via Major Helfers, direct to General Patton or his appointed deputy.

On 1 August, the Third Army was activated and so began the mad dash across France. Up to the end of September, there was a move to a new location almost every four or five days. Between landing in France and the end of hostilities, some nineteen moves were made, setting up the station each time. Upon arrival at a new location, the very first task was to site the radio truck and aerial in the most advantageous position within the confines of the area they were allocated, and establish contact with the control station at Windy Ridge. Once this was done, they could then pay attention to their own needs and find the most comfortable site for the tents, which always close to the vehicles. Upon setting up, and indeed at practically every change of watch, it was customary to request from Windy Ridge the number of messages waiting. Right up to within a week of the end of the war there would always be a backlog of anything up to about 80 awaiting attention.

The move to Beauchamps on 3 August, 1944 was the only time when the unit came very close to the front line. The move was not a long one, only about 36 miles. The detachment left Lebignard in the late morning, but did not arrive at Beauchamps until about 1800, having to wait on the way whilst German vehicles and tanks were bulldozed off the road. They also passed through territory where notices were displayed warning that roadside verges and buildings had not been checked for mines and booby-traps. It was during this move that they had to stop and contact Windy Ridge using the suitcase set [A Whaddon MkVII]. On arrival at the allotted field, there was a further delay whilst snipers were cleared out of surrounding trees and hedgerows. The two drivers were detailed to perform guard duty over the cypher and radio vehicles during the night.

That evening, there was an air raid in the HQ area whilst Wilf was on watch. The noise of the aircraft and machine-gunning was such that it was impossible to work and he took it upon himself to send the Q code signal for 'I am being bombed'. He does not know what response, if any, this brought from anyone at Windy Ridge. At about 2200 hours, there was the sound of a lone aircraft nearby and this was accompanied by anti-aircraft gunfire. Then there was silence and, after a few seconds, a loud explosion, as presumably the aircraft had been shot down.

A move at the end of September meant that they could now look forward to having a solid roof over their heads for the first time in months. It was getting a little cold sleeping in tents, and even working in the van at night. A Valor oil heater picked up earlier at an airfield was not ideal for use in a confined space. (Wilf had worked for the Valor Company before call-up and so he was well aware of its limitations!)

The move brought the unit to Etain in the Verdun region where they were billeted in the roof space of the village school. This lacked any access to outside light and the only illumination was provided by a small number of electric light bulbs. The operators were still working in the van, which was parked in the school courtyard.

Up to now, the latrines had always been of the 'pit and pole' types, (sometimes without the pole!) in fields, woods or alongside pathways and they anticipated better at Etain. Those provided for the school were not in use, however, and so theirs were holes drilled in a grassed area beside the school, but suitably secluded from public view by a strategically placed canvas screen.

The following move in the unit's travels was to Nancy, where they remained during a rather cold three months whilst the front remained static, This saw them established in French barracks (just vacated by the Germans)

Army Form C2133
(Pads of 200)

W/T RED FORM

Ref. No. ....
Page ....

Ship or Station	Set		Date		Operator's Remarks *
	Opr.		Time Ended§		Q.S.A. ....
	To †		Frequency and System		
	From†				

All before the Text

Text, Time of Origin, Signature, etc. Write ACROSS the page, code and cypher on alternate lines.

Do not use Left Margin.

G.03556/25. *Constancy and reliability of signals, quality of operating, interference, atmospherics, etc. †Name of Station ; if not known
Sta. 1/30. leave blank. §G.M.T. to be used on Home and Mediterranean Stations, other users indicate " Zone " Time employed.
Sta. 103/31.

Wt. 46073/898. 35,000 pads. 3/43. J.M. Ltd. 51-6289.

Red Form used for all signals in SCU/SLU traffic including incoming Ultra in the Rue du sergeant Blandin. Here, they were at last able to remove the gear from the radio truck and operate entirely from inside the building, where they had a room on the ground floor, virtually next to their sleeping quarters. The cypher section was also able to move in close to them.

Compared with their previous billets, the stay at Nancy was sheer luxury. They were within a short walking distance of the city centre, and with the enlisted men's mess a few yards away from their quarters, they no longer had to go tramping across two or three fields in all weathers to reach a marquee for meals. A film show was held most evenings in the mess, although it usually entailed sitting at the tables rather than in rows of chairs, to get a decent view. In the early evening, the tables in the mess were laid with tins of fruit juice ready for the next morning, and it did not take them long to get into the habit of tucking a tin or two of this delicacy inside their battledress jackets before the lights came on at the end of the show.

The official 'History of the SLU's' rightly lays much emphasis on the subject of security. At Nancy, a serious breach of security occurred. One of the RAF sergeants departed very suddenly and the operators were told that it was because he had sent a personal letter that had not been censored by Captain Hutchinson. In later years, Wilf discovered that this was not true. It transpired that a German message had been picked up from Bletchley Park which, when decoded, indicated that an allied air raid was expected in the Nancy area. Sergeant Jack White (RAF) of the cypher team had overheard this information being discussed with an American serviceman. The other RAF Sergeant was reported by Jack White for the breach and was quickly posted away.

Whilst in Nancy, the German Ardennes offensive commenced, and on 28 December, parts of the HQ moved up to Luxembourg City. The SCU/SLU remained operational in Nancy and did not move to join them until 10 January, 1945. At this point, changes were made which lasted about a week. Their station (Zeta) became non-operational and the communications and cypher sections split. The SLU part worked with another, located in a building belonging to Direction des Chemins de Fer A.L. adjoining the Pont Adolphe, and helped deal with Twelfth Army Group traffic.

The operators teamed up with the SCU/SLU serving XIX TAC. (The U.S. 19th Tactical Air Corps acting as the air support for the 3rd Army). For the whole of the European campaign the US 19th Tactical Air Corps SLU had shadowed 3rd Army moves, but for some reason had now arrived in Luxembourg before them, and 'set up shop' in a tent in the roof space of an engineering factory near the main railway station.

This arrangement lasted from 14 to 22 January. Wilf's unit shared billets with the SLU in the Hotel Select (which was not quite so luxurious as the name implies!) Their mess for this period was in the entrance hall of the railway station. They operated in letter code, which came somewhat hard to the SCU personnel who were used to figures. The RAF cypher people had a Typex machine also located in the roof space.

On 22 January, Zeta again became active as an independent station with Third Army and its own SLU. Located with the Army HQ, they operated from an up-market old peoples' home bearing the name Fondation J.P Pescatore, in the centre of the city. Billets for both Army and RAF personnel were at a school, the Ecole Industrielle et Commerciale 1907, some distance away, which they shared with American troops. It is interesting to note that the Foundation J.P. Pescatore building, a large and well appointed structure, had previously been reserved for occupation by SHAEF but with the onset of the Ardennes offensive it was considered too dangerous to move it there, so it remained in the rear, at Rheims.

After a week's leave, Wilf caught up with the unit and moved to Idar-Oberstein where he and his fellows were accommodated in German barracks, and again were able to operate with the equipment indoors. They left on the 3 April and subsequently spent brief spells in Frankfurt, Hersfeld, Erlangen and finally Regensburg, always occupying German barracks.

A Christmas card issued to all in 3rd Army, showing the great sweep by Patton across France. This came to a sudden halt at Christmas time with the German counter attack through the Ardennes.

Life in all these locations was very similar. There were occasional trips out in one of the vehicles, but non-fraternisation orders were in operation and for that reason they spent much of their time in barracks, exploring and seeing how their adversaries lived.

One thing that stands out in Wilf's mind was the journey through Mainz, on account of both the vast amount of bomb damage that had been done to the city and the fact that it was here that they crossed the Rhine. The crossing was made over a pontoon treadway built by Third Army engineers of the 80th Infantry Division, and at 1,865 feet it was the longest assault bridge put across the Rhine.

During the closing few days of the war, traffic from Windy Ridge dwindled so that the operators only took an occasional message, contact being maintained with quarter- or half-hour schedules. It is worth mentioning, in praise of the wireless equipment, that, although the station was operating 24 hours a day, seven days a week, for almost twelve months, Wilf cannot recall one breakdown – not even a valve filament burning out!

The unit left Regensburg, and the Third Army, on 17 May. That night, the men slept in the open alongside their vehicles, in a field in the edge of the Black Forest, before making their way to Nancy for a more comfortable night in a transit camp. After a day or two at SCU 8 HQ Versailles, the unit made its way back to Little Horwood and the prospect of many weeks at the tented camp at Nash, followed by two weeks in the Nissen huts at the rear of shops in Fenny Stratford, when the weather began to get cold.

During their time at the headquarters of the U.S. Third Army, the SCU crew saw little or nothing of General

Patton, despite working as they did with the HQ G2 Section, and usually being in very close proximity to his office. Two sightings Wilf recalls were when Patton was greeting or seeing off, first, the lady in charge of Red Cross women and second, Marlene Dietrich.

On 26 October, 1945, along with many others, Wilf left Bletchley for Southampton and a journey that over the next twenty months, would take him to SCU 12 Delhi, Indian Special Wireless Centre, Bangalore, and No.4 Wireless Regiment, Singapore, these last two on intercept work. But that's another story!

Headquarters
Third United States Army
Office of the Assistant Chief of Staff, G-2
A.P.O. 403

16 May 1945

Subject: Commendation.

To : Lt.Col. R.F.Gore-Browne,
c/o Air Ministry, A.I.1(c),
King Charles Street,
London. England.

1. It is desired to commend Special Liaison Unit/Special Communications Unit No. 8, commanded by Captain C.W.Hutchinson, 252161, Royal Signals, for the splendid performance of the duties of that Detachment while with the Third U.S.Army during the period 8 June 1944 to 16 May 1945.

2. This Unit at all times performed in a highly efficient manner and its spirit of cooperation reflected in its work. It was to me a great pleasure to have been associated with it. During the rapidly moving situations which characterized the operations of this Army, a great number of problems were presented in connection with this Unit but always they were successfully solved.

(Signed) "OSCAR W. KOCH,"
Colonel, GSC,
AC of S, G-2.

Commendation to the SCU/SLU based at 3rd Army

Epilogue

Early 1994, in the magazine After the Battle, Wilf saw a photograph, the caption of which referred to a George Patton Kimmins, living in Portslade, Brighton, who runs the George S. Patton Jr. Appreciation Society. He became a member, and very shortly afterwards received a telephone call from Helen Patton, one of the General's granddaughters who was organising the unveiling of a new memorial to the great man, in the orchard they occupied at Nehou on their arrival in France.

Helen Patton insisted on Wilf being there as he was the only person she could trace as having been present in that orchard in 1944. Wilf and his wife Barbara attended, staying with a local family for four days and were given VIP treatment during their visit. All of the original apple trees in the orchard had been grubbed up, except the centre one, which was left for the sake of nostalgia. The site is now known as Camp Patton.

Chapter 26

Martin Shaw's Story
Ashby's, Gees and Rockex at Hanslope

Martin Shaw was born on 11 May, 1925 in Halesowen where he still lives. He was educated at a technical college in the town and then went to work for Accles and Pollock, the precision tube makers in nearby Oldbury. He became a jig- and tool-maker and his work included tools for the manufacture of aircraft. In his spare time, he was a messenger for the Auxiliary Fire Service, and later joined the Air Training Corps where he studied Morse code and other RAF trades. He went for several interviews for aircrew duties but the only vacancies were for air gunners and he declined the invitation.

His work at Accles and Pollock was a reserved occupation at first, meaning that he could not be called up for military service, but in 1944, he volunteered for the army and was recruited into the Royal Corps of Signals.

Ashby's – the SCU basic training unit at Stony Stratford

Martin finally received his call-up papers for the army in December, 1944, a week before Christmas. He set out on Friday, 12 January, 1945, as instructed by the Ministry of Labour and National service to report to the Royal Corps of Signals No 1 SCU at Bletchley railway station. He was dressed in his best (and only) suit and raincoat. It was January and very cold. The bits and pieces he had been told to take with him were: soap, a towel, a pair of pyjamas and his shaving kit, all of which he had in a brown carrier bag. He had very little money and only a single railway ticket to Bletchley.

As he stood waiting at the bus stop opposite his home at 293 Stourbridge Road, Halesowen, for a 130 Midland Red bus to take him to Birmingham on the first leg of his journey, there was no one to wave him good-bye and he remembers thinking, 'I may not see that house or my folks again'. Once he got to Birmingham, he found there were no through trains to Bletchley so he had to change at Rugby, then catch a local train.

He eventually arrived at Bletchley railway station just before lunch time. There, an NCO in the Royal Corps of Signals, told the few young men leaving the train to go to the station café. This was just outside the station where another NCO was talking to other young men in civilian clothes. He told them to get themselves a cup of tea, as they might have to wait some time for transport to the camp.

After about half an hour, a lorry arrived and the men were told to climb on the back after the NCO first checked their names on a list. Martin was unaware at the time that he was only a few hundred yards from the now famous Bletchley Park.

Martin's group were unlucky, because when their lorry arrived it was half-full of coke and coal! There were no seats so they had to sit on the sideboards, and hang on tight. There was a canvas cover over the back to protect them a little from the elements, as it was very cold and wet. The lorry left the town behind and they were soon travelling down narrow country lanes. After about half an hour they reached an army camp. As they disembarked from the lorry they were told, 'This is little Horwood – Gees for short – a Royal Corps of Signals training camp.'

After a cup of tea in the NAAFI, Martin and his fellows were taken to the stores to be kitted out with their uniforms and equipment. The stores were in the old stables that had been whitewashed and fitted out as an equipment stores but when they went in, it smelt just as if the horses were still there. They were told later that it had been used in the past as sleeping quarters for earlier recruits. Each stall contained different items of clothing or equipment.

They started at one end with a large empty kit bag, and finished at the other end with a kitbag full of two of everything in the way of clothing. They were issued with all kinds of equipment, including a rifle and a bayonet, a 'housewife' (needle and thread), and a large clasp-knife which had a spike for getting stones out of horses' hooves. They were also given three somewhat smelly army blankets. They had to carry all this equipment to a waiting army truck and load it on the back. They placed their newly issued gear in piles, with various means of identification. Next, they were given an empty palliasse, which they had to fill with straw. They joked that the straw was probably left over from when the horses were there! Once filled, these straw mattresses were stacked on another lorry. Then they had to find their kit and load themselves on the lorries.

Being early January, it was now getting very dark even though they had the benefit of double British summer time all the year round. There were no lights as the black out was in force. The lorries drove round the country lanes for another half an hour before they eventually pulled up in front of a large house beside a main road. They were told this was Ashbys in Stony Stratford. The main road was the A5 Watling Street, then the main London-to-Birmingham road. They were to stay here for the next eight weeks doing their basic training.

The Guard Room of Little Horwood camp

They unloaded their kit and were ordered inside. With the full kit, they had to climb three flights of stairs to an attic where twelve of them were allocated a small bedroom. This contained six bunk beds, the lower bed being only about six inches from the floor and the top one about four feet up. Martin was lucky; being the first one in the room he had dropped his gear on a top bunk near the door. It was bitterly cold and the room was unheated.

The men were taken downstairs to the mess, bringing their new army knife, fork and spoon and new pint mugs with them. The mess was simply the closed off entrance hall of the house. It had a tiled floor and there was no heating here either so it was very, very cold. The front door had been locked and could not be used. They sat down to a late tea consisting of corned beef sandwiches and a pint mug of tea.

The next day was a Saturday but it was not a day off. They were roused at 6.30 a.m. while it was still dark and very cold. They washed and shaved in an out-house which had once been the coach-house and stables and was now converted into the ablutions. There were wooden washstands with rusty metal bowls in which to wash. They were lucky, however, in that it had hot and cold water, a large communal shower, and two flush toilets.

They went back upstairs, folded their blankets and put all their kit and equipment on the bed. Then there was just about enough room for the twelve of them to stand by their beds as they soon discovered they would have to do every day for kit inspection.

The men were still in civilian clothing and not yet officially inducted into the army. They went down to the mess for their first army breakfast which consisted of the normal army porridge with no sugar and watered down evaporated milk. (Martin cannot recall having any fresh milk during all the time he was in the army).

Ashbys in Stony Stratford used for basic training by recruits to SCU7 at Little Horwood. It was named after an army officer who set it up.

The porridge was followed by two sausages, a piece of bread and margarine and a pint mug of tea to finish with. Having been given their official new address, they were instructed to write home to say that they were fit and well and enjoying themselves!

At 8 30 a.m., two lorries arrived to take them back round the lanes to Gees. Martin recalls that it was an unpleasant feeling not knowing where he going or what he was expected to do when he arrived.

The men were assembled in the main orderly room at the camp. There were about twenty-four new recruits, all still in civilian clothes. They lined up before an officer sitting at a desk. He told them about the their enlistment and about King's Regulations, but stressed that the most important thing was the Official Secrets Act, a copy of which they all had to sign in front of him. They then 'signed on' and were now officially in the army 'for the duration of the emergency' but were informed that the 'emergency' could easily last longer than the war!

They were then given their pay-books which indicated that they were then members of His Majesty's Forces. This most important of documents was Army Book 64 (AB64), the soldiers' service and pay-book. It was a record of all of the soldier's personal details, clothing sizes, a will, next of kin, medical history, inoculations and army training. It also recorded all privilege leave so a soldier could work out when his next leave was due.

Martin and his comrades were also issued with an Army Pay Book Part Two, which recorded their weekly payments and any debts they incurred to pay for lost clothing etc. They were told that they were entitled to the magnificent sum of three shillings per day or twenty-one shillings a week. If however, a soldier decided to send seven shillings a week home to support a wife or parent out of the twenty-one shillings, the army would add another three shillings to it as a bonus, making it ten shillings per week. Most of the single lads agreed to this deduction, hoping their mothers would save it for them until they got out of the army. This only left them fourteen shillings to last the whole week. The Pay Parade was usually held on a Thursday at Gees.

They were all ready to return to Ashby's, but then they were told to line up in front of another door. As this door was opened, they saw that inside it was a medical room with several men in short white coats waiting for them with hypodermic needles in their hands. The young men at the front pushed back and those at the back pushed them forward again. No one wanted to be first in to face those medical orderlies with their syringes at the ready!

At their first inoculations, some men fainted and others were sick. It turned out that those at the front were lucky because the needles were sharp to start with but got blunter as they neared the end of the queue (there were no such things as disposable syringes in those days) because one needle had to do for the lot. The orderlies just kept filling up the syringes.

It took the recruits over a week to sort out their uniforms and kit. A kind lady further up the road would sew the Royal Corps of Signals flashes on their coats for sixpence. They had to exchange parts of their uniform with each other to enable them to match the different shades of khaki or to get them to fit. Those who were particularly short or tall and could not get the uniform to fit at all, had to visit the camp tailor at Little Horwood to have them altered. They had to pay a small 'fee' for the privilege.

The recruits were not allowed out of Ashby's on their own for over two weeks without an NCO being in charge. They were told they had to look smart in their uniforms before they could go out alone or they 'might disgrace the King's uniform'. Their hair was cut each week by Sgt. Varney who charged them sixpence a time. It was either that or go to the so-called 'Sweeney Todd', the camp barber at Gees and he only used clippers!

The men were required to make their boots shine like those worn by the NCOs. This meant they had to spend most of their free time burnishing the toecaps of their boots with spoons, and the back of a toothbrush, together with plenty of 'spit and polish'. Army regulations then specified that only dubbing could be used on army boots, but they soon found out that if you used it, you were unable to get any kind of polish on them at all. So to comply with the rules, a tin of dubbing was always put out for inspection, and the tin of black boot polish hidden away in their kitbags. This seems to have satisfied everyone concerned.

A rumour was put about, possibly by the NCOs, that there would be a prize for the best boots. One recruit even took his boots to the local cobbler in the nearby town of Wolverton and paid him to polish them on his polishing machine! They looked marvellous, and he did not reveal his secret until weeks later. The instructor picked him out as having the best boots, but his only prize was that he was given the job of cleaning and polishing the instructor's two pairs of boots!

The worst chore was peeling potatoes. Every night, six men would be detailed to do 'spud-bashing' as it was called. They met in the ablutions and from a large sack they had to peel enough potatoes to fill a large dixey but only had the army-issue clasp knives with which to do the job. Many had never peeled a potato in their lives before; they would chop off the ends and sides and finish up with small square pieces. More went into the pig swill bin than in the dixey. When they were eventually let out of Ashby's, Martin went to Hootons, the threepenny and sixpenny bazaar in the High Street, and bought a potato-peeler. It was much easier to use, much quicker, and wasted less potato. Within a week, most of the squad had bought one.

The next eight weeks were spent on basic but serious army training. Many of the men were not very fit and suffered quite badly. It was also so cold that everyone was afflicted with chilblains on their hands and feet. They were marched up and down Russell Street, which they nicknamed 'Agony Lane', doing arms drill every day, rain or shine, under Sergeant Martin, their strict drill instructor. Martin kept up his cross-country running and was allowed to go out training every Saturday afternoon in his army vest, shorts and plimsoles. As part of the recruits' final training, they had to do a timed 'one-mile run', ending in the Calverton road. Thanks to his previous training, Martin managed to win it.

Those of the men who were on the wireless operators' course had to do at least three hours of Morse each day in the wireless room. The course instructor was a CQMS (Company Quartermaster Sergeant) Cousins. He was an army signals man from the World War I. The men estimated that he was over sixty years old and later found out that he was actually sixty-two when he was teaching them in 1945. He had been in army signals and had re-enlisted at the age of sixty in order to train special signals operators.

After eight weeks, basic training was finally finished. The wireless squad passed out at 12 words per minute in Morse. They were ready for their passing-out parade and looking forward to the week's leave associated with it, before they had to report back to Gees to start the twelve-week advanced radio and signals training course.

At the end of this last drill, they marched off in formation, an officer and all the NCOs in front, turning left into New Street, then on to Stony Stratford's High Street. They 'marched to attention' (with rifles on their shoulders) up the High Street. On the footpath on the right hand side, at the junction with Wolverton Road, stood another officer at attention and he took the salute. It was March, 1945, and they were the smartest and happiest soldiers around on that morning. They marched the short distance back to Ashby's. Parked in front of the house on the main road, were lorries waiting to take them away on leave.

The SCU7 advanced Morse training course held at Gees, Little Horwood

After his leave ended, Martin returned to Bletchley railway station early on Sunday night. Nobody wanted to be back late as they had passes that had to be handed in by 23.59 at the guardroom at Gees, their new camp.

A lorry was waiting to take kit and men back to Gees but when the lorry reached the camp it was dark as the blackout was still in operation. They were directed to a large hut where they were told to put their kit on a spare bunk. This was to be their living accommodation for the next three months.

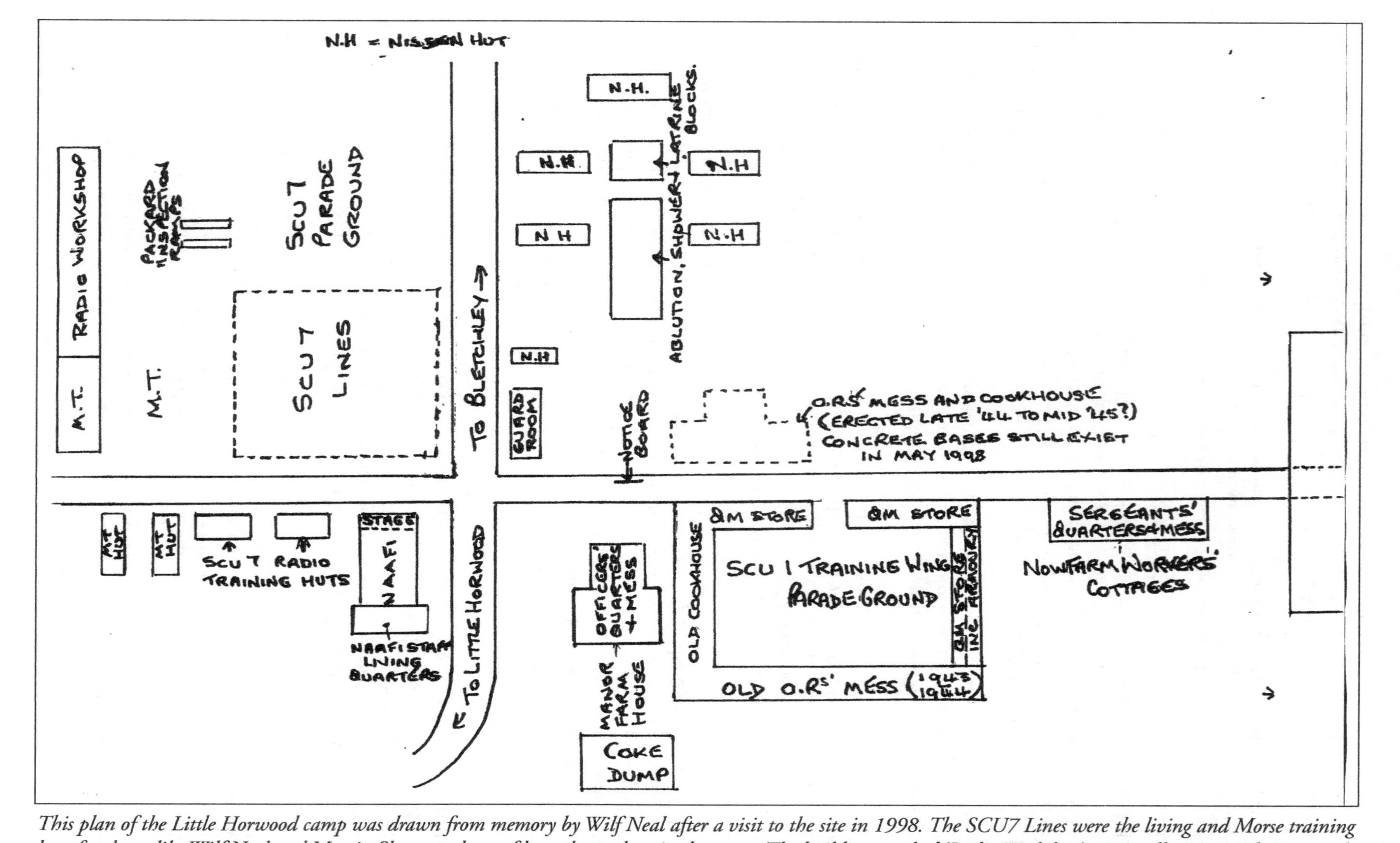

This plan of the Little Horwood camp was drawn from memory by Wilf Neal after a visit to the site in 1998. The SCU7 Lines were the living and Morse training huts for those, like Wilf Neal and Martin Shaw, and out of bounds to others in the camp. The building marked 'Radio Workshop' was totally secret and very strictly out of bounds, being the workshop where agent's wireless sets were made.

The buildings marked 'NH' to the right were Nissen huts housing the drivers, cooks, dispatch riders and other military staff of SCU1.

Names were checked against a list for new arrivals and they were told that these were the SCU7 lines by Corporal Samson, the NCO in charge, who was responsible for discipline in the hut. Samson slept in a single bed with a box-spring matress behind the door, while all the trainees had to sleep in tall wooden bunk-beds with wire mesh bases.

SCU7 was a self-contained wireless training unit. There was a unit orderly office and several signals and other training huts. The new brick-built ablutions block was quite modern with hot and cold water in the washbasins, flush toilets and a proper bathroom. These all had to be cleaned each morning by everyone in the squad.

Corporal Samson advised the men to get to bed early as the next morning they would be rising at 6.00 a.m. whilst it was still dark. He explained that they had to be up, dressed, washed and shaved, with their equipment laid out for inspection and the hut cleaned before going for breakfast. After breakfast, they had to clean the ablutions and do half an hour of physical training, before starting Morse training at 09.00.

Assembly took place in the large wireless hut and the trainees were introduced to the RAF officer commanding SCU7, Squadron Leader Matthews. The actual instructors were all army NCOs from the Royal Corps of Signals. There appeared to be one instructor for every two trainees.

Squadron Leader Matthews reminded them of King's Regulations and the Official Secrets Act they had all signed. He told them were special signals trainees and must not tell anyone else on or off the camp what they did. The SCU7 lines at Gees, although consisting of only a few huts, was out-of-bounds to the rest of the camp.

After introductions, training started straight away for the apprehensive recruits. They practised Morse all day and all were assessed for speed and accuracy. They were told that if they did not reach the high standard required in four weeks they would be posted away. Besides doing four hours of Morse each day, they had lectures on all kinds of radio equipment – receivers, transmitters, aerials, radio cars and radio vans.

Those that came up to standard by the end of the four weeks were allowed a weekend pass from the Saturday lunchtime until Sunday night. Several who did not reach the standard were posted elsewhere the next week.

On 7 May, 1945 there was an announcement to the effect that all personnel had been given two days' privilege leave to celebrate VE day. The camp virtually closed down for the two days. Many who could get back home and had the money did so, but a lot of the men went to London to celebrate. Only the lads from distant places such as Scotland remained behind to guard the camp, after being promised they would all get extra days' leave added to their next privilege leave.

The remainder were now doing advanced signals work. This entailed doing 'schemes' which meant going out in radio cars or vans, setting up mobile radio stations and manning them. The NCO instructor in charge drove two trainees out some 25 miles from the camp. They mostly used the A5 and went north from Bletchley.

The instructors all had their favourite locations. One was at Stoke Bruerne by the canal, another at Pxulspury, and another one at Pottersbury. When they arrived at the site, the station was set up. This involved putting up aerial poles, connecting them to the vehicle and setting up the mobile generator, receiver and transmitter. It was called a 'three man team'.

The driver, usually the NCO, was in charge and he was responsible for the guarding the station, while the two operators took it in turn to encode the outgoing messages and decode the incoming ones, whilst the other one sent and received them. Every message had to be in code, plain language was absolutely forbidden. The second operator had the job of encyphering and decyphering the messages. For this he used a special metal frame that had several long rubber bands on it containing the alphabet and these were moved round in a set pattern.

The SCU7 training room at Little Horwood was the 'Main Station' and the mobile units could work either to it, or between the other mobile stations that were out in the field that day. The operators were allowed to change frequencies but this could make it very complicated for young trainees. They could, and often did, get into trouble when doing it.

Martin was increasingly frustrated with an operator who kept changing his frequency and wanted him to put the other operator back on air, who was a friend by the name of Bill Winded. Instead of having the message coded, he did the unforgivable and sent in plain language the message 'Get Winded'. Back came a reply, again in plain language, 'No! – No! – No!' The instructor at the other end had been listening on a spare set of earphones and had heard him send the message. Martin did not realise the serious trouble he was in until he returned to the camp, where he found the NCO who had overheard the message waiting for him. He was marched to the office and in front of an officer, charged with 'indiscreet signalling by sending plain language over the air'.

He was told he would have to appear before the Commanding Officer the next day for sentencing. Apparently, all the frequencies used were monitored by a special security unit. The incident was considered a disgrace, as it had never happened in SCU7 before. Signalman Shaw would have to be punished as an example to the other trainees still on the course.

The next day, 15 May, dressed in his best uniform, a very frightened young Martin Shaw was marched in, under escort, to appear before Squadron-Leader Matthews, Officer Commanding SCU7. The NCO making the charge gave evidence but Martin had nothing to say in mitigation and pleaded guilty. The sentence was seven days CB (confined to barracks), one month's loss of privileges and immediate removal from the signals training course.

After the war in Germany had ended things slowed down. The new operators realised they were probably no longer required. More weekend passes were handed out and there was more time off. The radio course eventually closed down in August, 1945. It was the last one at little Horwood. No one in Martin's squad managed to complete the course and qualify for extra pay.

On 3 July, 1945, Martin was told to report to the transport office the next day at 0800, as he was being sent for a job interview at another Royal Signals unit at Hanslope Park. Transport had apparently been laid on to take him there. Next morning, dressed in best uniform with boots polished, he reported to the transport office. He was driven in style, in a small Austin pick-up truck, along country lanes which he had never seen before. Hanslope Park looked just like any other army camp but as they drove through the gateway, there was no guard on duty to check them and in fact, no one about at all.

The driver took him to some huts marked 'Workshops'. Martin went in and told the NCO there that he had a letter for Major Hawkes. He was taken to the Major's office and marched in. He stood to attention, saluted and handed the letter to the Major.

The letter apparently suggested that Major Hawkes, who was in need of skilled engineers, might make use of Martin, who was not going to continue as a trainee operator. The Major questioned him at length about his

---

Opposite:
An 'end of course' photograph taken at Ashbys. The CO of SCU7 wireless training wing at the time was Squadron Leader Matthews and he is seated centre. Second from right on the front row is WO2 (Warrant Officer 2nd Class) is Harry Cousins, the chief Morse instructor. Others include drill and weapons instructors and those on the course.

background, first at technical school, then as a bench hand fitter, and as a tool- and gauge-maker. He called in the NCO and had Martin taken to the workshops to make some test pieces for him.

The workshops were clean and very well-equipped and Martin made the pieces which were then taken to back to Major Hawkes. He examined them, nodded approval, picked up the telephone and spoke to the office at Gees. He then told Martin, 'You start work here on Monday morning'. Before Martin left to collect his gear from Gees, the Major stressed the importance of security at Hanslope Park, but did not mention the indiscreet signalling that had brought him there!

Martin returned to the hut at Gees and told his pals of the day's events and his pending departure to Hanslope. Little did they know that several of them, also with an engineering background, who had been wireless trainees on Martin's course, would be following him there over the coming weeks.

Working on Rockex at Hanslope Park

It was only one week later, after his interview with Major Hawks that Martin was posted to Hanslope Park. He left SCU7 at Little Horwood on the morning of Wednesday, 11, July, 1945. He had to say goodbye to the friends he had made during his first seven months in the army.

Martin realised they were nearing their destination because the fields were full of very tall wooden radio masts connected with a mass of wireless aerials. Telegraph poles lined the roadside. These carried telephone wires as well as heavy cables. His driver informed him that these cables were important land lines that carried all the important messages to and from Hanslope Park and Bletchley Park. He said, being a Royal Signals driver, he had to know everything – and young Martin, still only 20 years old, believed him!

They drove through the gates of the park leading to the mansion. The driver stopped outside a hut marked 'army stores' and unloaded the kit. Martin went into the hut and reported to a corporal. The corporal knew all about Martin, even the number of the hut he was to live in. He looked at Martin all dressed up in his best uniform, polished boots and blancoed gaiters, with his rifle slung over his shoulder. 'Put your kit here and those dirty blankets over there on the floor. We issue all newcomers with clean blankets and a fresh mattress,' he said.

He looked at Martin's rifle and said, 'we must get rid of that at once. They don't like those things here!' They walked through the camp until they came to another hut marked 'armoury'. There was an NCO inside, 'Not another one,' he said when he saw the rifle, 'I have a job to get rid of them now, no one seems to want them.' He took the rifle and gave Martin a receipt for it. 'Keep that safe' he said, 'or they will charge you for a new rifle when you leave the army.'

Martin was taken to Hut 3 which was to be his home for the rest of his stay at Hanslope. The NCO said it was an easy camp and, as long as he behaved himself, kept himself to himself, and told no one on or off the camp what he was doing, then he would be alright. Martin thought that would be very easy as he did not know anyone, did not know what he was doing, nor what he was expected to do – so he was most clearly not in a position to tell anyone anything!

At 1.00 p.m. Martin went to the workshops. The engineers were lined up outside and Lance-Corporal Willis took the roll call to make sure that every one was reporting for duty after lunch. When every one was accounted for, they went inside to work. He was handed over to Sergeant Haddon who was in charge of the machine shop. Sergeant Haddon said nothing to him, no words of welcome, no pep talk, but put him to work with Ralph West who came from London. They became firm friends from then on.

The main meal of the day consisted of beans on dry bread, as they had no means of making toast in large quantities. This was followed by a piece of bread, margarine and jam, plus the inevitable mug of tea. There would be nothing more for them until breakfast the next morning. The evening shift, who worked from

4.30 p.m. until midnight, had a half-hour break at 8.00 p.m., when they could go to the mess for soup, a crust of bread and a cup of cocoa. This meal was strictly for the shift-workers only, no one else was allowed in the mess. In the winter, when it was very cold, you needed the soup and cocoa to keep you warm until midnight.

Martin went back to Hut 3 to make his bed and meet his new hut mates. He was told that there was no NAAFI on this camp, only a regimental institute, which was run by a lady who lived in the village. He had not heard of one of these but found out that it was a canteen run by the army on army funds. A cup of tea was tuppence, and a glass of beer was sixpence – if you had the money – and if they had any beer. You could also buy razor blades, toothpaste, Brylcreem and soap, if you had coupons. He decided to go to bed early that night as he was completely exhausted. The first day at Hanslope Park had been a long and weary one. Before he got into bed, however, some one told him to be sure and leave his empty mug on top of the bedside cabinet.

It was a great surprise to be roused the next morning at 6.45 a.m., by a man with a large enamel jug who was pouring everyone in the hut a cup of tea to drink, as they lay in bed. Martin had never been given a cup of tea in bed in his life before – let alone in the army! It appeared everyone put sixpence a week into a kitty to pay a cook in the mess to make the tea every morning. The men in the hut took it in turns to get up early and fetch the tea, pour it out and return the empty jug to the cookhouse.

The 'tea in bed' arrangement continued all the time Martin was in Hut 3. He was told it had been organised by Lance-Corporal Willis so it was his privilege to be waited on by the men in the hut. He did not to have to get up to fetch the tea or pour it out. Such are the privileges of such exalted rank!

Martin was used to getting up at 6.30 a.m. every morning, so after the mug of tea he decided to get up and get dressed and go for a wash and shave. The washroom was an old Nissan hut with wooden stands fitted with hot and cold water taps. The washing bowls were made of steel and were rusty. The galvanised coating had worn off because they had been cleaned every week, for the past few years, with abrasive sand to make them shine for the weekly inspections.

Breakfast was the usual porridge with watered down evaporated milk, two sausages with a large portion of baked beans and a slice of bread with a portion of margarine and a mug of tea. As he was leaving the mess, there was a mad rush as the latecomers jostled in. They all grabbed a sausage and a piece of bread and the most important a mug of tea. It was obvious that these men had only just got up and it was the same each morning.

Martin went into the workshop, expecting to do a full day's work but forgot it was Thursday. Thursday was army pay day. After lunch, some of the men had to parade in an empty hut next to the cookhouse. It was some time before he realised that many of the engineers did not appear on this parade.

From the pay parade, with money in your pocket, you went into another hut to receive your weekly free ration of twenty Woodbines or a packet of ten Players cigarettes; there was no pipe tobacco. Once a month, they got a free sweet ration of hard-boiled sweets or a small block of chocolate. These were usually in sealed tins. After picking up the rations, they took them back to the workshops. The men who did not smoke, quickly sold their ration to the smokers.

It seemed strange, as over the next few weeks Martin noticed that many men could afford to go home at weekends. Some of them even seemed to have enough money to run small cars. Many had old motorcycles and everyone had a cycle.

During the first month, Martin settled down well in Hut 3 at Hanslope Park, making many new friends. He enjoyed the engineering work much more than the wireless and Morse code work at Little Horwood. At the end of the first month, he had a real surprise. Willis took him to one side and told him that he had passed the month's trial and was now on the permanent staff. He would be receiving a bonus each month which would make his army pay up to a total of five pounds per week, although he would have to pay income tax. He did not explain where the extra money came from and Martin did not ask. It was, of course, from the Section VIII account.

This extra money explained why so many men were able afford to go home most weekends, owned cycles and motorcycles, and why some even ran a car. No one ever spoke about this extra pay, and he did not tell anyone else, in case that they did not receive it.

Just before Christmas there had been good news for the ATS girls. They were told that they were going to be demobbed. They would be leaving Hanslope Park and would be moving to the 'Head office' in London where they would continue to punch master tapes for Rockex. It was good news for the ATS girls but not so good for their boyfriends at the Park.

In late January, 1947, there was an announcement that all the SCUs were being disbanded and shut down. This was the best news Martin ever had at Hanslope. He was told he could have early demobilisation if he agreed to return to Hanslope Park, and continue to work on Rockex as a civilian, irrespective of his demob number. Martin was in Group 64 and at the time, they were only starting to deal with Group 26.

The agreement was that the engineers would return and work for at least twelve months but they were warned that if they did not return after two weeks' leave, they could be recalled to the army. They were entitled to eight weeks' paid demob leave and as they were still in the army for that period of time. Everyone of them accepted the offer as a way of getting out of army uniform. There was no mention of the pay or conditions and nobody bothered to ask.

Martin returned to Hanslope Park after two weeks' demobilisation leave. He was now a civilian engineer but aware that he was still in the army and subject to King's Regulations for another six weeks. He soon discovered that he was now working for His Majesty's Government Communications Centre (HMGCC), Hanslope Park.

One of the first jobs given to the returning engineers was to remove the Weald wireless station which was situated in a field off a lane which lead from Weald to the A5 and Stony Stratford. Weald had been an important link to the SIS agents in Europe but now only the buildings remained. The men had a carpenter with them who told them that the buildings were urgently required to enlarge the wireless station at Hanslope Park to cater for the increased wireless traffic, arising from the closure of Whaddon Hall.

The men soon got organised and busied themselves brewing up tea, in the good old army way. Over the next few days the station, which had played such an important part in the Secret Wireless War, was largely demolished, although the brick building which housed the generator and in which the tea was brewed is still standing today.

Shortly afterwards the men learned they would be getting two weeks' paid holiday at the end of July. The bad news for some was that the production of Rockex would cease at Hanslope Park at the end of August 1947. The whole of the engineering department would be moving to London to the Palace of Industry at Wembley.

This move turned out to be a disaster for the Rockex project. Most of the skilled operators living in the north were forced to leave and only those living close to Wembley were able to move. After working on Rockex, the crew had to go home to their old firms or find alternative jobs.

In Wembley, new operators had to be found and trained and many of those that had moved decided to leave after a short period. The production of Rockex was affected for the next twelve months. Because of security, the production of Rockex later moved to Arkley near Barnet. Much later, the famous machine moved back to its birthplace at Hanslope Park. The design was probably improved and with modifications it remained in production for many years and was used in British embassies around the world until it was finally withdrawn in the 1980s.

Martin finds it strange that in the year 2002, information on the 'Rockex' is still on the secret list and most of the machines and handbooks have been destroyed. He points out there is one on open public display in the Military Communications and Electronics Museum in Kingston, Ontario, for all to see.

Chapter 27

Lawson Mann
One of the Fifty Despatch Riders at SCU1

Lawson Mann joined the army at the age of eighteen in January 1942, signing on at Shoeburyness Barracks for basic training in the Royal Artillery. After some training as a despatch rider he was transferred in June to Corfe Castle in Dorset, to join the Royal Welsh Regiment which, he tells me, had just changed over to handling artillery.

He was then posted overnight to SCU1 at Little Horwood on a War Office Transfer. There he was trained as a driver and had more training as a despatch rider. As he had been a carpenter before he joined up, the builder who was putting up the camp employed him to assist in the construction work. He built the asbestos huts near the canteen and did the brickwork for the hall, stage and dressing-rooms of the NAAFI canteen. He also built the accommodation for the NAAFI girls, and the manager and his wife.

Lawson Mann – one of our 50 despatch riders

He remembers that several D.R.s (Despatch riders), had to escort long timber 'drugs' from Woburn Woods where special 70-foot pine trees had been felled and loaded on the back of Bedford lorries. These were taken to Whaddon where they were spliced together, using special bands, to form 140-ft long wireless masts. The masts had rings at intervals to which wire was attached to anchor them to the ground.

The bottom ends were covered in metal that rested on concrete bases made in 40-gallon drums, sunk into the ground. A special lorry and crane were required to hold the masts in position whilst they were fixed with stays. Due to their sheer size, the masts were always assembled at the actual site of the wireless station. Lawson recalls that the motorcycle repair unit was situated at Fenny Stratford on the A5 but the lorries and wireless vehicles were repaired and maintained in the MT workshops at Little Horwood.

As soon as he was fully trained, Lawson was based at Whaddon Hall but was billeted in Ansell Road, Stony Stratford, with his wife Grace. When he went on leave, he always kept his motorcycle at Ashby's in Stratford. When at Whaddon, he worked from the two huts in front of the Hall (Main Line). His commanding officer was Major Saunders, who later disappeared in a Lysander on a return flight from France. Other officers were Captain Darling and Lieutenant-colonel Tutt who worked in the wireless station office. He also helped Sergeant Dickens who came from Wrexham who was in charge of the Teleprinter Room.

The teleprinters at Whaddon were in constant contact with the War Office, Foreign Office, 'Codes' in London, Forfar in Scotland and Bletchley Park. Lawson was assigned to the local runs and worked shifts all the time, first day 1.00 p.m. to 5.00 p.m., second day 8.30 am to 1.00 p.m. and 5.00 p.m. to 8.30 a.m. and the third day he had off. He later worked at all the radio stations in the area, Whaddon (Main Line), Windy Ridge, Upper Weald, Nash, Hanslope Park and Bletchley Park.

The despatch riders did a run to London every night, delivering messages to the War Office and to 'Codes' at Broadway. The riders would sleep at the Salvation Army Hostel in Birdcage Walk before returning to Whaddon the next day.

Lawson Mann

Despatch riders also did runs to Eastcote and Pinner in Middlesex, Knockholt in Kent and to Crowborough in Sussex. All of them had to be done during the blackout with very little lighting on the bikes, no maps or road signs. The bike had to be kept full of petrol and well maintained so that it did not break down.

On the night of 9 November, 1943, Lawson was returning in the dark down the narrow lanes leading to Whaddon, when he was in collision with two men pushing a cycle in the middle of the road. He knocked them both over and ended up in the ditch at the side of the road. The two men were Italian prisoners-of-war who were working at a local farm and were outdoors well after their curfew.

Luckily for Lawson, an army truck came on the scene and took the men away. By the time he returned to camp and reported the incident his superiors already knew about it. It appeared that the prisoners worked for a local farmer whose wife knew Richard Gambier-Parry!

Lawson was put on a charge which was reported in:
SCU1 Daily Orders Part II No. 131 3 December, 1943 (Issued 6 Dec. 43).

Item No. 3 – Stoppages of Pay.
1137528 Sign. Mann L J.
Placed under stoppages of pay of 6/6 (six shillings and sixpence) on 4 December, 1943 for:- Section 40 A.A. W.O.A.S. being involved in a traffic accident on 9 November, 1943. AB 64 Part 11 entry No. B6.

The general view was that this stoppage was unfair since the Italian prisoners should not have been out past their curfew, and especially without lights in a country lane. Some were quite incensed by it and Reg Cockayne, one of the operators, 'passed the hat round' and collected the then princely sum of £4. That was more than four weeks pay for a signalman at the time. So Lawson showed a profit from his escapade.

A group of despatch riders at their base at Little Horwood

It is extraordinary that Lawson Mann only saw the official Daily Orders recording his accident over fifty-six years later. This was a result of the sharp eyes of Martin Shaw who had spotted his name in the visitors' book at Bletchley Park, and remembered the name Lawson Mann from a copy of my own 'Special Enlistment' which I had sent to him earlier. This document is shown in Chapter 9.

Chapter 28

Bill Miller – Tea with the Germans

Bill has written his story entirely from memory as none of us were allowed to keep diaries – well at least not those on the same level at Bill and myself, though clearly John Darwin and others did!

Bill has tried to keep the story in chronological order but there may well be some events that are out of sequence. As he still has his original passport, however, he has a record of all his travels during the war. Whilst this account was originally written for his family, it offers an insight into the work of a wartime Section VIII operator in a neutral country.

There were two completely different ways in which the war was fought. First, of course, there was the combat between the armed forces, and then there was the secret intelligence war. The intelligence war was fought both in belligerent and neutral countries, and in the latter the enemies faced each other almost daily. Both sides even knew their counterparts by sight, but of course there was never contact between them. It might seem strange that whilst great battles were being fought on land, sea and in the air, in the neutral countries, intelligence officers of both sides frequented the same restaurants, cafés and cinemas.

Bill Miller was an active participant in the intelligence war throughout hostilities. His full story, which has had to be curtailed for the purpose of this book, ought to be published in full at some stage, or at least deposited with the Imperial War Museum.

As a boy, Bill Miller was fascinated by short wave wireless, then in its infancy. He learned Morse code, and spent hours copying ships' wireless messages and constructing simple short wave radio receivers. It was his hope that some day he would become an amateur wireless operator, and eventually open his own amateur wireless station.

Bill was eventually able to obtain the necessary qualifications but put this on hold for a couple of years, as he had other interests. The Munich crisis caused him to renew his wireless activities and when he registered for military service, at the age of 19, his wireless and Morse code qualifications were recorded.

Bill was called up for military service in December 1939, and reported to a town hall in London, where he was enrolled into the 1st London Divisional Signals of the Royal Corps of Signals. The recruits were taken to Eastbourne, where they were issued with two blankets and a groundsheet. They were then allocated a place, sharing with five others, in a bell tent. It was snowing, the ground was frozen, and it was very cold. Eventually they were accommodated in requisitioned hotels which had been stripped of furniture. They were issued with uniforms, and the next three months were spent in drill and military training to make them into soldiers. The military training took all morning, and the afternoons were occupied with training to become wireless operators. The recruits also had to learn Morse by flag and Aldis lamp.

Bill had a great advantage over his fellow recruits, since his wireless skills and Morse qualifications were in excess of the initial army requirements. Fairly soon afterwards, he was able to pass the proficiency tests, and was transferred to operational active service as a qualified wireless operator. With a sergeant in charge, Bill and two other operators were allocated a wireless truck, and formed part of a communications network, moving around Kent to establish communications at locations which would become Divisional HQ in the event of a German invasion.

Along with two other soldiers, he was sent to the Post Office Training School in Newcastle on a 'high-speed' telegraphy course. The instructors were civilians. To the soldiers' dismay, they had to learn to copy Morse by 'Sounder', in other words, they would not be able to hear any tone, but had to read Morse just by key clicks. It took some time to get used to this, but it eventually proved to be invaluable when receiving messages sent by unskilled agents. The course was cut short as a German invasion was feared, and the soldiers had to return to operational duties.

The day that changed Bill's life happened in November 1940. He was stationed in Charing in Kent, when he was approached by the section officer, Captain Palmer, and asked if he would be willing to have his name put forward for 'Special duties involving overseas service'. He was told nothing about what was entailed. For some reason Bill said 'yes'. He had been on standby for some time, and he and his fellow soldiers were fed up as the war was going so badly, and they wanted to do more to help.

Sometime later, a Captain Oswick visited the wireless truck where Bill was working, and told him that he was to go to the War Office in London and report to a Major (whose name he cannot remember) at a certain room number. Captain Oswick seemed as surprised as he was, and Bill was very apprehensive. He had never spoken to anyone as high-ranking as a Major, and wondered what it was all about. He only had one day's notice, so he set off after breakfast, with just a haversack in case he had to stay the night. Little did he realise that that breakfast was to be last meal he would ever have in the army!

On arrival at the War Office, he found another soldier waiting who had had a similar experience to his and who told Bill he was a professional wireless operator. The men were interviewed separately; Bill was told that he would initially based at a place called Bletchley Park. He had to sign the Official Secrets Act and he was to catch a train to Bletchley that afternoon, in the well-worn phrase 'where you will be met'. Bill asked if he could return to his unit to collect the rest of his kit but was told he should have brought it with him. Arrangements would be made for the kit to be sent on. He was not to have any communication with his former unit nor any of his former comrades. The officer who had interviewed both men separately then called them back in and told them that this interview had never taken place and that they had never met him! They were both mystified and did not mention their interviews on the way to Bletchley.

On arrival at Bletchley, there was, as usual, no sign of anybody waiting for them. A sergeant and another soldier were waiting in the booking hall, but they took no notice of the two of them. When everybody had left the booking hall, they asked the sergeant if he was waiting for them. He asked their names – which were on his orders – but said he had been expecting two civilians and he phoned for further orders.

They were initially taken to Little Horwood but Bill was later given a private billet, a bungalow, in Water Eaton Road, where the landlady provided him with a room, all his meals (at least two hot meals a day), and did his laundry. The soldier who had been with him was taken elsewhere.

By this time Bill thoroughly confused, since he had not the faintest idea of what he was there to do. Clearly, ordinary soldiers are not accommodated in private billets and he was only 21 years old at the time. The landlady told him he was from the 'Park' and introduced him to a rather odd kind of chap, also staying there, who was also from the 'Park'.

Bill still did not know what the job was, but he was to be interviewed by somebody they were referring to among themselves, as 'Pop'. 'Pop' was, of course, Colonel Gambier-Parry. Although he did not know it at the time, Gambier-Parry was Head of MI6 Section VIII (SIS communications).

Gambier-Parry finally revealed to Bill what he was there to do. Behind the colonel's desk, on the wall, was a huge map of Europe. The occupied section of France was clearly indicated by coloured pins, right up to the frontier of northern Spain. Gambier-Parry pointed with a cane to the frontier of Spain and occupied France and told Bill he was to proceed to San Sebastian, in Spain near the French frontier, where he would join the staff of the British vice-consulate and would perform the duties which would be allocated to him. His real job, however, was to assemble a wireless station and contact home base in the UK daily, at times to be arranged. None of the staff of the consulate, with the exception of the vice-consul himself, were to know that he would be operating a wireless transmitter – it was to be a clandestine operation. Bill was stunned when he was told where he would be going. It seemed to him that Gambier-Parry was pointing to the frontier of occupied France, and Bill wondered how he could get out of this – and go back to his old unit! But Gambier-Parry shook Bill's hand and welcomed him into the 'Firm' – which was how the SIS was known amongst its members.

Gambier-Parry told Bill that from that moment he was on the payroll. His salary would be £300 a year whilst in UK, and £400 a year when overseas. His pay as an ordinary soldier was two shillings a day – 14 shillings a week or £37 a year. To jump from a soldier's pay, to eight pounds a week, was simply staggering. His new salary, was more than most officers were receiving, and in 1940, £3 a week was good pay in civilian life. He only had to sign away his right to army pay – something he did gladly!

Bill then started his training. Firstly, he was introduced to the equipment he would be using. The receiver was a National HR0 communications receiver from America. At the time, it was considered the most sensitive and accurate wireless receiver in the world. The transmitter (a MkIII) was very low power – in the region of 30 watts – and made at Whaddon for this specific purpose.

For his training on the HRO receiver, Bill was taken to Whaddon Hall, where the receiving station was located in a small room on the top floor. There were a number of HRO receivers in pairs on benches around the room. Above each pair were notices indicating capitals of neutral countries – Lisbon, Madrid, Tangiers, Gibraltar and others he cannot remember. All the call signs were in figures, as were the procedure codes. The operators were civilians with the exception of a couple of army officers. There was a sergeant, but he was not an operator. Apart from him, Bill was the only other ranker and, of course, he was the youngest. The officer- in-charge was a Captain Saunders, the only person who knew where Bill had been assigned. Bill's training consisted of doubling up on stations being worked until they let him work a couple of schedules on his own, but under supervision. All telegrams were in five figure groups. There was a teleprinter in another room, and a despatch rider called in at times to fetch and carry telegrams.

Bill then started learning about codes and cyphers. This took place on the first floor of the mansion, in Bletchley Park. He was taught by a couple of nice young ladies who came from 54 Broadway. He knew this because when he had his final briefing at Broadway, one of these same young ladies gave him his tickets and some postage stamps to take overseas.

On his first day of cypher instruction, one of the ladies took him down to lunch. The dining room was on the ground floor, a beautiful room with panelled walls. There was an ante-room alongside where people waited to be called into the dining room, when spaces became available at the one long table. Bill felt very uncomfortable in his private's uniform, whereas all the others waiting were either civilians or officers. When the ATS girl at the door of the dining room called out 'two' they went in.

To his dismay, Bill found himself sitting next to a major who actually called him 'old chap', unthinkable in the army, but he soon learned that the uniform one was wearing meant nothing in Bletchley Park. Everyone was under the same rules: tell nobody what you were doing, and don't ask anybody what they were doing.

Eventually, one of the young ladies told Bill that it was time for him to 'get out of that uniform' and change into civilian clothes. He went home to get his clothes, but found that in the short time he had been eating army food (which was much more plentiful than he was used to), he had put on weight, and his clothes were very tight. So, they gave him twelve pounds and ten shillings, and told him to go to Northampton and buy some suitable clothes. He bought a nice suit, a couple of shirts, underwear and shoes, and to his mind the best of all – a beautiful raglan blue overcoat. This overcoat later caused a bit of a problem when he arrived in Spain. All these clothes cost just over ten pounds – clothes were not rationed then – and when he went to hand the change back- they told him to keep it. *'No receipts – no signature'*.

Bill then handed in his uniform, and all his army papers and identity tags, in exchange for which he was given a civilian identity card. He never wore uniform again.

This done, his training in codes and cyphers started in earnest. Messages were encyphered and decyphered, using 'one time pads' in conjunction with encoding and decoding books This system of coding, is one of the most secure system of coding even today. Its drawback is that it is manual and slow. The pads consisted of pages filled with five figure groups. To encode a word, one looked up the required word in the encode book. Each word had a four-figure group. These four-figure groups were written underneath the groups on the encode pad until the message was completed.

The upper and lower groups were added together, without carrying over any digit. The resultant figure group became the encyphered group. The reverse procedure was followed for decoding. Both sending and receiving stations had to have identical pads, the encoding pad of one being the decoding pad of the other stations (there is an example is in the appendix).

The second code he was taught, was to be his own personal code. This used a paperback Penguin book. One had to be sure, that the same book, in the same edition, was used by sender and receiver. No code books were used. The system of coding was to jumble up the words of the message in a very complicated way. The system used the page number, and a line number, to choose a couple of words or a phrase, from which a grid was formed. It is far too complicated to explain without an example, so in the appendix there is an example of how a message was encoded and decoded. Bill had to spend many days encoding and decoding messages using both systems, until his superiors were satisfied that he would be able to use what he had learned in the field. Only one person in the field held the appropriate copy of the Penguin book, the other copy was held at HQ in the UK. If he needed to send a message while overseas, and for some reason he was unable to get at the Penguin book, he was asked to select a short phrase which he could easily remember to be used instead of the book.

He was told that he would be going overseas about April 1941 and was given Christmas leave in December 1940. Before he went home he filled in an application for a passport, and had his photograph taken. However, he had only been on leave in Paignton for a few days, when he received a telegram recalling him to Bletchley.

It seemed that a complication had arisen about Bill being granted an exit permit to leave Britain. From what he was given to understand later, there was a law that meant that anybody eligible for call-up to H M Forces was not allowed to leave the country. He therefore had to re-submit his application for a passport but was told to leave the place of birth blank. Then a little later, he was told to put his place of birth as being in a Commonwealth country. It was only some time later that Bill was given to understand that the reason for this was that it meant that the country of his birth had the first call on his military service. At twenty-one he found this all very confusing. Moreover, 54 Broadway even told him the date he had arrived in UK and that his name was on the passenger list of the ship he had travelled on!

Bill was then told that a situation had arisen that required him to proceed overseas at once. This was on 16 February, 1941. Three days later, he reported to 54 Broadway, where he was given his passport, and his exit permit was issued on 18 February, on the same day as the passport was issued, with a diplomatic transit visa for Portugal, and a diplomatic entry visa for Spain, both granted on 19 February, 1941. Bill still has this passport, so can he be sure of the dates and it is illustrated.

At Broadway, Bill was given a final briefing and presented with the 'code' referred to earlier. This consisted of several pages of foolscap, on which were listed all sorts of German equipment, including tanks, guns and vehicles, together with lists of German military units and names of senior German officers. Each item on these lists carried a five-figure group. This was in case he had to send this information home by wireless, with no time to encode it. Thus he could send any of the above information if a situation arose in which it was required urgently.

Bill was becoming even more confused and a bit concerned wondering what he had let himself in for. He was told that he would fly from Bristol to Lisbon the next day. He would be travelling as a diplomatic courier, and would be taking all the equipment with him which would be classified as 'diplomatic despatches'. His wireless receiver was in a large wooden case, each nail head bearing a sealing wax stamp. A smaller case, similarly sealed, contained the transmitter. In addition, he carried a conventional diplomatic bag, containing the code-books, one-time pads and the lists mentioned above. He was sent out to buy five copies of the same Penguin book, but was told not to buy them in the same place. So Bill went round to various bookstalls and bought copies of *Poets' Pub* by Eric Linklater which happened to be the latest issue from Penguin publications on sale.

Firma del interesado,

William V. Miller

ALTA COMISARIA DE ESPAÑA EN MARRUECOS

N.o 61-

TARJETA DE IDENTIDAD

PARA

EMPLEADOS ADMINISTRATIVOS DE REPRESENTACIONES OFICIALES EXTRANJERAS

El Alto Comisario de España en Marruecos, CERTIFICA: Que,

Mr. William Victor Miller, súbdito inglés, es funcionario del Consulado General de S.M. Británica en Tánger —

Dado en Tetuán, a 6 de Marzo de 1944.

El Alto Comisario

Bill Miller's Spanish Passport

At Broadway they put one copy – his copy – of *Poets' Pub* in the diplomatic bag, and then sealed it. Bill was also issued with a courier's passport, a very elaborate item, in the form of a scroll, signed by the Secretary of State for the Foreign Office, and bearing a large wax seal. It was printed in English and French, and apart from the usual 'request and require', as on ordinary British passports, it said that W. V. Miller, holder of passport number 363041 was 'charged with despatches for His Britannic Majesty's Ambassador in Madrid'.

Bill was allowed to go home to say goodbye to his mother. She was naturally upset, since his brother was a prisoner-of-war, his two sisters who shared a house, had been bombed out, had lost everything and had been evacuated, and now Bill was going away. Sadly he never saw his mother again, she died in November, 1943.

On his return to Broadway, all was ready for Bill's departure. He was told that from then on his address would be 54 Broadway, Westminster, the headquarters of British SIS. He was given some UK postage stamps so that when he wrote home, his letter would bear a UK stamp and would be put into the diplomatic bag from Madrid to London. The letters would then be posted in London. Thus, it would appear to anybody who wrote to him that he was working in London. An army truck, driven by a sergeant, took him and all the baggage to Paddington railway station. They drove down the slip-road to the platform alongside the train, where the guard was waiting. Bill had a reserved first class compartment all to himself and the baggage. The rest of the train was packed, as was usual in wartime. Some time after the train was on its way, the Guard asked him if he would mind sharing his compartment with some officers. He agreed, of course, but received some odd looks from these rather senior officers.

Bill was met at Bristol, and taken to The Grand Hotel, a luxury hotel where a room had been reserved. It was rather late when he arrived. He noticed that people had still left shoes outside their rooms to be cleaned, even during wartime.

The people who had met him the night before came for Bill the next morning, and took him to Whitchurch aerodrome, telling him he would be travelling on Flight 777. On arrival at the aerodrome, he had to go through Immigration control and the Censor. All those who examined him were army officers and they seemed very suspicious of one so young, but his courier's passport got him.

Flight 777

All sea travel to Europe having ceased in late 1940, it was necessary to re-open the UK-Lisbon air link with whatever aircraft were available. KLM Royal Dutch Airlines, had some aircraft at the disposal of the Dutch Government-in-exile in UK. These DC3s were leased to the British government with their crews. The aircraft carried a maximum of twenty-one passengers and a crew of four.

The direct route to Lisbon could not be used as it passed over Britanny and would be in easy reach of German fighters based in France. So aircraft flying to Lisbon had to make a wide detour, flying westwards over the Atlantic before turning back to Lisbon. Bill was told there was a secret 'understanding' between Britain and Germany that these flights could continue unharmed because they carried prisoner-of-war mail, neutral diplomats and diplomatic bags, newspapers, etc. It was in Germany's interest, in fact, not to interrupt these flights, as UK newspapers and magazines on sale in Lisbon were a mine of information to the German Intelligence Service. Also, by using neutrals who were sympathetic to Germany and their diplomatic bags, they could infiltrate spies and use the diplomatic bags of neutral countries for passing intelligence information.

Passengers for Flight 777 waited at the Grand Hotel. Times of departure and arrival, and the name of airport of departure were kept secret from passengers. Passengers selected to leave were assembled on the morning of departure and then taken to the airport. Bill's late arrival with diplomatic priority seemed to upset things, as the weight of his luggage, including the heavy boxes, was well over the baggage allowance. This meant that one of the other passengers had to stand down and this caused some resentment. There were only about twelve passengers, but the vacant seats were piled with bales of newspapers, mailbags and all sorts of boxes.

The pilot briefed them, saying the flights left daily and would take about seven hours. He then told them about the route and said there were life jackets under the seats. There were also a couple of rubber dinghies on board – one was taking up quite a lot of space in the gangway. The pilot also asked his passengers to be diligent and keep an eye out for any aircraft that they might see from the windows. Naturally, Bill did not find this very comforting, especially as it was his first flight!

Bill arrived at Lisbon after about seven hours but there was nobody to meet him. There were no baggage-handling facilities such as we have today, all his baggage was dumped alongside the aircraft on the tarmac. After a long wait, the pilot himself took Bill to the British Embassy. The receptionist at the front desk did not seem too pleased to see him, saying 'You people keep everything so secret, nobody told us you were coming.' However they passed him over to the MI6 representative who took him to a first class hotel, the Grand Hotel Borges. He was to stay there until a passage was available on the weekly railway sleeper service to Madrid.

Bill was given some money and told he would be contacted in due course. Lisbon in 1941 came as something of a revelation to Bill. After the blackout, rationing and shortages of all kinds in Great Britain, here in Lisbon, there was no blackout, no war and the shops were full of fruit that had not been seen at home for a long time. He could not get over the brilliant lighting everywhere, the streets, the shops, crowds of people sitting in cafés and restaurants – in fact a country at peace. He just wandered around, taking it all in. A man from the Embassy took him out to dinner in a restaurant one night. During the meal, he muttered 'take a discreet look at those chaps over there', indicating a table. 'They're from the German Embassy'. These were the first of many Germans he was to eventually see.

Bill had arrived in Lisbon on 20 February, 1941, and left for Madrid on the 27 February, so he had seven glorious days of rest and relaxation. On the train, he had a sleeping compartment to himself, with all his baggage. On arrival at the Spanish frontier, he was a bit apprehensive. The frontier guards did not speak English, but his diplomatic visa gave him immunity so his personal belongings were not examined.

In Madrid, Bill was taken to the Embassy where he met his new boss, Commander Hamilton Stokes, the Chief Passport Officer and Head of British Intelligence in the Iberian Peninsular. Bill was lodged in a hotel called Nuestro Senora del Carmen which was located in the Plaza de Santa Barbara, quite near to the Embassy. The Passport Control Office was located on the first floor of a building opposite the Embassy. HS (Hamilton Stokes), told him that there had been a change of plan. Bill was to go to Bilbao and not San Sebastian. The reason was that San Sebastian was only a vice-consulate and so did not enjoy diplomatic immunity. Bilbao however, was a consulate, and although not having the same protection as the Embassy, would be more secure for him than San Sebastian.

Hamilton Stokes then gave Bill a full briefing. He already knew some of what he was told, but he now got the details. If the Germans entered Spain, with or without the permission of the Spanish government, in order to attack Gibraltar, there was only one place where they could cross the Pyrenees, i.e. near the coast of occupied France and Spain. The only road leading to it passed through San Sebastian and Bilbao. The French and Spanish used different railway gauges, so if they used the railway, they would have to change trains at the frontier.

We had agents in that area watching out for any German movements. At the first sign of any significant German troop movements indicating a move into Spain, Bill was to establish wireless contact with the UK base. His contact in Bilbao was a man named Vallet, who was attached to the consulate but was not there every day. He was some kind of pro-consul.

The UK had to know immediately if the Germans were to enter Spain and most particlarly if they did so with Generalissimo Franco's permission. Spain had a bomber aircraft base in Seville, so the Royal Navy had to be given the earliest possible opportunity to disperse the ships of the British fleet, anchored in Gibraltar, out to the open sea. This therefore, was the prime reason for Bill being placed in Bilbao, as a truly vital part of our intelligence.

In the event of very short notice, Bill would be able to use the lists of German units and similar information given to him at Broadway, as he would not have time to encode such information while operating the radio, and he had nobody to help him. His cover would be that he was a member of the consulate staff. It was emphasised that, apart from the consul himself, and of course Vallet, no one on the consulate staff must know that Bill had a wireless transmitter. So, in addition to his real work, he would have to undertake any consular duties, that might be allocated to him.

Madrid station would monitor his transmissions to London. Although it had the appropriate one-time pads, it did not have a copy of his Penguin book. Hamilton Stokes told Bill that he was not to frequent the same café or bar regularly, nor was he to associate with any girls, since wireless operators were prime targets for counter intelligence agents. At Bill's final briefing, Hamilton Stokes looked at Bill and said 'your overcoat'. At first he had been pleased that Bill was so proud of his new overcoat, but he later decided that the bright blue colour was too conspicuous. He was taken out to the Madrid equivalent of Regent Street where he was bought a gabardine raincoat and a rather large beret.

Bill was then taken to Bilbao by car with the courier on the weekly diplomatic bag run. Mr. Mclauren, deputy head of the Passport Control Office, was also with him. They stopped for lunch in Burgos at a large hotel which was virtually empty. Apparently all the hotels there had been crowded during the Civil War, as Franco established his headquarters in Burgos, but when the war ended, Franco and his staff and administration moved to Madrid.

When the party arrived at Bilbao, they dumped Bill, with all his boxes, on the pavement outside the British Consulate, which was located on a first floor above a Spanish bank. Since the road had a tramline, there was nowhere to park. Suddenly, Bill got the shock of his life. A couple of immaculately dressed officers passed him. Bill was certain they were German and thought the Nazis had already entered Spain. He was petrified. They might not have been Germans but they equally well could have been, as sometimes senior German officers came to Bilbao for conferences.

A couple of consulate staff came down to help Bill to get the boxes upstairs. They wanted to know what was inside, so Mclauren told them 'stationery'. They then took Bill to a boarding-house, the Pension Allegue, located in the Gran Via, quite close to the consulate. Nobody there spoke English, but he managed to register and left his passport for the Spanish police. He left his suitcase on the bed and returned to the consulate.

He was first interviewed by the consul, a Mr. Graham, who had apparently been resident in Bilbao for many years. His first words to Bill were not exactly encouraging. He told him that he had expected an older man, as all British subjects of military age had been evacuated 'a long time ago' and Bill was only twenty-one. He told him that he was not concerned about his own safety, but it was for the sake of his staff. He reminded him of something he did not want to be reminded of – that Spain had a military government and that all wireless transmission was strictly forbidden. Bill subsequently found out that Mr. Graham had never wanted a wireless operator in the consulate, and had to have his 'arm twisted' before he agreed.

Graham also reminded Bill that he would be operating clandestinely, again something of which he did not want to be reminded. He added that the consulate would not be able to help him, if he was detected. Not much of a welcome.

The consulate staff consisted of a pro-consul named Richardson, a young man about his own age, Pat Dyer, a Spaniard and Mr. Vallet whom he believes was also a pro-consul, who came into the office three or four times a week. It was Vallet, of course, who was the only other person who knew Bill had a wireless and who, Bill assumed, would give him any orders. It was all rather vague as to exactly what action he should take in the event of Germany entering Spain. All he was told was to use his own initiative.

Bill was shown the room in which he could set up the equipment. They called it 'the archives', a dusty room which had obviously not been used for years, full of ancient files and record books. There were cupboards along one wall, and Bill decided he could fit all the equipment in one of them, so that when he was not operating, the cupboard door could be locked and there would be no sign of any wireless equipment.

An aerial had already been erected, the excuse being that the consulate needed it for a wireless receiver in order to receive the news from Great Britain. This solved a problem, as Bill could not very well go on the roof to set up an aerial. The building was quite a high one with a central well, down which ran all the electric cabling for the building. The aerial down lead came down this inner well. This was technically not very suitable for transmission, but there was nothing Bill could do about it. He had to connect and disconnect the aerial every time he used the equipment, and transfer the down-lead from the kitchen (the adjoining room) to where the receiver was located.

The archives were out of bounds to the other staff and only Bill had the key, but if, for any reason, anybody did get in there would be no sign of his equipment. He had to establish contact with control station in Great Britain three times a day at different times, in accordance with a signals plan. Once contact had been established, he merely exchanged signal strength with the station, then closed down. Thus he was only on the air for a few seconds. The reason for this was because the Germans, just a few miles away in occupied France, had radio detection equipment so the shorter the time he was operating, the less chance there was of being detected.

Bill was not sure of the actual location of the home station to which he worked. He did not use the operating procedures used at Whaddon or its call signs. He used the call sign PNO and used normal commercial procedures, so that it would appear that he was an ordinary commercial station. But since all radio transmissions were monitored, and he doubts whether this fooled anybody for long. He never sent telegrams, and was only to do so in an emergency. So, all the time he was in Bilbao, he was only transmitting for a few seconds at a time.

On his return to the boarding-house, Bill found that his suitcase had been broken into, no attempt even being made to disguise the fact. He pointed this out to the owner, but the man just shrugged his shoulders and said something Bill did not understand. He reported for duty the next morning, and when he told them about the suitcase, they said they would arrange for a move. He spent the first days unpacking and installing the equipment. He was able to keep the codes in the consul's safe. He had to do a bit of juggling to modify the aerial down-lead, in order to make it compatible for the equipment.

When everything had been installed, Bill made his first contact with the home station. He was delighted when the signal was reported as QSA5, the top grading. He had a signal plan that indicated the times of contact, as these varied every day. He then had to perform a series of frequency tests, to establish the optimum frequency for day and evening.

Bill was then informed of the office duties he was required to perform. Fortunately, they were very light, as he could not speak Spanish. There was very little consular work involving British subjects, since all that were left were one or two elderly ladies, and certainly no tourists. Occasionally some refugees from France turned up and were interrogated, before being passed on, but he knew absolutely nothing about this, and nobody spoke about it.

Bilbao was a port and, at the time, a very important and busy one, with regular sailings to and from other neutral countries, and especially South America. Bill was told that he would have to assist with the 'navicerts', documents issued by the Royal Navy and distributed by British consuls. All neutral ships were liable to be stopped by the navies of the belligerent countries, to ascertain that they were not carrying goods to or from enemy sources. Companies exporting goods from neutral countries had to have a 'navicert' for each item of cargo. They obtained a navicert from the British consulate on which they had to declare the origin and ultimate destination of the cargo, together with all the appropriate invoices and receipts, with an exact description of each item. When completed, the Navicert was submitted to the consulate, and this is where Bill came in. The consulate had copies of the Ministry of Economic Warfare 'Black List', which was frequently updated. The names of all the firms indicated on the navicert, both of origin and ultimate destination, had to be compared with the list and if any of these names appeared on the Black List the navicert would not be accepted by the Consulate.

Ship's captains did not want their ships stopped, so they insisted that all items of cargo on the ship's manifest had its navicert. The ship's manifest was submitted to the consulate, and if the consul was satisfied that all items of cargo had the appropriate navicert, he signed the manifest. If not, the refused to sign. If a ship sailed without a signed manifest, London was informed.

Bill also had to decode the telegrams the consul received from time to time, mainly notices to mariners about location of minefields and buoys. He did not handle any of the consulate's telegrams from commercial sources, but he also had to do any odd job they gave him. So what with all that and the wireless, which was his real job, he was kept fairly busy.

Ships used to visit Bilbao regularly from South America, some of them carrying Germans returning to 'fight for the Fatherland'. Bill happened to be in the port area when such a ship came in. For the first time, he saw Germans in uniform. They were Hitler Youth, wearing their brown uniforms with swastika armbands. They stood in the port square, directing coaches full of Germans from the ship, indicating the route out of the port that led to the road to France. It seemed strange to him to be watching these Germans, with nobody taking any notice of him. This was in early 1941, when the British were fighting Germany all alone.

Throughout Bill's time in Bilbao, the war was not going very well and the Germans were expected at any moment. He was always 'looking over his shoulder' and often wished he had someone to talk to about what he was doing. He felt very alone at times. Wireless contact was becoming difficult and he had to spend more time 'on the air' than was advisable. Bilbao was surrounded by iron ore mines and mountains, neither of which are conducive to good radio communications.

Bill well remembers the fateful day that the Germans sank HMS Hood, the largest British battlecruiser. The local newspaper published an aerial photograph of the Gran Via (Bilbao's main road, on which the consulate was located, and superimposed upon it an outline diagram of the Hood, so that people could see how big it was. The paper also continually published maps showing how well the Germans were doing. Crete was one example, he remembers. None of this did Bill's morale any good.

The local cinemas often showed old American films, with English dialogue and Spanish subtitles. On one occasion, when Bill went to see such a film, there was a long German documentary called *Victory in the West*, a German film about Dunkirk, with actual film taken by their planes whilst attacking our troops on the beaches of Dunkirk. Later, when the Bismarck was sunk, however, it only merited a few lines on an inside page of the local press.

Then, one day in May, 1941, Bill received orders to pack up all the equipment and return to Madrid. No reason was given and neither Vallet or Bill could understand why, as they were still expecting the German arrival. So, he said goodbye to Bilbao and never returned.

On arrival in Madrid, Hamilton Stokes told him that there had been a change of plan. Bill was to stay in

Madrid, not to work on the wireless but to concentrate exclusively on cyphers. He was thus kept very busy encyphering and decyphering top secret telegrams. He worked in the Passport Control Office, which although part of the embassy, was located in a building opposite. He therefore made frequent trips backwards and forwards to the Embassy wireless room, to deliver or collect telegrams. As he did so, Bill used to get some odd looks from a group of escaped British prisoners-of-war awaiting for repatriation. They probably wondered what a young Englishman was doing in Madrid who was apparently not serving in the forces.

It was around about this time that the Spanish government became concerned about the large increase in the number of staff in the various Embassies, and issued an order restricting staff increases to a certain percentage of pre-war levels. When Bill's passport was sent in for his resident's visa to be extended, the Spanish government refused permission. Instead they issued a 'Permission to leave Spain ' stamp in his passport.

So Bill had to leave Spain. It was very likely that the Spanish authorities knew what he was doing in Bilbao and this was an opportunity to get rid of him. But to return to U K was not easy as there were no flights or ferries. Gibraltar was in a military zone, so was also out of bounds. There was only one place he could go, and that was back to Lisbon. But his time limit was running out and he could not wait for the weekly sleeper train so he had to go by air. The problem was that the flight was part of the Berlin – Paris – Barcelona – Madrid – Lisbon commercial service. At Barcelona, a Spanish crew took over. Although Bill felt uneasy about it, the embassy assured him it would be all right.

The plane was a German Junkers, not very big by today's standards. The fuselage looked like it was made of corrugated iron. There were single facing seats on both sides of the aircraft. When they showed him to his seat, he found he was sitting opposite a rather stout,old gentleman who had his feet underneath Bill's seat, and made no effort to move them. He ignored Bill's greeting (by this time his Spanish was quite good, as he had had private lessons in Bilbao and Madrid) so Bill had sit with his feet in the aisle which was most uncomfortable.

When the plane took off, the stout gentleman went into the pilot's area and stayed there for most of the flight. There was no division between the pilot's cabin and the passengers. So, Bill put his feet under the empty seat opposite, and when stout gentleman returned to his seat, he had to sit with *his* in the aisle. After landing, Bill walked behind the man as they were leaving the plane but he was held back when they approached the door. The stout gentleman stood at the top of the gangway; below him, stood a crowd of people who were waiting for him – all with their arms raised in the fascist salute! He responded.

Bill was later informed by the embassy representative sent to meet him that the stout gentleman was Franco's brother, the new Spanish ambassador to Portugal.

Bill arrived in Lisbon on 17 July, 1941 and reported to the British embassy, this time to his own department, where he was told that he was on stand-by awaiting orders from London. and was to report to the Embassy every day.

Bill spent seventeen days in Lisbon and had a great time. He often thought of home where they were enduring wartime conditions, while he was enjoying himself in a neutral city, with no rationing, no blackout, and few other restrictions. He could even walk around anywhere without having to 'look over his shoulder' as he had had to in Spain, especially in Bilbao. In Lisbon, the shops were full of all the goods that were unavailable at home and, unlike Spain, food was plentiful.

Although Bill did not have to work whilst in Lisbon, he visited the wireless room in the embassy and met the two operators, Les Teal and a man called Arrowsmith. They were both considerably older than Bill.

After his daily report to the Embassy, he wandered around Lisbon, sometimes with Embassy staff who were off duty. At the end of July, he received orders from London. to proceed to Tangiers, and report to a Colonel Ellis.

Bill Miller in Tangier 1941 – 1944. A picture taken for identification.

Bill had heard vagely of Tangiers but he did not really know where it was. He left Lisbon by the Portuguese airline, on 4 August, 1941, arriving the same day. Little did he realise that this would be his home for eight-and-a-half years.

TANGIER The International Zone of Morocco

Before World War II, Morocco had been divided into three zones. There was the French zone whose capital was Casablanca, the Spanish zone whose capital was Tetuan. The small international zone of Tangier that had been created in 1920. The Sultan of Morocco, the titular ruler of all the Moorish people, who had no jurisdiction over Morocco as a whole, had his palace in Rabat, in the French zone.

When World War I ended, it became apparent that the tip of Morocco was a very strategic location with regard to the Straits of Gibraltar. Britain, through its base in Gibraltar, controlled the Straits and thus the entrance to the Mediterranean. Any power occupying the area facing Gibraltar could greatly reduce its role in controlling the Mediterranean. That is why the victorious Allied powers decided in 1920 that the tip of Morocco, bounded by the Atlantic and the Mediterranean, should become an international zone, with the small town of Tangier, as its capital, and should thereafter become known as the Tangier International Zone of Morocco. Tangiers was to be governed by a 'Committee of Control', initially consisting of the Minister ranking Consul-generals of France, Spain, Italy and Britain. An International Administration was set up, together with International Police and Courts of Justice.

The population of Tangier was very cosmopolitan. Apart from the indigenous Moors, there was a large Spanish community, together with French, French-Moroccans, Italian and other European nationalities. There was a small British community and several British companies. The premier hotel, The Minzah, together with the Villa de France and the Consulate Hotel, were British-owned, as was the bus company and several import and export companies.

Children in Tangier assumed the nationality of their parents. There were three official languages – Arabic, French and Spanish. There were two currencies, French Moroccan francs and Spanish pesetas. It was a free currency market with no restrictions; any currency in the world could be bought and sold at the numerous exchange booths, as well as at the French, Spanish or British banks. Tangier was a free port and no visa was required to enter it. Consequently, numerous refugees from occupied Europe – mostly Jewish – had found their way to Tangier. Some of the Spanish residents were refugees from Franco's Spain, who had fought against Franco in the Spanish Civil War.

Tangier had three Post Offices – French, Spanish and British. British stamps were sold, overprinted 'Tangier', at the British Post Office which also performed all the usual functions of the post offices in Britain, including operating the Post Office Savings Bank. Francs and pesetas could be used at any shop or café, at the daily rate of exchange.

In June 1940, taking advantage of the fall of France, Franco sent in Spanish troops to occupy Tangier. Franco claimed that the object of the occupation was to 'protect Tangiers' neutrality'. Despite Tangiers' strategic location, and the obvious British interest, there was little the British could do about it, as our troops were fully occupied in fighting the Germans on our own.

Whilst the Spanish occupation was recognised as de facto, Tangier still retained its international status in many fields, due to pressure from the United States and Great Britain. It continued to have three official languages, French and Spanish currencies, British, French and Spanish post offices, a free port and free, unrestricted currency exchange. Spanish, French and British newspapers were available.

Tangier soon got the reputation of being a city of wine, women and song and, of course, it was a clearing-house for all the intelligence services – and for contraband and smuggling.

Franco sacked all the international administration staff and the International Police, and Spain took over the government of Tangier. For the first time, the Germans were allowed to open a legation, even though there were no German residents nor did any German ships call at Tangier. So, it was obvious that virtually all the staff at the German legation were spies. This was nothing unusual, as most of the other Consulates and Legations were similarly staffed, including our own.

This was the position when Bill arrived in Tangier. As had so often happened before, there was no one to meet him, even though he was acting as a diplomatic courier and was carrying a diplomatic bag. Bill had no money and all he knew was that he had to report to Colonel Ellis. So, he hailed a taxi and asked to be taken to the British Consulate-General. When he arrived, it was closed, for lunch! The guard took him into the residence, however, where he met the British Minister Mr. (later Sir) Alvery Gascoigne who took delivery of the bag, gave him money to pay the taxi and telephoned the Villa de France hotel to reserve a room. Gascoigne told the taxi to take him there and said Bill should return to the consulate after lunch. He did so, and finally reported to

The British Consulate in Tangier during the War

Colonel Ellis, who proved to be a retired Indian Army officer of considerable personal wealth, the owner of Ellis & Company a large import and export company, which also owned the local bus company.

Ellis also happened to head the British Secret Intelligence Service, covering French and Spanish Morocco, and other parts of North Africa. His cover was that he was the Press Attaché of the British Consulate-General. This was unusual, as consulates do not normally have press attachés, but the *Tangier Gazette*, the British newspaper in Tangier, was owned by the British Government. Thus Bill became a member of the staff of the press attaché at the consulate.

The *Tangier Gazette* was published daily in French and Spanish editions, with an English edition once a week. It was the means of circulating our war news and political propaganda. Copies of this newspaper were widely, though unofficially, circulated throughout French and Spanish Morocco. All British Intelligence for Morocco and French North Africa was controlled from Tangier. The offices of the Press Attaché were housed in the annexe, a rather a large house. The basement and ground floor were occupied by the administration and editorial staff of the *Tangier Gazette*.

BBC news broadcasts were monitored by shorthand-writers, and translated into French and Spanish. A daily press release and political commentary were transmitted in Morse code from Great Britain, and formed the contents of the paper. The upper floor however, was occupied MI6 staff, and was out of bounds to the locally employed *Tangier Gazette* staff. Intelligence reports and information were collated and all telegrams were encyphered and decyphered. A considerable amount of telegraphic material was sent or received by its wireless station. All telegrams were in cypher – telegrams en clair were never sent or received.

The station was located within the residence of the British Minister, which was located inside the compound of the consulate. Whilst the German and Italian operators made no attempts to conceal the aerials of their radio stations, the British had to make do, with a simple inverted 'L' aerial slung between two buildings within the compound so that it was out of sight to the public. Although the Spanish authorities obviously knew of the existence of the station (which was officially clandestine), no action was ever taken to stop its operations. The staff had to be careful, however, as the Spanish administration were very pro-German.

Naturally, they quickly got to know their German and Italian counterparts by sight. All consulates and legations flew their national flags every day. The largest consulate was the French, located in the centre of the residential and business sector of the town. The Germans had taken over a palace overlooking the main marketplace. The American legation and our consulate were situated nearby, with the Italian legation more or less between them. All the consulates and legations were within easy walking distance of each other. It all seemed rather strange that, with the war raging, Bill could see the swastika flag flying whenever he passed the marketplace which was practically every day.

After his initial interview with Col. Ellis, Bill was taken to the MI6 office on the first floor of the annexe and introduced to the staff. There were not many. Neil Whitelaw was Colonel Ellis's deputy, there was another man whose name Bill cannot now remember, and a couple of women. Later, the number of staff increased considerably. Colonel Henderson, Major Thompson, Captain Kane, Jimmy Ware, John Machin and another woman joined the embassy but these staff increases took place over time. Initially, there was some uncertainty as to Bill's exact duties as he had originally come to them from Madrid as a cypher clerk but London said he should assist John Hankey, the wireless operator. As he was an experienced cypher clerk, and the post needed an extra one, he was asked to do both encoding and wireless operating. He much preferred coding as he had been doing this all the time he was in Madrid, whereas he had not carried out operating duties – sending and receiving messages – for about a year, and felt he was a bit rusty. All the time he was in Bilbao he only spent a few seconds 'on the air,' to establish contact with UK; no messages were ever passed.

It was decided therefore, that John Hankey (who was not a codist) should take care of the morning wireless schedules, and copy the evening press broadcast, while Bill was working on the cyphers. In the afternoons, Bill worked the wireless schedules. Apart from the daily schedules with UK there were also had bilateral wireless contacts with Madrid, Lisbon and Gibraltar, and later with a couple of agents in French Morocco. Bill was certainly kept busy, encoding telegrams in the morning then transmitting them in the afternoon. The telegrams received in the afternoons had to be decyphered in the evening and the following morning.

Unlike Bilbao, which was a one-off, Tangier was part of a wireless network that involved UK and some embassies in neutral countries. There were not many stations in the group during 1941/42. All telegrams were encyphered using one-time pads consisting exclusively of figures, as were the operating procedures. All call signs were three-figure groups, the middle figure indicating signal strength. Thus for example, if Tangiers' call sign was 501, the operators at the home station called Tangier '501' until contact was established, at which time they substituted the middle zero for a figure denoting signal strength, on a scale of 1 to 5. If they could hear Tangier at, say, strength 3, they changed the Tangier call sign to 531. Tangier would do the same so they always knew what reception was like at each end.

Everyone at the Tangier consulate-general was kept very busy. There were no such things as set working hours, it was a case of all hands to the pump. Consequently, everyone working in the office was expected to do other jobs.

The first extra job Bill was requested to do was to look after the *Jean Bart*. *Jean Bart*, it turned out, was a French battleship. The *Jean Bart* and the *Richelieu* were the two newest French battleships, and were considered key elements of the French Navy. On the capitulation, both sailed out of St Nazaire; the *Jean Bart* made its way to Casablanca. Although not yet completely seaworthy, its 15" guns were fully operational.

In World War II, a battleship was the ultimate weapon at sea because, even if it never left harbour, it held up ships of the opposing navy, which had to be kept in readiness, in case it ever put to sea. Thus its presence and state of readiness had to be monitored at all times, usually by the RAF. In this case, however, it was not easy to do so, because Casablanca was on the Atlantic coast of North Africa, and thus a considerable distance from the nearest RAF base. However, we had an agent in Casablanca, with whom we were in wireless contact.

The *Jean Bart* was anchored in Casablanca, but at times it moved around in the harbour. When it did so, our agent in Casablanca would send a diagram of the harbour, indicating the exact position of the *Jean Bart*, sometimes on an odd scrap of paper. Whenever the ship moved, he would send another diagram, usually by courier. There were no copying facilities in those days, so Bill was given the job of tracing these diagrams whenever they came in. One of the tracings was sent to Gibraltar, the others to London. There were also some notes with the diagrams, but these were in French so he could not understand them.

Thus, the Admiralty always knew exactly where the *Jean Bart* was, and its state of readiness to put to sea. There was the danger of the ship being taken over by the Germans, even though this would be in contravention of the armistice agreement.

Monitoring the *Jean Bart* was Bill's job right up to the North African landings. When he later visited Casablanca, he was told the *Jean Bart* proved its guns were operational since it engaged in a duel with an American battleship, resulting in the *Jean Bart* being beached.

Outside the consulate-general, the staff had to be very vigilant. Since the Spanish authorities were very pro-German, they had to make sure that they never got into compromising situations. They were under constant surveillance, and although they got used to it, they could never allow themselves to become complacent.

Some time in 1942, another operator joined the consulate-general from Gibraltar Roy Perry, was originally from Great Britain, and he and Bill became great friends. Roy also undertook cyphering duties, after he had been trained, in addition to his wireless work. In fact, it must have been around that time that the staff became so busy that they had to employ a member of the local British community to assist with the cyphers. Her name was 'Teddy' Dunlop, and she was the wife of the local British doctor. This was rather surprising as most locally engaged staff were not allowed anywhere near the cyphers. However, it turned out that she was a member of the local network. In fact, Teddy stayed with them throughout the war, and eventually became Head of Station. Some time after the war, she moved to Lisbon in the same capacity.

Bill was asked to perform another unusual job. Some time after his arrival in Tangier, Neil ('Willy') Whitelaw, asked if he would be willing to assist with the 'boat operations'. Owing to the 'need to know' requirement, only three or four of the consulate-general staff were involved in these operations, and they had to swear that would never discuss or mention them, in or out of the office. Willy lived in a house owned by Admiral Gaunt (who was in Great Britain at the time). This house was situated at the end of a long, narrow road, that lead from a square on the outskirts of the town to rocks high above the sea shore where it ended. There was no beach at this point, so the spot was never used by the public. It was from Gaunt House that the 'boat operations' were conducted. They consisted of a clandestine boat service, operating between Gibraltar and Tangier, using a fishing vessel. All the runs were at night. The fishing boat sailed from Gibraltar, after being loaded with the appropriate cargo, to a designated section of the shore situated at the end of the garden of Gaunt House, and some adjacent properties that had been acquired by the British.

The fishing boat waited off shore, while a rowing-boat was despatched from the boat to a prearranged point on the rocky coast, where a reception party waited. This was a difficult operation, because there was no communication with the fishing boat; exact time of arrival was very uncertain, and invariably a long wait was

Gran Café de Paris in Tangier, a gathering place for all nationalities and where German, Italian and allied intelligence services often enjoyed meals – at adjacent tables!

involved. Sometimes the operation had to be aborted for various reasons – if the sea was too rough for the rowing boat to approach the shore, for instance, or if there were Spanish patrol boats in the vicinity.

In addition to those in the Passport Control Office, a few members of the local networks formed the team involved in these operations. One was a Monsieur Bals, a member of the French resistance, who owned a bar-restaurant opposite the Annexe. Another was George Greaves, the Daily Express correspondent.

Unloading was not easy in the choppy waters. All sorts of boxes were unloaded and Bill remembered one case which was dropped into the sea, since it contained a suitcase wireless receiver/ transmitter, which they had to dry out. On one occasion, there were several cans of petrol roped together. Another time, a number of vehicle tyres, also roped together. No one was sure who the tyres were for but thought they might be for the local buses, because Colonel Ellis owned the bus company. Once or twice they sent passengers across to Gibraltar in the boats. They were refugees and others who for some reason, had to leave Tangier. There were also some agents ultimately destined for Great Britain or France via Gibraltar.

Among the other duties Bill was called upon to do was the training of a couple of agents in the use of suitcase wireless receiver/transmitters. Teaching the operation of the wireless set was fairly simple, but, teaching the Morse code was a different matter. One such agent was a Spaniard, codenamed BODO. He was a lighthouse-keeper, due to work in Tangiers should the need arise. Training took place in Gaunt House, but he could not attend regularly. Technically, he was quite good, but Morse was a slow process. He must have done a lot of homework, as in the end he did not do too badly, but, of course, he was very slow.

Another agent, was from French Morocco. He was a Moslem, and spoke only Arabic and French, which was a bit of a problem, as Bill's French was poor. As with BODO, use of the equipment was easy enough, but again Morse was difficult. They used to meet at the local tennis club, where the manager was a member of their network. It was very hard-going, training this agent.

Towards the end of 1942, Bill and his fellow workers were instructed to maintain continuous radio watch with Great Britain for a few nights. They would receive a one-word telegram which it had to be delivered immediately to the Minister, Mr. Avery Gascoigne, no matter what time of night it arrived. Bill was on duty when the telegram arrived, and it indeed consisted of one word. Bill thought the word was 'Torch'. He actually had to query the word, as he had never before received a plain language telegram. A check revealed that 'Torch' was correct. It was around midnight or the very early hours of the morning, and he had to go through to the Minister's bedroom (the wireless room was in the Residency), and bang loudly on the door to wake him. Bill gave him the telegram when he came to the door. He immediately knew what it meant, dressed quickly, dashed out to his car and drove off.

The next morning, we learned that Allied troops had invaded French Morocco and our Minister had to go immediately to the Spanish High Commissioner in Tangier to assure him that the Allies had no designs on Spanish Morocco, and would observe and respect Spain's neutrality. This was great news for the wireless room staff as all the intelligence work they had been doing had come to fruition.

One of the first things to happen after Casablanca was occupied, was to re-open the British Consulate in Morocco. As no King's Messenger Service had yet been organised for Casablanca, the diplomatic bags were sent to Gibraltar, and thence to Tangier and Bill and Roy then took it in turns to act as couriers to Casablanca.

They travelled by Jeep all through Spanish Morocco with a driver and escort. Roy went first and Bill made the second trip. On entering French Morocco, they were stopped by a black American soldier who told them he had never seen an Englishman before. In Casablanca, he had the pleasure of meeting the agent, whom he had been working to by wireless, and who had been the man who sent the diagrams of the *Jean Bart*.

In November, 1943, Bill's mother died, and he was granted ten days compassionate leave. There was, of course, no commercial transport that he could use to travel home, so the consulate-general made an arrangement for him to travel on an RAF flight from Gibraltar. Bill crossed over to Spain in transit to Gibraltar on 16 November, 1943. He had to wait several days for a plane, eventually leaving Gibraltar on 22 November 1943, arriving at Bristol the following day. The plane was a Warwick, and as he was the only civilian on board, the passengers gave him several strange looks, as they were all serving officers. All wore heavy overcoats or flying suits, but all Bill had were his civilian clothes, plus a pair of flying boots which he borrowed. They travelled all night and, sitting at the rear of the aircraft, he was nearly frozen stiff. It was a long flight and sadly he arrived too late for the funeral.

On his return to Tangier, he had to obtain an exit permit to leave the UK, which took time. For the return flight, also on a Warwick, Bill was given a flying suit, some sandwiches and a flask. He eventually returned to Tangier on 23 December, just in time for all the Christmas parties.

In the summer of 1943, another event occurred that changed Bill's life. He and Roy were walking down to the Old City and had to cross the 'Socco Grande'. This was a large open square, where the market was held. On market days, there were story-tellers, snake-charmers and, of course, every kind of local merchandise but on this particular day, the square was empty.

Bill glanced across to the far side of the square to the German Legation, to see the swastika flag flying. It was then that Bill glanced across to a café at the entrance to the market place, which had an adjoining tea-room. Although they knew most of the bars and cafés in Tangier, they had never been in that one, one reason being that it was frequented by Germans from the legation opposite. The café had tables with built-in chessboards which seemed popular. As for the tea-rooms, what red-blooded male in his early twenties was really interested in tea rooms!

It was at that moment that Bill was stopped in his tracks. He felt he had never seen such a beautiful girl as the one who was then glancing through the window. He pointed her out to Roy and suggested they go inside. Roy demurred, indicating that the place was full of Germans. On their return, however, the pair of them went into the tea-rooms. Sure enough, there were a couple of Germans having coffee and cakes. They looked at one another, each knowing who the other was, but then ignored each other.

Bill's eyes and thoughts were on the girl. She was well-dressed and extremely beautiful. There was a cashier at the entrance to the tea-rooms, and the girl and the cashier were talking in French. Bill was disappointed as it seemed to be a French-speaking establishment and although he could speak a little French, he was more fluent in Spanish. She came over to their table and asked what we wanted, in French. As the Germans were speaking loudly in German, he thought he would do the same, and ordered something in English.

Naturally, he had been warned to be very careful about talking to women. It is oldest trick in the world of intelligence to ensnare men with pretty women. So Bill and Roy sat in the tea-room they tried to make out her nationality. Whatever their nationality, people in Tangier were either strongly pro-Allies, or pro-German.

Bill began going to the tea-rooms regularly, often with friends from the office. At various times, they heard her speaking, French, Spanish and Italian. Whoever she was, for Bill, it was a case of love at first sight. Eventually, he asked a friend who she was and was told, 'Oh, that's Ramona'. The friend added that she was Spanish, and a member of the pro-Allies Spanish community. Bill asked to be introduced and, to cut a long story short, eventually they were married.

Although throughout the war, there was never any open unpleasantness between the representatives of the nations at war in Tangier, there was one most exceptional incident. The Germans were monitoring and reporting the passage of British convoys through the Straits of Gibraltar. These convoys were vital as Malta and the war in the North African Desert depended on the supplies they carried. The Germans had acquired two or three properties on the coast of Spain in the vicinity of Algeciras, facing Gibraltar. It was from these buildings that the convoys were being detected. In daylight, ships could easily be observed, but although the convoys passed through the Straits at night, the Germans were still able to detect our ships. Although the truth was not known in Tangier, Bletchley Park had decrypted messages that revealed that the Germans were using infra-red equipment operated from these properties to monitor the shipping. It seemed that they could even detect the class of ship from the heat radiated.

At the same time, the Germans had acquired a house in Tangier, on the opposite side of the Straits, for the same purpose. From sources within the local British network, the British had found out that the Germans were installing what was thought to be wireless equipment. In actual fact, it was infra-red equipment. The house was situated on top of some cliffs overlooking the Straits. From this vantage point, ships could easily be monitored.

It was suggested that the Royal Navy shell the properties in Spain, but the British ambassador in Madrid vetoed the idea, as it might have given Spain an excuse to enter the war. The British and the Americans made strong representations to Franco to order these monitoring stations be closed down. It was not certain whether or not Franco actually knew about them. The existence of these stations was a breach of international law and of Spain's neutrality.

Franco depended upon America for oil and grain, so he decided to accede to the Allied requests and ordered the closure of these stations. For some reason, however, the German house in Tangier required a different approach. London decided that direct action should be taken against it.

One of the local British spy networks had been requested to survey the house. They reported that, owing to its vulnerable location, a suitably placed explosive charge could destroy it. Some time prior to all this, a new man had arrived in Tangier from Gibraltar. His name was Wharton Tigar, although no one in the office knew it at the time, he was a member of SOE and became the organisation's representative in Tangier. Tigar set up his office in the room next to the SIS wireless room. Bill and his colleagues had been eyeing this room themselves for use as their own wireless station, and had received permission to move, but Wharton Tigar beat them to it. The SOE keep themselves to themselves,, and although Tigar met the other office staff many times they never knew what he was up to. He acquired an assistant in his office, a local British businessman. The office knew this man very well, but they had had no idea that he was an SOE agent. Nor did they know at the time that plans were being to plant a bomb at the house.

One day, all the staff were instructed to make sure that they were seen in public, for the next couple of nights or so. This was no problem, as after dinner in a restaurant, they could end up at one of the many nightclubs.

On this particular night, Bill was in a nightclub – The Lido – with friends. The first thing he and his knew of anything being wrong was when they left the club in the early hours of the morning. There were police and soldiers at the door, checking the identities of everybody as they left. The next morning, they learned that the German house had been blown up and had collapsed into the sea below.

The Germans were furious and immediately accused the 'English' of bringing bombs over from Gibraltar in their diplomatic bags. The Spanish police rounded up several suspects, including some members of the local network, but everybody had an alibi. They were told to be on their guard, however, as the Germans certainly would not take this lying down. True enough, shortly afterwards, the Germans took their revenge.

The Gibraltar–Tangier ferry service was served by two vessels, a salvage vessel called the *Rescue* and, when it was not available, a much smaller vessel called the *Alert.* They invariably carried a number of diplomatic bags and mail bags for the British Post Office. Unclassified diplomatic bags travelled without a courier and were met by elderly member of the consulate staff named Da Silva, who had been hired locally. He also manned the

A pause on the courier run between Tangier and Casablanca in 1943

consulate's telephone exchange. Confidential diplomatic bags were always carried by a courier who would take them directly to the Consulate upon arrival, leaving Da Silva to deal with the rest.

One day, the unclassified bags were lined up alongside the taxi-rank as usual and Da Silva was arranging for transport to the consulate, when there was a terrific explosion. Da Silva was killed, as were five Moslem taxi-drivers. Colonel Ellis was certain that the Germans had planted a bomb on the quayside. The Germans claimed, however, that this was proof that the British were bringing bombs over in the diplomatic bags and took full advantage of the propaganda opportunity to stir up the local Moorish population against the British. The dead taxi-drivers were buried the next day. The Moslem Cemetery was quite close to the Annexe, where the staff were all assembled with all the window shutters closed and the outer gates locked and bolted. The funeral cortège passed the Annexe whilst the staff watched through the shutters. There seemed to be hundreds of mourners chanting in Arabic. The bodies were shouldered on litters.

After the burial, a mob surrounded the Annexe, shouting and throwing stones. There was no sign of any police or soldiers and Colonel Ellis had ordered Molotov cocktails to be prepared. These were lined up on the floor, under the veranda shutters. The staff were ordered to use them should any attempt be made to enter the building. Looking back, it was a mad idea as these petrol-bombs could easily have set the building alight.

The military eventually arrived and cleared the crowd away. It was quite late when the embassy staff ventured to leave the building, but there were still plenty of police around. They had to be very careful, as the Germans were claiming the staff of the consulate-general were responsible for the death of the taxi-drivers. A source at the Spanish High Commission reported that a list of suspects was being prepared and people might be hauled in for questioning. Colonel Ellis headed the list – and Bill was on it as well.

Colonel Ellis took up residence in the consulate residence, which had diplomatic immunity. Bill was sent to Gibraltar on the next courier run. He returned the following week by which time things had quietened down and Colonel Ellis was back in his own home.

The Head of SIS outside Great Britain was invariably the Chief Passport Officer. The Passport Control Office was part of the Embassy, although not always physically attached to it. All MI6 staff had a number, the first two digits of which would be that of the host country. The head of MI6 say in France, would have been 45000, in Sweden 36000. In Sweden, for example, the CPO's number would have been 36000, his deputy 36100. Office staff started at 900, wireless operators at 950. Thus in Stockholm, the first Section VIII wireless operator would be 36950, and the second 36951. In correspondence, internal minutes etc., names were never mentioned, only the individual's number. Tangier was 56 land and when Bill moved there he became 56952. All communications with London, were addressed to 'X W' and went through Main Line at Whaddon Hall. Apart from his breakfast, Bill usually had meals with a colleague or colleagues from the office. For dinner, they usually frequented various cafés or restaurants. For some time, four of the consulate staff would lunch at an

Italian restaurant called The Esquardilla. Even though the Italians were on the Axis side, it was a very popular restaurant, mainly because it sold books of seven meal vouchers for the price of six. So consulate staff ate there regularly for about two or three months. This was the time of the Desert War. The British always sat at the same table; four Germans from their Legation also lunched there regularly, usually at the next table. There was never any unpleasantness, and the only sign of anything out of the ordinary, was that when a British staff member had good news of the Desert War or any other victory, a bottle of wine with a 'label' on it was ordered, as opposed to the carafe of wine supplied with the meal. The British then drank a little toast among themselves.

Oddly enough, the Germans did the same whenever they appeared to have some good news. Looking back, Bill thinks it might seem rather childish. Only a relatively short distance away British and Germans were killing one another with impunity.

As soon as the war ended, Bill applied to XW for permission to marry Ramona. This was necessary because she was a foreign national. Her life history had already been checked, and of course all those at the Passport Control Office and the consulate knew her. They supported his application to marry and permission was granted. They were married on 11th August – first at the British Consulate-General Tangier at 11 am by the Consul, Mr. Dundas. After the ceremony, the Consul's wife produced champagne, and the Mr. Dundas presented Ramona with her new British passport. Then, at 5 pm they had a second wedding at the French Church – Jeanne D'Arc.

In March 1946, Bill had a letter from Head Office, telling him to return to Britain in order to go through the official demobilisation procedure. When he had handed in his army papers and uniform at Bletchley Park in 1940, he had wondered about it at the time as it seemed an unusual way to leave the army – which he genuinely thought he had. After all, he was a civilian, with a passport, and on the payroll of MI6. As far as he was aware, he had had nothing whatsoever to do with the army for years.

Ramona and Bill flew to Lisbon, where they were met by a young woman from the local MI6 office. Hotel accommodation was arranged for them and an onward flight to Northolt. Bill had to report to a demob centre at Guildford. It was all rather complicated. Like every man being demobbed. Bill was put into Z reserve, which meant he had to have a uniform. As he did not own one, they had to first issue him with a uniform, and then a complete civilian outfit. Bill received several odd glances at the demob centre, as he was the only one dressed in civilian clothes!

Bill found things had changed quite a lot since he had left Bletchley Park in February, 1941. He was now attached to a department based at Hanslope Park which he had not seen before. After about six weeks in Britain, the couple returned to Tangier in May via Lisbon and found the office was still quite busy.

Roy Perry was relieved shortly afterwards by Bob Cree, who arrived with his wife and baby son. Bill bought his first car, a brand-new Austin 8, in 1947. Bill stayed on until he was finally relieved in 1950. He and his wife drove home by car. One thing Bill has never understood, and has never found out, was why he was kept in Tangiers for so long. Other operators went out to Tangier, did a couple of years or so, and returned home, but some reason that never happened to Bill.

George Mooney took over from Bill in 1950 and Bob Cree took charge of the station. It took Bill and Ramona and their baby son a couple of weeks to get back to UK as they travelled at a leisurely pace. It was extremely cold, especially crossing the mountains – they were held up in Madrid as the mountain passes were blocked by snow – and they had no heater in the car.

When Bill telephoned to report his arrival in the UK, he was informed that Tangier station had closed just after he had left. He continued to work as an wireless operator but now with the newly formed Diplomatic Wireless Service.

Chapter 29

Phil Luck, Operating with Black Propaganda

Chapter 18 described the very successful Black Propaganda operation which was carefully monitored by Gambier-Parry who took a great interest every aspect of the work. This is the story of Phillip Luck, one of the engineers supplied by Section VIII, based at a Black Propaganda station.

Phil was born in 1925 in Northampton and educated locally. His close friend, Peter Hill, introduced him to wireless as a hobby, and they built battery-operated portable wireless sets in a shed in the garden. In 1941, they answered a BBC advertisement for personnel and after a while, they were interviewed at the Overseas Broadcasting Station at Daventry by its engineer in charge, Mr. Burkinshaw.

Phil was surprised to be offered a post at Daventry, whereas his friend Peter, who was probably more qualified, was not. However, they discovered that Peter had a cousin with 'connections' at the Foreign Office and wanted to introduce him into SCU1. Shortly afterwards Peter was employed at Gawcott.

In 1941, Phil began working for the BBC and attended the training school at Maida Vale Studios in London, obtaining his City and Guilds Parts 1 & 2 qualifications in Telecommunications. He then went to work at the BBC's Borough Hill, Daventry transmitter station, operating and maintaining the high-powered radio transmitters of the Overseas Service of the BBC.

At first, Phil worked on 'Sender 3', an interesting transmitter built by Metropolitan Vickers which, in its final stage, had valves that could be disassembled long before the transistor had been invented. This meant the valve could be taken apart, a new filament or grid installed, the valve reassembled and the air removed with a vacuum pump. The valve would then be good as new and the transmitter would be back 'on the air'. The whole operation usually took six or seven hours and was always performed on the night shift. Later, Phil worked on numbers 9 & 10, the latest Marconi equipment with water-cooled, final stage valves. He put 110kW into the antenna arrays hanging from the 500-feet high steel masts.

Later, SCU1 needed a replacement for a man who had unfortunately been killed at the Potsgrove station due to an equipment malfunction. Peter mentioned Phil's name and gave brief resumé of his experience to date. He was interviewed by telephone by Mr. T. Pryke who was the Engineer-in-charge (EIC) of the Gawcott transmitter station, and told 'I will be in touch again'. Shortly afterwards, Phil was Bletchley bound!

He remembers waiting outside Bletchley railway station on a bright and sunny day in December 1943, having been told he would be met by a Lieutenant Fuller. The road outside the station was a cul-de-sac which was full with uniformed personnel coming and going, and there were a large number of vehicles so Phil had no idea how he and Lieutenant Fuller would make contact as they had never seen each other before. The meeting time had been set for 13.30, and as he waited on the kerbside service and civilian people approached him with questions about being met and about transport.

[These famous words: *'Travel to Bletchley station, where you will be met'* were said to hundreds and hundreds of new recruits heading for BP or the various Section VIII units in the district. At times, however it left many as 'waifs and strays' lingering forlornly at Bletchley railway station!]

Eventually, at around 15.30, Lieutenant Fuller arrived and introduced himself and they drove off in a camouflaged Hillman Minx car. They travelled to Buckingham, and then to Gawcott, to premises in a field approached by a gravel driveway from a country lane. The buildings were surrounded by four-foot-high barbed wire fences. When Phil left the car to open a five-barred gate, he noticed that the top strand had insulators between it and the wooden posts. Electrified fences? He had never seen anything like this before but the next four years were to be full of many more new experiences.

Lodgings were found for Phil in Gawcott village with a nice elderly couple, the husband being a retired carpenter and undertaker. Their home was comfortable but, as was quite normal at that time in the country, it had no mains electricity. They had oil lamps for lights and coal fires for heating. Fortunately, there was hot and cold running water at the wireless station, and most of the men shaved there.

There were two transmitters, both manufactured by RCA (Radio Corporation of America). They were housed in two buildings, one of brick and one of wood that stood about 300 yards apart with a small brick building midway between them. The central building was where the incoming telephone lines terminated, and also served as an office for Lieutenant Fuller. He was not permanently stationed on this site because he was also responsible for the Potsgrove station. When Fuller went on leave, Second-lieutenant Ainsley was in charge.

Near to each transmitter, there was a brick powerhouse containing a diesel-driven generator. One was a large single cylinder Rushton with a flywheel eight feet in diameter. Since it was a single cylinder machine, the flywheel needed to be set in a certain position after use, with a crowbar, ready for starting next time. This was done with the aid of compressed air from a steel pressure vessel which had a small petrol-driven compressor to replenish the air supply. The other generator was powered by a six-cylinder Crossley diesel. It was also started by air with a similar system installed to replenish the air-bottle. Both machines were water-cooled. The maintenance on these diesels was carried out by a three-man crew, lead by Sergeant Bill Heatley.

The wooden transmitter hall was known as 'Geranium' and the brick building as 'Gardenia'; Phil worked mostly at Geranium. Apart from the materials of which the buildings were constructed, they were identical in layout. Both had the normal wartime brick-built blast walls surrounding them. After entering through the sliding door you were in a cloakroom-cum-locker-room which led via a sliding door into the operations room which contained a desk complete with a control console and two record turntables.

Facing this position was the front panel of the RCA 7.5 kW air-cooled transmitter, beautifully finished in two shades of grey enamel, with stainless steel horizontal strips running across the whole front. At the right hand side of this was a sliding access door leading into the 'works', although limited access could be gained to some units through two hinged doors on the front panel. All the doors had safety interlocks that would cut off all power and automatically ground the power-pack condensers for the protection of personnel. It was alleged to be the failure of this system at 'Poppy' which resulted in the unfortunate death of a young engineer named Locke. A manual earthing-rod was provided as back-up and Phil was careful to use this, even if the automatic system indicated 'Safe'.

On one side of the control room, there were a kitchen and toilet, and on the opposite side a workshop equipped with hand-tools with a bed folded into the wall for use if the station was cut off by snow drifts or for other reasons. The last room housed a large oil-cooled transformer. where the incoming high-voltage electricity supply was stepped down to 220 volts and an auto-transformer dropped this to 110 volts, the standard for U.S. wireless equipment.

The 'crew' at Gawcott. Phil is on the right with Tommy Pryke, the engineer in charge, standing beside him.

From the first day, wearing his civilian clothes, Phil was under instruction at Geranium, working on the day shift from 0800-1700, understudying the regular shift operator. The shifts ran 1700-2300 and 2300- 0800, in addition to the daytime shift, with two days off a fortnight.

The operating staff were a varied collection. Phil at nearly 19, was the youngest and the oldest members were in their early fifties. Some were civilians and permanent employees of the GPO, others were RAF on secondment and the others SCU1. Sergeant Major Tommy Pryke ('call me *Mr.* Pryke please'), was responsible for the running of the station, assisted by his deputy, Staff-sergeant Russell 'Rusty' Coleman.

Geranium had 'Tony' Anthony and Jack Fiefer, ex-GPO men wearing the uniform of the Royal Observer Corps, as well as Fred Higgs from the RAF and Charles Tucker and 'Ashie' Ashworth who, like Phil, were in army uniform. All worked under the direction of SCU1.

Gardenia was staffed by Reg. Turner ex-GPO (an ex-merchant navy warrant officer), Harry Oakley RAF, 'Butch' Butcher and Peter Hill from SCU1.

Also on the site, and working the same shifts, were members of the local Home Guard – Mark Bennett, Jim Prickett (who was also landlord of The New Inn in Gawcott), Leonard Jerrams and the Thorpe brothers, Fred and Bert, always known as Darkie and Ginger. Darkie was the carpenter/maintenance man and the others were employed keeping the buildings clean and the fences in order.

The transmitting station at Gawcott

Peter Hill at the control desk – note the large turntable on the left

On his birthday in February, Phil attended Whaddon Hall for enlistment. He was 'sworn in' holding a Bible, and given one shilling, plus army number 2602925 with rank of Signalman and an Army Pay Book, AB64 – intended as a record of army service. Then he was sent to the Quartermaster's Stores at Little Horwood where he received two battledress uniforms, shirts, trousers, socks, boots, greatcoat and a forage cap (surely, Phil thinks, the most stupid headgear ever devised), together with Royal Corps of Signals shoulder flashes.

His AB64 was endorsed with the words 'Special Enlistment 2 Special Operations'. No military training was given and he was nervous that he would fail to respond in the correct manner if challenged by a military policeman or an officer. This had obviously happened previously because he was later provided with a telephone number for just such an emergency. It was a Whitehall number, linked to SIS HQ at 54, Broadway in London. Personnel in his position were warned not to avail themselves of this facility unless absolutely necessary.

Phil was soon taking his place as a regular shift watchkeeper at 'Geranium'. Consignments of large shellac discs were delivered twice a day by civilian drivers, usually in a black Hillman Minx car. These records were made in the studios at Milton Bryant code-named Simpson, by teams of foreign nationals and even some captured U-Boat Commanders who had been 'turned' by skilful interrogators into becoming traitors. These people all worked under the strict supervision of a specialist language editor from scripts prepared at the Foreign Office.

The editor sat with his headphones on, the script in front of him, and his finger holding down a switch. Whilst this was depressed, the recording machine functioned, but one wrong word, deliberate or accidental, and he would release the switch. This whole operation was the brainchild of Sefton Delmer, the journalist.

The programmes covered the following languages: German, French, Polish, Czech, Slovak, Norwegian, Danish, Dutch and Flemish. Unlike the BBC which strived to be truthful at all times, these broadcasts had no such reputation. The sole interest of Black Propaganda was in furthering the Allied cause by sowing alarm and despondency amongst the enemy. Black Propaganda was deemed to justify the deception, whatever the cost.

Upon the arrival of the car with its cargo of recordings, the civilian driver would be allowed into the outer cloakroom but no further, and the sliding door to the control room had to remain closed. The box containing the records would be signed for and the previous day's box handed to the driver, for return to Simpson. Each disc had a numbered card with details of the frequency of transmitter, antenna array and the time (or times) at which the programme had to be broadcast.

The discs and cards were marked with numbers such as F1, F2, F3 & F4, a series of G's 1 to 10, then P's, Y's, H's and W's. Some were only transmitted for a few days while others had regular daily schedules for months and even years. The F's were directed at France, G3 was for the German Afrika Corps in Tunisia and G9 was for Occupied Europe, the Atlantic area and Norway/Denmark.

The first task for the operator was to sort the cards into transmission times and then prepare the discs. These were very thin shellac discs about 18 inches in diameter, and were played at 33⅓rd rpm, from the centre outwards towards the rim. A mark had to be made with a wax crayon in the centre, near where the grooves began and the pick-up arm stylus was positioned on this starting point The operator then ran the turntable, listening on headphones, and counting the number of revolutions from the start mark to the first word of the programme. This number was written on the disc. The object of this care with the timing was to simulate a live broadcast, by fading in the programme using the control on the console, at the exact start of the speech or music, without any hiss caused by the stylus running in an unmodulated groove. The control desk was provided with two turntables for lengthy speeches, or so that the output from one turntable could be broadcast while the other discs on the adjacent equipment were being prepared for transmission.

The transmitter was always warmed up 10 minutes before air time by applying power to all units except the final one, then two minutes to transmission time full power was applied, and the RF stages tuned. Then the operator went back to the control desk, started the turntable and fed in the audio, exactly as the second hand on the chronometer reached the twelve. To monitor the programme for any distortion, there was a Hallicrafters Communication Receiver which also received the BBC 9 o'clock news and even the popular BBC radio comedy programme *It's That Man Again (ITMA),* when time permitted.

The various language discs contained information for broadcasting to resistance groups in occupied Europe and morale-building speeches from their leaders-in-exile in Great Britain. Twice a week, the P1 contained a Roman Catholic church service. Suggested sabotage methods and targets were also included, including non-violent ideas such as frequent absence from the workplace to visit the toilets so as to cut productivity. Anything judged to disrupt the German war effort was acceptable, it did not have to be a spectacular affair like a train derailment.

The station broadcast mainly in the evening and at night, due to movements of the Heavyside and Appleton layers of the ionosphere. This enabled the skip distances to be adjusted to give maximum strength signals into the areas of Europe being serviced. The day shift was therefore engaged mainly on maintenance tasks, valve changes, cleaning of insulators with methylated spirits, running the diesel generators on a bi-weekly basis and repairing any storm damage to the aerial systems. Fences damaged by cattle, owned by the land owner, Mr. Benny Warr, also needed frequent repair. Phil soon realised that it was to prevent damage from cattle that the fencing was electrified and, not as he had first assumed, to deter the intrusion of humans!

On one occasion, Gawcott was visited by two RAF officers who came to check the mast heights. They had recently moved into the aerodrome at Finmere with a squadron of Mitchell light bombers, and needed to brief their crews on the location of local hazards, as no warning lights were displayed on tall structures during wartime.

The secret nature of the operations had been so successfully instilled into the group, that these quite official visitors to the station were treated like enemy aliens. They were made to remain outdoors whilst Tommy Pryke was summoned by bicycle from Gardenia. It was decided that he alone could decide if their questions could be answered without contravention of the Official Secrets Act. The information was finally imparted and the two gentlemen were then escorted to the public road – as if they were plague-carriers! Contact from the outside world was clearly discouraged.

After several months of busy routine, Phil was transferred to Potsgrove, virtually a replica of Gawcott, located just off the Woburn-Hockcliffe road, two miles south of Woburn and, again, very isolated. He lived inside the grounds of Woburn Abbey at the home of the Duke of Bedford's farm manager, where the garden centre stands today. His task at Gawcott was to help in the preparation of a mobile transmitter and audio studio, for when the Second Front was established. The equipment, of pre-war Italian manufacture, had been captured by the British Eighth Army during the campaign in North Africa. It was to be installed in four large British Army trucks to form a completely self-contained unit.

At the time, Phil did not know what it was for, but it is now known to have been a medium-wave station that was to follow the invading armies through France and into Germany. It was expected that the retreating German army would destroy local wireless stations as they went and that is exactly what happened. The mobile unit would allow 'local' broadcasting to continue, in addition to the transmissions made via Crowborough, Gawcott and Potsgrove.

The trucks were for the transmitter, audio studio, diesel generator set and crew accommodation, with the antennae was housed in a trailer. The crew would need to be very versatile, capable of driving heavy lorrries, operating and repairing all the equipment. Knowing how to erect a circus 'big top' would be a definite advantage if they were to deal with putting up the antennae.

The Engineer-in-charge of Potsgrove's 'Poppy' and 'Pansy' transmitters, was Norman Bowden. Phil's colleagues included Jack Morton and Percy Jones of SCU1 and Bill Womack from the GPO (supposedly in the Royal Observer Corps). Les Button and Bert Weatherhead came from the RAF. Again, all these men, from different military backgrounds, were under orders solely from Section VIII.

After several weeks of intense activity, the job of building the mobile unit was completed but such was the rapid advance of wireless technology that it was never used for its intended purpose. Phil returned to Gawcott where he found big changes. The Geranium transmitter was no longer using records but was linked by landline direct to the studios and broadcasting a 'German Forces Programme' called 'Soldaten Sender Calais'.

This pretended to be located in Calais in Northern France, but after the liberation of Calais the call-sign was changed to 'Soldaten Sender West'. The broadcasts were clearly aimed at all branches of the German military with segments addressing the Luftwaffe, Kreigsmarine and units of the SS and Wehrmacht.

The bulletins did not contain good news for the listeners in the German forces; they reported recent German losses of planes and ships, including U-Boats, tanks, men and territory, and described the serious situation on the home front. There were also reports of the damage caused by the previous night's bombing raids by the RAF on major cities in the Reich. Individual factories would be named in the news bulletins, and street names were listed, naming the shops and house numbers destroyed or damaged, estimated casualties and so on.

At the time, Phil wondered how this was possible, and thought that if this were just guesswork then the credibility of the broadcasts would be short-lived. Phil knows now that all the information came direct from the German authorities' damage reports to their own HQ's sent in code on their Enigma system. This was being intercepted, of course, by the Y service (or by SCU3 at Hanslope), decyphered by Bletchley Park, then made available via PWE at Gawcott!

The news bulletins were then formulated (with the aid of RAF photographic reconnaissance pictures), by a team of psychological warfare experts, into the reports broadcast to the German armed forces to sap their morale and cause maximum distress. The accuracy of the stations information was frequently confirmed by listening German service personnel, by letters from their homes in Germany and telephone calls to their families. All this added to the authenticity of the broadcasts and increased the listening base.

The news slots were strung together with hours of the best music broadcast by either side during the war. Every taste was catered for but mostly it consisted of the Big Band music of Glenn Miller, Kay Kaiser, Tommy Dorsey, Joe Loss and Geraldo. It even featured a regular spot with a German Forces sweetheart (a Tokyo Rose in reverse), with her signature tune of Lili Marlene, and the Andrews Sisters further contributed to this honeyed, seductive flood of sound.

Then, just as the German serviceman was enjoying the programme, would come the news that his home town had been blitzed the previous night and the allied advances on both East and West fronts were continuing apace. The German listeners were constantly reminded of some of their leaders' broken promises, especially the boast of Hermann Goering: 'No enemy aircraft will ever fly over Germany, or you can call me Meyer!' After a long damage report, the announcer would ask '*Wo ist Herr Meyer heute*?' ('Where is Mr. Meyer today?'). This programme was transmitted from Poppy at Potsgrove, and also by the huge, high-powered transmitter called Aspidistra at Crowborough, which had been built underground by Section VIII.

In June 1944, it was decided to move all personnel from their scattered civilian billets into a new purpose-built camp at Tattenhoe, on the road from Buckingham to Bletchley, about a mile and a half from Bletchley Park. This was an unusual experiment, because although the new camp housed various ranks of all branches of HM forces, plus a number of civilians, rank meant nothing here. Everyone ate in the same dining-room, and not in segregated messes as is usual in military camps.

The accommodation was fairly basic. Each hut had a central corridor with cubicles on either side measuring about ten feet square. There were solid walls on three sides and the entrance was covered by a black curtain. The furniture consisted of a single iron bedstead with a straw-filled mattress, pillow and blankets; later on, sheets and pillow-cases were issued. A wardrobe completed the furnishings. The heating was by electric tubular strips but was inadequate and this was later changed to a steam system fed from a large centrally-located solid fuel boiler.

Phil's previous life of virtually solitary existence was transformed. He was suddenly pitched into a social whirl when off duty. There was a bar, a games room and a quiet/reading room well supplied with books and the daily newspapers. All this was within walking distance of two cinemas and Bletchley railway station for when the seven-day leave period came around every three months. There was a noticeable improvement in morale as a result of the move.

Personnel from many SCU stations were housed at Tattenhoe, but such was the secrecy instilled into all concerned that none of them ever knew what the others were doing, nor did anyone volunteer information concerning their activities. Phil says his mother, father and brother never knew what he was doing.

With the installation of the landlines, work in the transmitter hall at Geranium had lost the previous busy period of record preparation, the many changes of frequency and antennae couplings. The work now consisted of monitoring the outgoing programme for any distortion, and making certain that every available kilowatt of power was reaching the antenna array by retuning to correct for any slight frequency drift.

The Gardenia transmitter was now employed on other work. A new process was installed as the station switched to forwarding Ultra traffic directly to SHAEF (Supreme Headquarters Allied Expeditionary Forces). It consisted of a machine fed by a punched tape which ran at high speed and was transmitted in the normal amplitude modulated way. The tape arrived neatly wound on reels but as it was run through the machine it spewed onto the floor and needed to be constantly cleaned up and disposed of before it took over all the available space.

The tape transmission itself soon became obsolete. The addition of new filter equipment to the incoming landlines meant the machine could be coupled directly to the lines. The machine, called the Hellschreiber, transmitted at a fantastic speed and its output was read by another machine which printed characters onto a paper tape with a scrolling motion. At the time, it was at the cutting edge of technology.

VE Day, 6 May, 1945 arrived with great celebrations. Phil remembers he was off duty that day and invited to go to Little Horwood, lured by talk of a big feast with free beer, A group of the men, including Bill Heatley and Sergeant Steadman, set out from Tattenhoe only to discover upon their arrival that only a few tired sandwiches remained and all the beer had gone!

The defeat of the Axis powers in Europe silenced Soldaten Sender West which left only one of the four transmitters gainfully employed as the traffic for SHAEF continued. People were now talking of demobilisation but the Japanese remained to be defeated.

Every serviceman's nightmare at the time was to be posted to the Far East. They had all read and seen newsreels of the war in the Pacific, of the fanatical defence of those remote islands by the Japanese and the high cost in Allied lives in this theatre. Accepting the moral aspects involved, Phil regarded the atomic bomb as a miracle, which may have saved a million casualties.

The idle transmitters were put on a 'care and maintenance' basis while someone tried to devise a use for the organisation. First, they said goodbye to all the GPO men, who returned from whence they came and resumed their civilian employment.

Eventually, Phil was 'demobilised' – as far as he could be from non-military employment – and he returned to civilian life. He changed employment completely, never again working in the wireless world.

Chapter 30

Tom Kennerley, Training Agents in China

Soon after I started writing this book, I was told of a meeting of Diplomatic Wireless Service (DWS) pensioners that was held every quarter at the Conservative Club, Stony Stratford. David White kindly invited me to attend and I was hugely pleased to meet several members of wartime SCU units.

They had gone on to work for the DWS, and included several friends from that time, amongst them Ken 'Zook' Howarth and Tom Kennerley. Sadly, both have since died. Tom agreed to write his story for the book but unfortunately did not finish it. I have pieced together this chapter from the information he gave me, from our conversations at the DWS meetings, from my fond memory of him and from personal knowledge of the events he describes.

Tom was a much travelled and experienced wireless operator when he left Singapore for the UK in 1947 and it is a great pity I cannot relate more of his undoubtedly interesting story.

After undergoing initial military training at Markeaton Park, Derby, Tom was posted to the 5th OTB – Operators Training Battalion – at Slaithwaite, near Huddersfield. It was from here, after only four weeks, that he was posted to 'B' Group (SCU7), at Little Horwood. This was simply due to the fact that he had some knowledge of the Morse code, but no knowledge of army wireless procedure. Incidentally, like a number of others who joined SCUs, he had his first experience of Morse code whilst in the Boy Scouts.

Tom Kennerley in Calcutta.

After undergoing the rigorous Morse training at SCU7, Tom eventually passed his B3 tests at the second attempt (the first time he had omitted to press the Morse key whilst attempting to tune the MkIII transmitter). On passing his B3 test, he was transferred to 'A' Group at Windy Ridge, and billeted in nearby Jubilee Hall.

Tom worked for a while at Windy Ridge, where he estimated that between fifty and sixty men would be on duty over a 24-hour period. After a while there, he was sent to work at Main Line station near the Hall itself, where he used AR88s and HROs, on the various types of the traffic it handled. Tom spent a short time at Main Line, before being posted to SCU 11/12 in Delhi but very soon was off again, being sent down to Calcutta. There, he lived at 36, Ballygunge Park, and worked in our wireless station, which was then at 28, Queens Park. This building housed the regional offices of ISLD (Inter-service Liaison Department – a 'nom de guerre' for SIS). At Queens Park, he relates that he had the pleasure of working with Ted Mesquita, Bill Beveridge, Dan Paul and Cliff Jacobs, ex-Burma PTT (Post, Telephone and Telegraph) and all very experienced operators.

From the station in Calcutta, Tom worked in to Delhi, Colombo, Kunming and Kweilin, using HROs, MkIII transmitters and others, the largest being 100 watts. After a few months in Calcutta, he was sent to Kunming in China to take over the training of agents operating in southern China. He was flown there by the U.S. Air Force in one of the notorious 'Over-the-hump' flights that traversed vast areas of uninhabited jungle and high mountainous regions.

Ultra was being fed from British SLUs in Kandy, Delhi and Calcutta, to General Slim of the 14th Army in Burma and to General Stillwell, commanding the U.S. forces in Burma and Southern China. An SLU was also in place in Kunming; amongst other advantages, it was able to feed Ultra to General Chennault, who was in command of the U.S. Air Force in China and Burma in support of Generalissimo Chiang Kai-Shek.

Tom as duty officer SCU11/12 wireless station at Dhakuria, Calcutta

The British operators were also able to forward Japanese naval Ultra messages, providing Chennault with great advantages in planning operations, such as directing his bomber force on to Japanese convoys, and other vital targets.

The wireless unit in Kunming was run by Section VIII, and indeed for a period, I believe the cypher work was done by them as well. The SCU representative in Kunming was Derek 'Jock' Leslie who ran the Main Line station. As soon as he arrived in Kunming, Tom set up a training-school in an outhouse of the SLU, and found he was responsible for some twenty would-be agents. He described them as 'a motley crew.' After being brought up to the required level of efficiency, these gentlemen were equipped with two large suitcases, bought at the local market. One case contained a B2 transceiver, and the other a car battery drained of its acid, along with a couple of bottles containing the acid, safely packed in straw and paper.

The agents, along with their two suitcases were then taken to the nearby U.S. air base and parachuted into their allotted area. It was then Tom's job to set up a listening watch and await their appearance on the air. Some made contact, and ominously, others did not!

'From the front – Tom Kennerley, 'Yorky' Byford and Jock Leslie, on duty at Dhakuria, using AR88s.

Tom's son Mike sent me the pictures shown in this chapter but the one of the operators in front of AR88s was already in my possession, since I was there when it was taken. The shot of the aerial array at Dhakuria was made by me at the request of several photographers who wanted a picture of the camp in its setting beside the lake. I climbed the mast with three cameras and took this shot for them. I had worked on these 100ft. I P&T (Indian Post and Telegraph aerials here, and at Dum Dum. You can see how slender they are with only short pegs jutting out for one's feet and hands. It is a little daunting when you first attempt it!

After a few months, Derek Leslie was recalled to Calcutta. This entailed Tom becoming responsible for both agent training and Main Line operation. Of course, with the Main Line, he was assisted by three Chinese operators using MkIII transmitters and HROs. From here, they worked back to Calcutta and onwards to Kweilin, Tunki and Nanning.

The war against Japan finished soon after the dropping of the atom bombs on Nagasaki and Hiroshima. Towards the end of November, 1945, the office was closed down. All the wireless equipment, arms and ammunition were packed up and returned to the SCU 11/12 stores at Calcutta.

Before leaving, Tom had to get rid of the stock of hand-grenades because he was ordered not to return them to Calcutta. He decided to bury them in the now disused air raid shelter in the garden, and then collapse the roof down on them. He placed the grenades in one corner, the fuses in another, and told a coolie to start the work. Clearly the coolie misunderstood and to Tom's horror, started the job by trying to hit the fuses with his spade. Tom was fortunately able to catch his arm and stopped him in mid-air!

On returning to Calcutta a week before Christmas 1945, Tom Kennerley and Bob Beadle were told that they were being sent to re-open the station at the British Legation in Bangkok. When Tom and Bob arrived in Bangkok, they doubled the number of the entire staff, there being only a cypher officer and a chargé d'affaires. During Tom's year there, he saw the number of staff grow considerably. Their job in Bangkok was to work into Singapore, using an AR88 and a Mk33 transmitter run from a 350 Onan generator.

I will let Tom finish the story in his own words: '...It was from Bangkok that I returned to the UK in January 1947, through Singapore, where I was met at the airport by Geoffrey Pidgeon, who later in the day bought my sports jacket for ten or fifteen Straits Dollars. We next met in 1999, at a DWS reunion, and he had the audacity to ask for his money back as the jacket was now worn out. I had to tell him that the guarantee had expired some years previously.'

Chapter 31

Edgar Harrison
Wireless Operator to Winston Churchill

This is the story of my friend Edgar Harrison who enlisted in the Royal Corps of Signals on 29 April, 1929, and was transferred to MI6 (Section VIII) in January 1940. His story starts with his background and early training.

Edgar was born on St. David's Day in 1915 in the hamlet of Energlyn in the parish of Eglwysian.The nearest town with a railway station, some two miles away, was Caerphilly. His birthplace had neither electricity nor gas, and water came from a well in the adjacent hamlet of Pwllypant, reached via an unmetalled road.

Edgar was the eighth of ten children – seven boys and three girls. His schooling started in 1919 at the Hendre school at Penyrhoel, a walk of over two miles. When his hamlet became part of the Urban district of Caerphilly, he moved to the Gwyndy school in the town.

In 1926 he successfully passed a scholarship exam to grammar school with the likelihood of university to follow. This was 1926, the year of the General Strike, the depths of the Great Depression. With his father out of work, it was decided that the family could not afford the clothing and other expenses involved in a transfer to a grammar school. So Edgar continued at Gwyndy school and before he was fourteen he started working underground in the local coal mine, the Windsor Colliery at Abertidwr. There were two shifts, 0600 to 1400 and 1400 to 2200. For the early shift he had to get up at 4.00 a.m. and walk the three miles to the colliery so as he could start work on time.

At the age of thirteen and still at school, Edgar had sat a competition set by the War Office for apprenticeships in selected Corps of the British Army. The competition was open worldwide to boys of British parentage. The examinations took place at central army establishments and consisted of two English papers, two Mathematics papers, and a paper each on Geography, History and General Knowledge. The entrance fee was 5 shillings, a huge sum for working-class boys in those days, and non-refundable!

Edgar came twenty-third out of the 150 places available but only 12 places were on offer for entrance to the Royal Corps of Signals). So, at the tender age of fourteen and one month, young Edgar arrived at Catterick to commence his training as an Operator Signals in F Company.

At the end of his four-year apprenticeship which included eighteen months at the Liverpool General Post Office, Edgar had acquired the necessary proficiency in the fields of line, wireless and visual telegraphy, the installation of telephone exchanges and maintenance and repair of the associated equipment.

The telegraphy test was severe. He had to pass the Post Office grade known as SC&T (Sorting Clerk & Telegraphist) and achieve thirty words a minute. In the final test, he was required to receive 30 telegrams in

30 minutes and then return them in the same timescale, without making a single, solitary mistake! He was also taught the craft of laying cables, erecting aerials and a chance came along to take (and enjoy), an equestrian course. Those on the four-year apprenticeship were expected to reach a standard of education equivalent to today's 'A' level.

Having successfully completed his apprenticeship, Edgar was posted to Tidworth and for a few months served as an operator/gunner in the armoured vehicles of the 11th Hussars. In the autumn of 1933, he was selected to become a member of His Britannic Majesty's Foreign Office W/T station (F.O.W/T) at the British Legation, Peking. Subsequently, Edgar travelled extensively in China and did some covert work for the British military attaché. Part of this work involved plotting the Japanese diplomatic wireless network in China and Manchuria. (Examples of Peking callsigns are shown in Appendix. 5).

Edgar Harrison, Royal Corps of Signals

After completing Equestrian course at Catterick Camp in 1932

Returning to Britain in 1938, Edgar attended a six-month teleprinter course at Bristol Post Office, before being posted to Catterick as an OCTU instructor. It was from this position he was ordered to Bletchley Park in January, 1940, and thence into Whaddon and Section VIII.

At Whaddon, Edgar was accommodated in the Dower House and assisted 'Spuggy' Newton with the installation of radio equipment in vehicles, Packards and Gin Palaces. He also assisted Charles Emary in the training of an agent destined to serve in Norway. Apart from breakfast at the Dower House, he had all his meals in Whaddon Hall, sharing a table with, amongst others, Bob Hornby, Percy Cooper, Ewart Holden, 'Polly' Pollard, Wilf Lilburn, the butler and the housekeeper. The food was excellent. Occasionally he visited Nash where an SIS station was being run, at the time, by Bert Gillies.

In late March or early April 1940, Edgar accompanied the agent trained by Charles Emary and himself to northern Norway. After landing the agent, they quickly returned to England and to Whaddon. Edgar's return coincided with the start of heavy wireless traffic to and from Brussels where Section VIII's operator, Bungy Williams, required assistance if he was to cope with the volume.

Shanghai signal section 1937. Edgar is standing sixth from right.

At very short notice, Edgar was sent to Brussels to assist V. F. 'Bungy' Williams in clearing the backlog of signal traffic and keep pace with the wireless workload, until the German attack in May. Major-General Mason-Macfarlane was head of the military mission in Brussels and it was he who told the men to close down the station and make their way as best they could to a Belgian or French channel port.

Original transmitter HX8 in Peking. Installed by the Royal Navy and taken from the first RMS Mauretania in 1931.

The roads were crowded and, from time to time, came under enemy fire. During this chaotic journey, Edgar and Williams lost each other, and Edgar made his own way to Dunkirk and subsequent evacuation. Williams never made it to back home and Micky Jourdain recorded against his name in the Code Book 'lost at sea, 28 May, 1940.'

Arriving back at Whaddon, Edgar worked for a while at the new SIS station at Nash and relates that it gave him great satisfaction to make the first contact with the agent they had taken into Norway. They had almost given up hope that he would appear 'on air'.

At this time, soon after Dunkirk, there was real fear of an attempted invasion by German forces from across the Channel. Faced with this threat, a decision was taken to base mobile W/T stations at the headquarters of Southern, Northern, Western, Scottish, and Irish Commands at Wilton, York, Chester, Edinburgh and Belfast. These mobile wireless stations were intended to handle the Ultra traffic that was just beginning to come through more regularly from Bletchley Park and any other high-level communications that might be required. The wireless equipment was fitted in Packard saloons in the grounds of Whaddon Hall (See Chapter 17).

An underground W/T station was also installed in London's St James' Park. Edgar was selected to take his Packard to Western Command at Chester where he installed the station initially in a field near to the H.Q. and the village of Eccleston. He later managed to move it to a more secluded site nearby. While Edgar was based in Chester, an exercise was held involving all the Commands that was designed to test the ability to handle wireless traffic at the highest level, and it proved most successful.

Although Edgar was comfortable and happy at Chester this was not to last. In September 1940, a Packard arrived quite unexpectedly, containing a passenger named Hunter Anderson who had orders to replace him. In short order, Edgar found himself on his way back to Whaddon in the same car that had brought Hunter to Chester. On arrival at Whaddon, he was billeted in nearby Stony Stratford over the weekend and told that 'Pop' (Brigadier Richard Gambier-Parry) would see him on Monday. What a momentous meeting this turned out to be!

Edgar was astonished when 'Pop' told him, 'I could have chosen Spuggy Newton for the task,' bearing in mind the huge importance attached to Spuggy within the unit. 'Pop' then went on to say that his brother, Major-General Michael Gambier-Parry, was in Athens as Head of a Military Mission and he needed his own communications link to and from the UK. Edgar's job was to fly to Athens with sufficient equipment to install both a base station and a mobile station and make and maintain contact with Main Line, Whaddon and the Section VIII station at Cairo.

When asked when he would be expected to leave he was surprised to be told 'tomorrow'. 'Pop' thought his services would be required for about twelve months. After thanking Gambier-Parry for placing such trust in him, Edgar went to the Main Line Station to see Charles Bradford and agreed a Signal Plan with him, including frequencies. From there, his next stop was the workshops where he went through the list of all the equipment selected by Spuggy Newton, for the new station in Athens. The major items were a MKIII transmitter, an HRO receiver and Tiny Tim and Onan generators, plus the usual ancillary equipment. All this gear was packed and loaded on a 30 cwt truck, ready for the off.

Edgar was supposed to fly by Sunderland flying-boat from Pembroke Dock but this was changed to Plymouth, from where he took off about midnight for Gibraltar. There was a two-day layover in Gibraltar and it was there that Edgar was asked what he knew about gunnery. The powers-that-be were pleased when he told them of his time as an operator/gunner with the 11th Hussars. They then explained that an extra passenger had to be put on the plane and this could only be achieved if the RAF gunner was off-loaded, so Edgar became the gunner for the flight as far as Cairo. Between Gibraltar and Malta, he fired a few practice rounds. There was a two-day stopover at Cairo before flying on and landing safely at Piraeus.

Here. Edgar was met by Colonel Casson, a sergeant who was his Greek interpreter and transport for the wireless equipment. In a very short time, all of them and equipment were at the Grande Bretagne Hotel where the Military Mission was based. Edgar had been allocated a room large enough to set up the station, a comfortable bed and en-suite facilities. Gaining access to the roof, Edgar could see it afforded him ample space on which to erect an aerial and within 24 hours he was in contact with the Main Line station at Whaddon Hall.

Having been introduced to Major-General Gambier-Parry, to whom he passed on best wishes from 'Pop', he asked if it might be possible to obtain a vehicle similar to a Packard which could be converted to a mobile W/T station. Gambier-Parry said he would do what he could and much to Edgar's surprise and delight a Packard was delivered to the hotel. By this time, he had made himself known to the PCO (Passport Control Officer), at the Embassy and to 'Curly' Meadows, who ran the Mainline W/T station there. Between them, they found a place to park the Packard and Edgar used his spare time to convert it into a mobile W/T station. A fortnight after arrival in Athens, he drove the Packard about twenty miles out of the city and demonstrated its usefulness by contacting Main Line, Whaddon and Cairo, and clearing all outstanding wireless traffic. Colonel Casson was most impressed!

So, from that time onward, the two stations operated efficiently in and out of Athens. Edgar also lent a hand from time to time to the Main Line station in the Embassy by helping to clear backlogs of traffic to Whaddon and Cairo, mostly the latter. All this relative calm was to change dramatically in April 1941, when the Germans attacked through Yugoslavia and overran Greece. Edgar was at Yoannina when he received an order to make for Athens at all speed. It was a terrifying journey as the road was under almost constant aerial attack. Arriving in Athens, he found the city strangely quiet – it had been declared an open city. Orders left at the Grande Bretagne Hotel instructed him to report to Army HQ at Omonia Square.

Upon arrival, Edgar contacted Windy Ridge and for twenty-four hours was able to pass Bletchley Park traffic to the Signals Centre. Then chaos set in. H.Q. and the Signals Centre disappeared, and Edgar was left to make his own way, as best he could, to Port Argos/Nauplion.

After destroying all the W/T equipment, he scrounged a seat on a train bound for Patras which took him as far as Corinth. There he left the train and another awful journey commenced. No commissioned officers were about and so Edgar, with the rank of Sergeant, found himself responsible for looking after an increasing number of other ranks. They expected him, as the senior NCO present, to get them to Port Argos and this was accomplished though sadly, not without casualties, as they were constantly under air attack. Edgar regards this episode as much worse than the retreat from Dunkirk.

Edgar found that at Nauplion, there was a better measure of command and discipline. He reported in with his motley crew of 'Odds and sods' and was told to march them the miles along the railway line to Port Argos, then to cross open ground to an inlet where a ship was at anchor which would take them to Crete. Needless to say, they reached the inlet in double-quick time and indeed there was a ship with its gangway down, ready to receive them – or so they thought.

As Edgar approached the gangway, a shot was fired over his head and an 'Aussie' voice shouted 'the first Pommie bastard who steps foot on this gangway gets it'. The group retreated in haste. Much later, they learned that the ship was reserved solely for Australian and New Zealand troops.

By now, they had again been spotted by dive-bombers and fighter planes, and were subjected to terrifying attacks. They were caught on open ground with no time to seek cover and suffered more casualties. After the attack was over, Edgar collected his dwindling crew in the corner of a small field in which broad beans were growing. They picked the beans and used them to make a meal of sorts.

They trudged back along the railway line to the railway station at Nauplion to find a large number of troops being rounded up into some sort of order by an officer and some Warrant Officers. There was talk of taking a train to Kalamata and, as Edgar had more than a passing knowledge of railway signalling and locomotives, he volunteered his services. An oil-fired engine was available and, in very short time, a fully loaded train of about eight coaches chugged slowly out of the station. Edgar travelled in the engine-driver's cab to ensure that 'staffs' were properly exchanged on the single-line track to Kalamata.

Although they were frequently buzzed by enemy aircraft, no attack was made on them and they made good time to what is now the old station adjacent to the port. In great haste, Edgar and his assorted crew made for the cover of the large olive groves at the back of the beach. Word reached them that, as soon as darkness fell, they were to be evacuated from the port by destroyers. The Navy duly arrived and they waited their turn. At

3.00 a.m. they were still waiting in the queue when the ships had to leave port so as to have time to reach the open sea before enemy aircraft found them.

Returning to their olive groves, they snatched a little sleep to be awakened by gunfire and enemy aircraft dropping mines in the harbour. The Germans claimed the town and demanded surrender. The commander at Kalamata, Brigadier Partington, would have none of this. He gave the order to 'fix bayonets', then came the order to charge. Casualties were sustained, including a New Zealander who was later decorated with the Victoria Cross for his valour, but at least Kalamata was temporarily cleared of the enemy.

Enemy reinforcements arrived and the British troops were forced back into the olive groves where they were constantly under fire. Not long after dusk, they observed signalling by lamp and Brigadier Partington shouted aloud for a signaller. Edgar was quickly by his side, a hand-held torch was given to him and contact was made with HMS *Hero*. His signals to *Hero* were first to inform HM ships that the harbour was mined and unusable for evacuations and secondly to give the number of troops to be evacuated – thousands of them.

HMS *Hero*, HMS *Kandahar* and the other destroyers replied that every effort would be made to get as many as possible off in the six hours or so of darkness. Edgar has no idea how many were actually evacuated, certainly many hundreds, but in the early hours he received the last signal from *Hero*: "Must leave now, good luck and God be with you". Thus ended the evacuation. Brigadier Partington said he would be surrendering to the Germans at first light (29 April) and the British were to retain the present positions.

Edgar, however, had no intention of surrendering. He crept off into the darkness and hid in caves for some time, until he found a small boat complete with oars. After obtaining food and water from some kind locals, he set out to sea with little or no knowledge of boats or navigation.

Days later, an eagle-eyed watch on HMS *Kandahar* spotted him and he was hauled aboard. *Kandahar* took him to Crete, and after spending a while there, he went on to Alexandria, where, to his surprise, he was met by Colonel Casson. Casson warmly welcomed him, gave him some money, arranged his train journey to Cairo and Inter-services Liaison Department (ISLD, a cover name for SIS abroad) and a reunion with the Main Line Station. A spell in hospital followed, as he had been slightly wounded and had contracted sandfly fever. Amazingly, Edgar discovered that his brother, Wallace, was a patient in the same hospital. Wallace was also an operator in the Royal Corps of Signals.

Once he was fit again, Edgar returned to duty at ISLD. His role now consisted of training agents who would eventually be dropped into Crete, Greece, Yugoslavia, Romania and Italy. He also trained Arab agents who would be infiltrated into North Africa and occasionally, he helped out at Main Line. It was a comfortable life and he got on well with Wing Commander 'Jock' Adamson, the commanding officer. Unfortunately, Edgar was not able to wallow in the comparative luxury for long. In August, he was sent to Kanka for parachute training and on his return was asked to 'volunteer' to be dropped into Yugoslavia. General Mihailovic was desperate for communications equipment and ISLD had agreed to provide it, as well as an officer to set things up.

In the fullness of time, Edgar and his wireless equipment were duly dropped, and a network was created that allowed Mihailovic to be in touch with Cairo and London, plus a limited internal system. Edgar's time in Yugoslavia was rather hairy as the internecine strife between the Serbs, Croats and Slovenes made life quite terrifying at times. He was delighted to return, by sea, to Egypt and the sanity of ISLD. Whilst in Cairo, he shared a billet with Mike Vivian, (the son of Valentine Vivian the Deputy Chief of SIS), and they became fast friends.

In October, Edgar was asked to become a member of the Way Mission to the Caucasus. This mission came into being to facilitate the distribution of Allied supplies to the Russians, that were being ferried into the Soviet

Union at the time by going all round the Cape of Good Hope. Head of Mission was Colonel R. Way, R.A. (Royal Artillery), supported by a major and three captains, all Russian-speakers, and a cypher clerk.

Edgar's task was to set up and maintain communication with Baghdad, Cairo, Kuybyshev and Whaddon. He was also expected to train a cadre of Russian signallers to fit British equipment into their vehicles, armoured and otherwise, mostly the latter. His own equipment was a MkIII, an Eddystone receiver and an agent's suitcase set. He left Cairo in early November, joined up with the other members of the mission, including Bruce Lockhart in Iraq, and then flew to Mosul in an ancient Wellesley bomber. Awaiting him there was a Russian-built Dakota, and some two hours after arriving at Mosul the unit was on its way to Russia.

Until then, the mission had not been told where they would be based in Russia. There was talk of it being Tiflis, the capital of Georgia. Sooner than expected, they prepared themselves for landing and touched down at Nakhichevan in Azerbaijan, where they spent the night locked up in a hotel. It was made obvious that they were unwelcome. Taking off the next morning, another short flight took them to Erevan, Capital of Armenia, where they spent the day and the next night. They found themselves the object of great interest to the locals, who were anything but friendly towards them, as they walked around the town, so they were glad to get back to the safe haven of the hotel.

Thankfully, a short flight next day took them to Tiflis, where transport awaited them at the airport to take them and the equipment to the Hotel Tbilisi which was to become their HQ. Edgar was allocated two rooms, a bedroom complete with toilet facilities and another in which to install the wireless station. The roof of the hotel afforded good space for an aerial and by midday the day after arrival, he was in contact with Baghdad, Cairo, Whaddon and Kuybyshev. This last was the acting capital of Russia. Most government departments, foreign embassies and missions were sent there, in case Moscow should fall.

The story of Edgar's six-month stay in Russia could be the subject of a book in itself. Suffice to say, he completed all the tasks expected of him, and as far as Section VIII was concerned, it was a successful mission.

Incidentally, the Bruce-Lockhart who accompanied them was the son of the Bruce-Lockhart who was mainly instrumental in spiriting Lenin back to Russia in 1917. Edgar returned to Cairo by rail and road – Tiflis – Tabriz – Kazvin – Zenjan – Teheran – Basra – Baghdad – Damascus – Jerusalem – Cairo. He offloaded the radio equipment in Baghdad where 'Curly' Meadows was in charge. They had not met since the short stay in Crete. An operator named Rice was running the station at Basra. Neither Damascus nor Jerusalem had Section VIII stations at this time. The trip from Tiflis to Cairo took three weeks and finally he was greeted at Cairo railway station by Mike Vivian.

After the travels and events in Russia, it was good to be back at ISLD. Edgar returned to his job of training agents and helping out by putting in time at Main Line. Not long after returning to Cairo, a retreat by British desert forces and the loss of Tobruk, led to a real fear that Rommel might capture Alexandria and Cairo. A decision was taken to move GHQ, including ISLD, to Jerusalem.

What was to become of the agents' station and training school? At that time, the agents' station was in contact with agents in Crete, Greece, North Africa and Yugoslavia. Permission was sought and given for ISLD to set up a station at Kufra Oasis where the Long Range Desert Group (LRDG) and Popski's Private Army were based. Edgar took a suitcase set to Kufra, flying there via Wadi Halfa, and soon found that a station in the Fort would be able to handle all the agents presently worked by Cairo.

Incidentally, the Fort was garrisoned by a company of the Royal Welch Fusiliers seconded from Khartoum. This was the regiment in which both Michael Gambier-Parry and Richard Gambier-Parry had served during most of World War I. Edgar signalled ISLD that Kufra would be an ideal place for the agents' station and when he returned to Cairo, two MkIIIs, HROs, an Onan generator and other equipment were already packed and ready for transportation to Kufra. Before flying back to Kufra with Henry Poole, he witnessed the departure of the Section VIII station on their journey to Jerusalem. The agents' station remained at Kufra until well after the victory at the second battle of El Alamein and it more than fulfiled the highest expectations.

In due time, after the Eighth Army had reached the Tunisian border, Edgar took the agents' station to Apollonia in Cyrenaica (now in Lybia) and set it up at a superb site overlooking the sea. By now it was the station responsible for all the Italian, Cretan, Greek and Yugoslav agents. Edgar knew most of these agents, having trained them in Cairo, with the assistance of Roy Peacock and Tommy Morton.

Major Potts, who had taken over from Jock Adamson, arrived unexpectedly in Apollonia in July 1943. Edgar was instructed to return to Cairo and then fly to Algiers where he would work in association with Tommy Morgan. Our Section VIII station there was eventually to take over responsibility for the Italian agents and it ended up in Bari under 'Red' Marshall. In the meanwhile, Edgar had a short but hairy stay on Sicily before returning to Whaddon in October, 1943.

During the remainder of 1943, Edgar liaised with Major Rooker in testing small transmitters and transceivers across the Bristol Channel. Rooker was at Porlock and Edgar was at Port Eynon, between Swansea and Tenby. The New Year saw a large intake of RAF personnel from Chicksands. A new SCU was formed, known as SCU7, under Sqadron-leader (Dickie) Mathews, to train these airmen in SCU expertise and Bill Wort, Dusty Miller, Bill Murray and Edgar were sent to Little Horwood where the training took place. By D-Day, their training was complete and the RAF operators were earmarked for various SLUs.

After D-Day, Edgar took a number of the airmen to France and Germany where they performed very creditably. When the ceasefire came, he was at Halten, near Hanover, with General Simpson's 9th U.S. Army with his SCU/SLU, which included his brother, Wallace.

Edgar in 2002, just prior to delivering the oration at the commemoration service of the Battle of Kalamata 28th-29th April, 1941.

Back at Whaddon, 'Pop' told Edgar that he had been recommended for Staff College. Edgar was delighted as he had always wanted a career in the Royal Signals. After successfully clearing the preliminary hurdles, he was extremely upset when he failed his medical. He asked to be demobilised, but this was refused and he was 'Detained Operationally Vital' (DOVd).

Edgar's first job was to go to the SOE station at Poundon near Whaddon, take it over from the SOE, demob all personnel and get rid of all the equipment. This was a mammoth task. When it was almost complete, Spuggy Newton took over and Edgar returned to Whaddon where he was appointed a permanent member of Courts Martial. He performed this job until, at very short notice, he was sent to Tokyo. This was the start of his career in the newly formed DWS, and twenty-one years later, he became its Principal Signals Officer.

Last, but by no means least, I must report on Edgar's duty as wireless operator to the Prime Minister, Winston Churchill, at four Summit Conferences. It is well known that Winston Churchill had implicit faith in the communications capability of Gambier-Parry and his Section VIII team, to the extent that they alone provided the highest level of communication for the P.M. when he travelled abroad.

All four of the conferences at which Edgar Harrison was responsible for the wireless traffic, followed a similar pattern. He was expected to maintain full secret communications with say, Cairo and London, by means of just an agent's suitcase set with an aerial, running on a 12-volt battery. This is Edgar's own account of the Conference held at Mersin in Turkey:

'At very short notice I found myself flying in a Dakota to Turkey. Arriving at Mersin, I was taken to a train standing in a siding where I had been allocated a whole compartment. Slinging an aerial across to a nearby signal gantry, I quickly established contact with Cairo and Whaddon and exchanged traffic during the two days there.

'On the evening before flying to Nicosia, the PM said he would not leave until we had a weather forecast from Cairo. I told him this would come in at 0800. Within minutes of passing the forecast to him, the train moved off, giving me no time to dismantle my aerial. We had another conference at Nicosia before flying to Cairo. Before leaving Egypt for the UK, and learning that I was not to go with the party as I was based in Cairo, the PM personally thanked me for my services and took a parcel home for me'.

After the war he continued in DWS, becoming its Principal Signals Officer. Edgar Harrison has been awarded the OBE, the Military Cross, and numerous other decorations, reflecting his great service to Section VIII, and to his country.

Chapter 32

Steve Dorman: the V2 and Ascension

Thinking back to my years at Whaddon Hall is like looking back at childhood summers, they were all sunny. Whilst that cannot possibly have been the case, my memories of the people at Whaddon is certainly rosy. One of the men I remember best was Steve Dorman. who joined Section VIII just after me in September 1942, but was a few years older and more technically qualified.

Steve had been interested in wireless since childhood and was a VI whilst at Queen's University, Belfast. He was also a member of the university cadet corps, and lectured the Home Guard on wireless. When he joined Section VIII, he developed into one of the unit's most outstanding engineers and designers, being in R&D from the very start. Notwithstanding the difference in our ages and position, he was always very friendly towards me, the most junior engineer around.

When David White asked me to write this book, the first thing I did was to find out who, were 'still around' those from those far-off days. I was delighted to be put in touch with Steve and in early 1997, Jane and I went down to see him at his home at Looe in Cornwall. We spent a few days there and he started me seriously believing I could put together a history of the unit.

I suppose we talked for about four or five hours about Section VIII each day and I recorded some two hours of the conversation on tape. On our last night, Steve and his second wife, Lesley, came to our hotel for dinner, where he presented me with a pair of wireless pliers that my father had issued to him as part of his tool kit on his arrival at Whaddon. These are now in a frame in our hall, together with the background story.

All his life, Steve kept a daily diary. The following is an extract from 1942 when he was about to join Section VIII, though he did not know the true nature of the organisation at the time. The year 1942 was clearly one of the most significant of his life. Following an interview with Dr. C.P. (later Lord) Snow of Lord Hankey's Science Council, he was directed towards science research, rather than a commission in one of the technical branches of the armed services.

At the end of the 1942 summer term at Queen's University, Steve spent the long vac training on an intensive wireless course at King's College, London University, which had been evacuated to Leicester. Because of his work as a VI (Voluntary Interceptor) with the RSS, his clearly outstanding wireless knowledge, and no doubt his connection with Dr. Snow, he was approached by Section VIII.

4 September 1942
'... telephone message from Lieut. Cmdr. Cooper concerning a research job ... arrange to visit him tomorrow'.

5 September
'Catch the 8.36 train to Bletchley, have a very satisfactory interview with Cmdr. Cooper and return to Leicester in good time. I think I have got a wonderful job and am very lucky'.

Tel.Anstey 245.

14, Branting Hill Avenue,
Glenfield,
Leicester.

WAR OFFICE.

21st July, 1942.

The Warden,
Knighton Hayes Hostel,
6, Ratcliffe Road,
Leicester.

Dear Madam,

Mr. Dorman, who is attending a special Radio Course at University College, wishes to utilise some of his spare time performing some highly confidential work of National importance on behalf of this Organisation.

Absolute privacy is essential and I shall be very grateful if you will grant the use of a room for this purpose. I understand from Mr.Dorman that he has discussed the matter with you and that such a room is available.

Yours faithfully,

Lieut. R. Signals
"M" Regional Officer, R.S.S.

WP/GEM

A letter to Leicester University from RSS asking for 'absolute privacy' for Steve (to continue his VI work) whilst studying just prior to joining SCU1.

17 September
'Arrive at Bletchley at 11.45 where a R.Signals car takes me to S.C.U.1 at Whaddon. I am shown round the place by Mr. Durbin and try to remember a lot of people's names. We play cricket at lunchtime. I am to stay in the Bull Hotel at Stony Stratford tonight. Telephone home'.

18 September
'Very interesting day studying 37, a transmitter-receiver. I am put in a billet with Mrs. Trimmer, at 4 Bedford Street, Wolverton. Not a very wonderful house but a very happy family with 3 grown children away and Peter, the cat. I have a small bedroom without a light but plenty of room to work downstairs. (No bathroom, but a zinc bath in the kitchen filled with hot water from a coal "copper" once a week)'.

[I was already at Whaddon when Steve joined and we were still making the MkV – a quite bulky suitcase agent's set. I believe these were the very early days of the MkVII which was still in the design stage. It is possibly an early version of that set he refers to in the diary. Of course, Steve had a great deal to do with its later development into our most important transceiver, the 'Paraset'.]

26 September
'We worked a 6 or 6½ day week. I travelled the 6 miles to work either with a lift from John Harding, or on the R. Signals bus, or later on my cycle. I used up my clothing coupons assuming I would soon be enlisted in R.Signals. On the 26th I bought trousers, 26/3d. shoes 21/0d. belt 3/6d. and a Royal Corps of Signals Brooch for Molly [His first wife]. There was a cinema a few minutes walk from my digs and most weeks I went twice, cost 1/6d. A Capt.Tucker joined our group on the 30th.'

[This was John 'Tommy' Tucker, with whom I worked in 1945 on the construction of the new Calcutta transmitter station at Dum Dum].

2 October
'I finish my prototype MkIIIB transmitter and carry out an analysis of its harmonic output'.

[This is remarkable. Is Steve claiming to have produced a new version of our standard transmitter after only fourteen days? He later went to live with Katie Weld, a cousin of his mother, who lived in nearby Woburn Sands].

7 October
'Catch the 8.36 to London. Buy books. 1942 ARRL Handbook. Wave Guides, Lamont. Radio Designer's Handbook, Radio Data Sheets, Camm. Try to get Blakey's book but it is out of print. See Vic Oliver and Celia Lipton in 'Get a Load of This'.

Spent £2.13.1½d. Train 11.0d. Books 30.0d, Lunch 2.0d, Tea 1-0.½d, Tubes 4d. Street Guide 6d, Illustrated [magazine] 4d. Hippodrome 5.0, Stamps 4d, Postcards 3d, Stamp pad 1.3d, Ink 6d. Comb 1.0d'.

[This attention to detail is typical of the man; the records he has left include a lifetime of diaries and I have no doubt his research notes at work were equally detailed.]

In early 1943, he records a covering note:
'I was working at SCU1 Whaddon which was under the direction of Brigadier Gambier-Parry. The head of our section was Lt. Cmdr. W.P. Cooper. The section staff comprised Alec Durbin, Bert Mason, John Shears, Ted Oliver, Wilf Lilburn, Alfie Willis, Capt. Tucker, Mac Hawkins (Mac) and Dennis Smith.

'The adjoining workshop was run by Charlie West, and the small 'production' shop by Hugh Castleman. Maurice Winnie was our test pilot who flew the Anson or Wellington aircraft when we were test-flying equipment. Others at Whaddon included Major Sharpe, Wing Commander Adamson, Commander Langley, Colonel Lord Sandhurst and Major Crocker.'

22 April, 1943
'Enlisted by Major W.G. Duncombe-Anderson this afternoon, going on to Horwood to be issued with uniform and various items of kit.

6 June
'To work this morning. Put Bill Reynold's 'Echo' radio together again. Back and change before lunch. [Bill Reynolds was the first carpenter at Whaddon who made the wooden cases for our MkIII transmitters; he was followed by Bill Barnes.]

8 June
'Letter from War Office indicating I should attend an Officer Cadet Training unit prior to commissioning. I feel this would essentially parallel my STC training at Queen's and would waste time.'

14 June
'Leave this morning on a long cycle run, Bedford, Cambridge and beyond, Royston, Shefford, Woburn Sands. Cycle first part with a soldier and second part with a geography mistress from Presteign, Wales. Cover 137 miles. Tea with Katie at 5 o'clock and back to Wolverton well refreshed'.

[Steve frequently records cycle rides of 130 to 150 miles both in the Whaddon area and at home in Ireland – he simply comments 'well refreshed'!]

22 June
'Sell my Triumph motorcycle on the 22nd. Buy a 1938 350cc P & M Panther model 85 EXU 150 for £34-10s from Pride & Clark in London on the 27th. Bought some small prints of paintings at the National Gallery. Had a letter from Ruth to say she has started work at Haselmere'.

9 October
'Commenced a three-day War Office Selection Board at Watford. This involved written papers, practical work including initiative testing, and an interview by the Selection Board'.

13 October
'Home on leave. Find Jack Smith [a fellow VI in Belfast] has been issued with a MkV transceiver, an early equipment built at Whaddon'.

1 November
'I am working on the Mk119 equipment. [I do not recall this model] Spend half the day mending my motorbike which broke down'.

11 November
'At last, I am commissioned 2nd/Lt in Royal Signals. I am very thrilled. I wore my father's Sam Browne belt which he had worn through the first World War. Thought, – I must be a very good officer. (I should have been, as over the years I had accumulated far more training and military experience than any of my associates who I think, without exception, had never handled military equipment or operated radio equipment in the field.)'

[I should point out that we had quite a number of World War I 'veterans' in the unit. Men like Gambier-Parry himself had fought in the trenches, John Darwin had front line experience in the Guards, and later joined the Royal Flying Corps. Jack Saunders joined us from the Navy and had been in World War I and we had others with a similar background.

Then, there were those like Edgar Harrison who was a real trained soldier in every sense of the word. However, it is certainly true that Steve had more army training, as a university cadet in the STC, than most of those working around him in the workshops at Whaddon, but few of them would lay claim to military expertise!]

17 November
'Train to London & go straight to Moss Bros, Covent Garden, to get "Kitted Out". Bring everything back except my service dress greatcoat and boots. Total costs £44.2s. 9d. Get a very nice trench coat with a fleecy wool interlining. Sew on pips in the evening and sort things out'.

18 November
'My first time at work wearing officer's uniform. I am introduced to the Mess by Cmdr. Cooper and am pleasantly surprised by the lack of formality. [That lack of formality was never entirely understood by Steve who had expected a more military approach from his fellow officers.] Letter from Dad and "Sam Browne" belt'.

29 November
'To Parish, Bedford to have an identification photograph taken'. [a further note says six enlargements were made and one is reproduced here – 'Parish' being the photographer.]

2 December
'Receive 100 cigarettes from Metropolitan Vickers and later a parcel with 2 pr. socks, 1pr. mittens and a "jolly fine" scarf from the M.V. Girls Knitting Guild. On the 10th I had a long letter from Ruth apparently advising me how best to "win" Molly. [Molly became his first wife.]

30 December
'Home on leave. Have a walk with Mum after lunch. Visit Molly at 8.30 and give her a bracelet of threepenny coins for Christmas.and my photograph. Her cousin, June Anderson [later Molly's bridesmaid], was there. Stay for supper. Play gramophone and piano'.

Steve's diaries continue through 1944 telling of his work at Whaddon and of his personal life. Later in 1944, Steve records:

'My mother's cousin, Katie Weld, with whom I had been living at Woburn Sands, died about a week before I became engaged on operation "Silent Minute". I had a few days to sort out Katie's house, organise a relative, 'Tiny' Townsend to help and move my own non-essential kit into store at Whaddon. Because of the secrecy then surrounding the operation, I kept a separate diary called "Operation Silent Minute".'

Gambier-Parry gave Steve one of the most important tasks ever undertaken by the unit. It concerned the threat posed to the country by Germany's V2 Rocket. As usual, Steve kept a diary but it is just as well that he concealed it at the time!

A few years ago, Pat Hawker wrote an article for a wireless magazine about the attempt to divert the rockets by the use of wireless and corresponded with Steve about it. The following is largely taken from that article, and the quotes in italics are direct extracts from Steve's notes of the operation.

Steve Dorman – just commissioned on 11th November 1943

Operation 'Silent Minute' – The Secret Story of the Vain Attempt to Jam the V2 Rocket

There is an ancient Chinese proverb that 'Success has a thousand fathers, failure is an orphan'. Most of the secret Intelligence successes of World War II have by now leaked out. There remain buried, however, some remarkable engineering feats among the failures, not least, I might suggest, the construction by Section VIII of a 75kW VHF jamming transmitter. It was built in less than six weeks at the Aspidistra site in the Ashdown Forest near Crowborough, East Sussex.

Although in terms of casualties and damage, the V2 rockets launched by the Germans between 7 September, 1944 and 27 March, 1945 proved less serious than the V1 pilotless flying bombs, they were militarily even more significant in paving the way for the post-war development of intercontinental ballistic missiles with nuclear warheads and the concept of total mutual destruction. Had it not been for the many problems of technology that delayed the operational phase (some 65,000 design modifications were made to the V2), the combined effects of the V1, the V2 and the unfinished long-range V3, with which it was planned to hit New York, might have saved the Nazis from defeat. There is ample proof that in 1944 the threat was taken extremely seriously by Churchill and the chiefs of staff, once they had become convinced that the many Intelligence reports could no longer be dismissed as fantasy or deception.

From 1942 onwards, such reports steadily increased. They came from SIS agents and Allied Intelligence and Resistance networks, prisoner-of-war interrogations and the bugging of senior POW officers at the special centre on Ham Common, near Richmond, Surrey; from the Photographic Interpretation Unit (PIU) at Medmenham, and from occasional Enigma and hand-cypher decrypts, augmented by interception of the German radar unit plotting test firings in the Baltic of V1s and V2s.

All this activity, some initiated by Dr R. V. Jones, SIS's Assistant Director of Intelligence (Science) – ADI(Sc), was brought together and analysed by him. These partial but not always reliable sources gradually added up to convincing evidence of German progress in rocket weaponry technology together with a vague idea of the basic specifications.

It was unfortunate that for many months, Churchill's own scientific advisor, Lord Cherwell (Professor Lindemann) continued to dismiss all reports as an elaborate deception by the enemy. He refused to believe that the Germans had developed a long-range rocket capable of delivering a worthwhile warhead on London.

The story of how Dr Jones, alerted by The Oslo Report of November 1939, found out about the work on V-weapons being carried out at the German Versuchsstelle Peenemünde research station (cover name Heeres-Artillerie-Park (HAD) has been well told in his book, *Most Secret War* and in others such as Brian Johnson's *The Secret War*. But there remained vital missing data, and some wrong assumptions. For example, right up to the launch of the V1 (FZG76) flying bombs in June, 1944, it was believed that they were powered by rockets rather than, in reality, by low-cost pulse jet engines with their distinctive put-put sound.

'All dressed up' – Steve Dorman told me he persuaded these chaps to dress in their best uniforms and go to a photographers in nearby Buckingham for this picture. Perhaps Steve was trying to make these fine engineers appear more soldierly but in my years at Whaddon I never saw them dressed in this way. I suggest they would have been very reluctant sitters! Left to right: Charlie West, Alec Durban, Alf Willis, Douggie Lax, Hugh Castleman with Steve Dorman standing to the right.

Fortunately, enough was known about the V1 to enable increasingly effective counter-measures to be planned and then put into effect. Just how effective they became is shown by the fact that, on 28 August, 1944, of 94 V1 launches, 65 were destroyed by anti-aircraft guns (ack-ack), aided by micro-electronic proximity fuses and American SCR-584 gun-laying radar, 23 by RAF fighters (mostly Tempests, as these were faster than Spitfires and could overtake the V1) and two by the balloon barrage. So only four out of the 94 got through to the London region!

In total, of the 8,617 V1 flying bombs launched from sites in France, 2,340 reached the London Civil Defence region, killing 5500 people and seriously injuring about 16,000. Horrific as such figures now appear, it should be remembered that at the height of the Blitz, in the month of September 1940 alone, over 5,500 people were killed in London alone, with a total of over 6,900 in the UK. A total of 6,300 people were killed in October 1940, and 6,475 (4,934 of them in London) were killed in April, 1941.

As intelligence accumulated, a massive RAF raid was mounted on Peenemünde during the night of 16–17 August, 1943 by 600 aircraft, 40 of which were lost. This seriously disrupted the work on V-weapons and caused much damage, but only one of the senior scientists (the prime target) was killed and many bombs fell on the associated foreign workers camp, killing a number of those who had been supplying information to British Intelligence.

The raid caused German rocket R&D to be largely transferred to Blizna, Poland, however, where it was soon identified by the Polish Home Army and from Enigma traffic. The major production plant for V2s was built in the cave complex in the Harz Mountains near Nordhausen, which was invulnerable to bombing, despite the plot of a post-war film about 'Operation Crossbow' which was pure fiction.

Intelligence networks in the occupied countries were asked to report any information on possible preparations, including rocket launch sites. This resulted in several valuable reports, and Dr Jones singles out for special praise those of 'Amniarix' (Jeannie Rousseau, later the Vicomtesse de Clarens), who worked for the Germans as an interpreter and who was able to pass reports to SIS via Switzerland where there were three SIS stations under diplomatic cover that were in radio contact with Whaddon. There were also, inevitably, a number of misleading reports based on rumour rather than fact.

Much to the annoyance of Dr Jones, the task of assessing the feasibility and progress of the V-weapons was entrusted to a new 'Bodyline' (later 'Crossbow') operation under Duncan (later Lord) Sandys, MP, Minister of Works, He set up a number of committees to which were recruited, amongst others, British 'rocket experts' who had no experience of intelligence work.

The V2 Threat

The rocket, known initially to the Germans as *Aggregat 4* (A4) and later as *Vergeltungswaffe 2* (V2), was code-named 'Big Ben' by the British. In June 1944, after the first V1 flying bombs had been launched at London, the Prime Minister formed the 'Crossbow Committee of the War Cabinet', but became alarmed and annoyed when Dr Jones revealed that stocks of the V2 appeared to have reached as many as 2000 (this was later shown to be an accurate estimate). The PM, shocked at this unexpected information, was even prepared to consider announcing that an attack on London by the V2 would be met by the release of poison gas over German cities.

One reason for government alarm was that for many months it had been estimated that V2 warheads would contain at least seven tons of high explosive. Similarly, the size and weight of the V2 had been over-estimated in many reports, despite what later was proved to have been accurate information from several of the best agents. The higher estimates were generally accepted, due to the continued insistence by Lord Cherwell that, even if such a rocket existed, the weapon could not possibly justify the effort being put into it by the Germans if it carried only the same explosive power as the V1 (about one ton).

The Crossbow rocket experts continued to insist for several months that a rocket would need to be very large to reach London. They assumed it would be powered by a solid fuel such as cordite, not realising that the German scientists had developed a far more powerful and lighter fuel, a mixture of liquid oxygen and alcohol. They also believed wrongly that the rocket would need to be launched from an elaborate 'projector' when, in fact, the Germans had developed a mobile *Meillier-Kipper* vehicle that brought the rocket to a concrete pad on which the vertical launching gantry was sited.

As late as 27 July, 1944, Sir Alwyn Crow, Controller of Projectile Development, was still stressing that 'the smallest rocket capable of a range of 100 miles with a five-ton warhead would not weigh significantly less than 33 tons, or have a diameter significantly less than 5ft'.

It seems, once again, to have been Dr Jones who had the foresight to recognise that such a horror weapon would have great appeal to Hitler almost regardless of the potential damage it could cause. Hitler had, in fact, given it high priority and personally ordered the construction of the unnecessary launch sites protected by some 3ft-thick concrete housing that had been detected and bombed by the RAF. In practice, the operational launch sites were virtually undetectable from the air and could be changed frequently, just as for the Cruise missiles eventually developed by the Americans with the help of German rocket scientists. If, by June 1944, intelligence rather than the rocket experts had a realistic and remarkably accurate assessment of most aspects of the V2, detailed information was lacking on how it was to be guided to the target area. Although it was correctly estimated that accuracy would be about a ten-mile circle around the actual target, this would still be sufficient to place rockets consistently within the Greater London area.

Radio Control

Agents had reported the V2 as being 'radio-controlled'. This could mean a complete tracking, guidance and ranging system or a tracking system that would enable the Germans to confirm where it fell (such a tracking system had been carried on many V1 flying-bombs), or conceivably just an initial radio control system that would shut off the fuel supply some 60 seconds after launch. It was known that the V2 guidance systems had been a source of difficulties and delays, and that some radio-controlled signals believed to be associated with the V2s had been detected at Beachy Head on about 27 and 50MHz. But more detailed information was urgently required if radio counter-measures were to be planned that might have some chance, no matter how slight, of successfully intercepting the missiles.

The Germans gradually developed a considerable number of different radio guidance and tracking systems for their rockets, although the majority of operational V2s depended on an inertia-navigation gyro platform called *das Mischgerat* that was impervious to radio counter-measures. Then in June–July 1944, two fortuitous events seemed to offer some reason for optimism. On 20 May, the Germans fired a V2 with a dummy warhead from Blizna that fell into the River Bug some 80 miles north-west of Warsaw. It was found and then hidden by members of the Polish resistance until the Germans gave up the search for it. Polish Home Army Intelligence had earlier set up a secret operation to try to seize one of the trial rockets being fired from Blizna before it could be recovered by the Germans.

The Polish government in exile, whose intelligence service was second to none, arranged for Professor Greszkowski of the Warsaw Polytechnic to examine the radio equipment. Warsaw, in turn, radioed London HQ and arranged for the rocket remains to be collected by an RAF aircraft from Brindisi, Italy (the aircraft nearly got stuck in the mud on the landing strip). It was then brought to London for further examination at RAF Farnborough. It seems likely that the recovered radio equipment was sufficient to show that while radio was used to shut off the fuel some 60 seconds into launch, using frequencies of the order of 50MHz (even this was later automated), the main guidance system appeared to be gyro-controlled, rather than radio controlled.

Misleading Intelligence

On 13 June, a V2 (possibly an A4b), without a warhead, launched from Peenemünde, went off course. As sometimes happened, it exploded in mid-air and the remains fell on Sweden. On 31 July, two tons of the damaged components arrived in the UK, having been 'bought' from the Swedes by the British in exchange for the promise of some mobile radar units. Although not appreciated at the time, this rocket was, in fact, an experimental hybrid version that differed significantly from the production V2. It seems to have carried the sophisticated Wasserspiel, FuMG406 radio-guidance system developed for the Wasserpiel ('Waterfall') surface-to-air missile, but the system was so badly damaged that little detailed information could be obtained from it.

The FuMG406 radio guidance system used frequencies of the order of 125MHz and 156MHz. It was tentatively assumed from the examination of the badly damaged V2 pieces at RAF Farnborough that, the weapons might be fully guided by radio after all. This would make them vulnerable to radio counter-measures, such as a high-powered jammer transmitting on suitable frequencies. It was presumed that any guidance system would come into operation only once the rocket fuel was switched off after the initial 'silent minute' of the launch. In this phase, the rocket ascended almost vertically from the platform to which it had been transported on the specially designed mobile launcher from an underground storage facility.

Countermeasures

On 5 August, 1944 the Air Staff prepared a 'Memorandum on Counter-Measures against Flying Rocket Attack' for the Crossbow Committee of the War Cabinet. The main text of this document can be found at the Public Record Office, Kew under PREM 3/111. The memorandum reported that five main chain radar stations between Dover and Ventnor were being equipped with special devices for detecting missiles at heights of between 5000 and 50 000ft (in practice, this height proved insufficient since the V2 soared to about 75km); additionally, radar installations at Martins Mills near Dover and at Pevensey were being modified to target greater heights.

The memorandum added: 'In addition to the normal RAF 'Y' radio interception services which maintain a continuous watch for enemy radio transmission, a special organisation to listen for rocket transmissions has been set up. This consists of coast stations extending from Lowestoft to Southbourne with a control centre at Beachy Head. With radio direction-finding equipment, continuous watch is maintained on wavelengths which it is considered are likely to be used in the rocket control. Information obtained from these stations will be used to assist in determining the area from which enemy ground radio control is operating. An organisation for flash spotting and sound ranging has been set up'.

According to Brian Johnson in his book 'The Secret War' (published by BBC Publications in 1978) by 15 August, the listening watch around 27 and 40-60MHz made use of some 60 Hallicrafters S-27 VHF and RCA AR-88 HF receivers and the pre-war BBC television transmitter (which had been used for jamming during the earlier Battle of the Beams) was ready to provide a jamming signal around 45MHz.

It was stressed to the Crossbow Committee in early August that: 'Radio counter-measures against the rocket are dependent on the availability of detailed technical information about the rocket radio equipment employed by the enemy. Only when further information is available will it be possible to say whether radio counter-measures can effectively interfere with the enemy's radio system.

'The present indications are that until more complete details of the enemy's rocket radio equipment are known, effective radio counter-measures against the rocket may not be possible. We believe, however, that it will be possible to interfere to a considerable extent with any radio ranging devices employed by the enemy to give them information as to the fall of the rocket.

'The most urgent need in planning and providing the radio counter-measures is the fullest intelligence on the radio control of the enemy rocket and technical details of the radio equipment employed in carrying out radio control. The urgent need for this information is known to all the appropriate intelligence authorities, and every effort is being taken to obtain the necessary information, particularly from the rocket equipment obtained from Sweden and Poland. Aircraft fitted with radio jammers may be required to counter the enemy control of the rocket'.

Jamming Requirements

The construction of a VHF jammer directed against a Wasserspiel-type radio-guidance system clearly needed to be more powerful than any previously built. It became the responsibility of Section VIII at Whaddon to provide such a jammer. What happened is revealed here for the first time in notes written by the late Steve Dorman, shortly before his death. He was one of the Whaddon engineering officers concerned:

'I attended a briefing given by Brigadier Richard Gambier-Parry at the beginning of August 1944. Intelligence had become available that a German rocket weapon, to be known as the V2, was being prepared for launching against the UK and might be guided by a radio signal between about 120 and 180MHz. Our task was to identify and jam the guidance signal before the weapon reached the South Coast. It was estimated that a jamming power of about 50kW would be necessary. There was no known equipment in the UK or USA capable of providing this facility but it was wanted without delay'.

Section VIII had good liaison with the British radar research establishment (TRE) at Malvern, with the BBC, with Post Office Research at Dollis Hill and with RCA Laboratories in Camden, New Jersey. A suitable building was hastily constructed on the Aspidistra site at Crowborough, Sussex. As well as an adequate mains supply, it was supplied with a 1 megawatt, 3000hp, 16-cylinder Crossley Premier diesel engine and alternator as part of the 'Aspi 1' 600kW medium-wave installation. [Aspi 2, 3 and 4 were 100 and 50kW U.S. General Electric HF broadcast transmitters installed at the site after completion of the 600kW MF transmitter.]

'Within a week, arrangements had been made for RCA to send over water-cooled triode valves, a pair of which should be capable of delivering 75kW RF power over the required VHF band. We obtained access to a site on Beachy

Head for receiver and transmitter remote-control facilities. Mervyn Wells of the Post Office set up the control link to Crowborough. Rowland Lees of TRE was seconded in order to use his expertise of VHF antenna systems. Louis Varney joined us temporarily from Hanslope Park (SCU3).

Bob Peachy saw to administration. Bernard Walsh (the SCU catering officer), owner of Wheeler's Oyster bars in London, arranged accommodation for and feeding of the team. To this end, he commandeered one floor of the Beacon Hotel, Crowborough. Overall management and site facilities at Crowborough were provided by Harold Robin and Ronnie Watton. They took over a small engineering works at Crowborough to supplement the small workshop already on site. The design and construction team was mainly from the Whaddon and Hanslope Park laboratories and workshops'.

In view of the urgency of the Task Z/Silent Minute project (according to information becoming available to British Intelligence, the V2 attack could be expected to start in September 1944, one month after the initial briefing of the Whaddon team by Brigadier Gambier-Parry), the work had to be planned to a time-scale of weeks, rather than the months or even years needed for a peacetime project of this magnitude. It was decided that the fullest use should be made of conventional designs, using readily available materials and straightforward workshop facilities.

'We tried to avoid problems such as the machining of difficult materials to close tolerances and to manufacture, for example, several resonant circuit components of slightly different dimensions while the machines were set up, rather than spending time on more precise designs or risk having to re-set the machines should the component not cover the required frequencies'.

An immediate problem was the requirement for a 100kW, 7kV DC anode supply. The BBC was 'obliged' to provide the Ministry of Information with the transmitters that had recently been installed at Woofferton, near Ludlow. Within a couple of weeks, they had been dismantled by an RAF technical team and brought successively on a low-loader to Crowborough where they were re-assembled. For the heart of the jammer, two powerful, water-cooled RCA 880 valves were used for a balanced resonant-lines power oscillator, the coolant running through resonant lines made from copper tubing about 2 inches in diameter. Frequency tuning was effected by sliding bridges driven by reversible DC motors on the grid and anode parallel lines.

Some idea of the urgency in which Task Z/Operation Silent Minute was tackled is given by the fact that the team worked around the clock in two shifts, 9.00 am to 9.00 pm and 9.00 pm to 9.00 am. At times even these 12-hour shifts were extended. One Whaddon engineer recalls that he once worked from 9am to midnight and then next day 9am to 9pm – 27 hours out of 48!

At the beginning of September, several doodlebugs (V1s) came down in the neighbourhood (one close enough to blow an engineering officer off his motorcycle as he was travelling from the Beacon Hotel to the site). It was therefore decided that a second jamming transmitter should be built as an emergency reserve.

Capacitor Failures

Once the first jammer transmitter had been completed, it needed to be run up gradually to full power. When this was attempted, new problems soon began to show up. The valve grid-coupling capacitors repeatedly failed due to the very high RF current passing through them. An attempt was made to build suitable capacitors from polystyrene. but without much success. Eventually, the unorthodox solution found was to use the grid-to-filament capacitance of type 889 high-power (50kW) transmitting triodes as used in Aspi 1 – probably the most expensive capacitors ever!

Another problem arose from the huge currents circulating in the resonant lines coupling with adjacent circuits. The DC supply lines had a nasty habit of glowing red, and the fluorescent lights in the building were as bright when switched off as when switched on!

Nobody stopped to think of the potential radiation hazards to themselves. An unanticipated problem was also caused by the 600kW radiated from Aspi 1 just a few hundred yards away. When a 1 x 1/16-inch copper earthing strap was bolted to the Aspi 5 antenna feeder, the induced MF current caused it to glow red and then melt! The jammer ran with a DC supply to its anodes of 90kW with an estimated continuous-wave RF of 75kW, derived from calculating the energy absorbed by the cooling water and subtracting this from the 90kW input power.

At about 10.00 pm on the evening of Friday, 8 September, a phone call was received to say the first V2 had been launched at the UK a few hours earlier (it landed in Chiswick at 6.34 pm, killing three people including an infant girl in her cot). The Whaddon engineers rushed to the site and Aspi 5 was successfully powered up within minutes under remote control from Beachy Head. There were further alerts in the following days but it would seem that few V2s were detected before they came hurtling down from the stratosphere. Even if the warnings had been given in time, Aspi 5, working on the frequencies of the experimental hybrid A4b/Wasserspiel system, would not have affected the missiles.

With the first transmitter running well and the second nearing completion, the Section VIII engineering team began to break up and return to Whaddon to continue with urgent work in connection with the war in the Far East. The team had successfully implemented 'Silent Minute' within less than five weeks and had the system working by the time the V2 attack began although, unfortunately, the equipment had no effect at all on the gyro system.

A 125MHz radio-guidance and tracking system may have come into use for some of the later V2 rockets launched in 1945, by which time all attempts to jam the V2 missiles from Crowborough or Alexandra Palace had been abandoned. Over 80% of those launched at the UK used the Mischgerat gyro system for guidance, accepting that the accuracy of the rockets was limited to a radius of roughly 15km of the target.

By 27 March, 1945, some 3600 V2 missiles had been launched (mostly from sites in Holland). About 1400 were directed towards London, 1265 towards Antwerp, 537 towards Norwich and towns in south-east England, and relatively small numbers towards other towns in France and Belgium. About 6000 were manufactured and the attacks would have continued had it not been for the final Allied advance into Germany.

'Flying Gasholders'

Of the rockets directed at the UK, some 300 went astray or exploded en route, 1057 exploded on the mainland, 517 in London and 537 in 11 other counties. They killed 2754 people and seriously injured 6523 people. This was less than half the casualties caused by the doodlebugs but the V2s were formidable terror weapons. If you heard the explosion, it meant you had survived! You never knew when one might arrive. It was not until 8 November, 1944 that first the German media and then the British media finally admitted the existence of the V2 attack. Previously, rumours had been spread that the sudden explosions were due to gas-holders blowing up spontaneously, provoking many ribald remarks about 'flying gasholders'.

Many Londoners found it difficult to excuse a memorable government whammy. This was the press conference called by Duncan Sandys on 7 September, 1944, just one day before the first V2s came hurtling down from the stratosphere. Relieved that all the V1 launch sites in France had finally been overrun, he claimed: 'Except possibly for a few last shots, the Battle of London is over'. This view reflected the general euphoria of the moment that the war in Europe was about to end and the belief that V2 rockets would be unable to reach London from Holland. It was also expected that Holland would soon be liberated by 'Market Garden', an operation that successfully secured three important bridges over the Rhine but found the vital Arnhem Bridge 'a bridge too far'.

The single worst V2 incident was in Antwerp, which suffered no fewer than 1200 V2 attacks, more than twice the number that fell on London. A cinema was hit and 240 servicemen and 250 civilians were killed. In central

London, 110 people died when a rocket came down in Smithfield Market off Farringdon Road; and on 25 November, Woolworth's store in New Cross Road, Deptford was hit, killing at least 160 people. The V2 attacks lasted seven months; one of the two final rockets reaching the London area on 27 March, the last day, fell in Stepney, killing 134 people.

First Launching

Pat's own memories of the V2 are less dramatic. The first successful operational launching of the V2 was not of the missile that exploded in Chiswick in the evening of 8 September, 1944, but one that landed in the outskirts of Paris some 12 hours earlier. Comfortably still asleep at a SIS/SCU9 base in the pleasant 16th arrondissement, he was awoken by a distant but unusually loud double-explosion. Since there were no further explosions he returned to sleep. It was not until much later than he realised that the explosion had come from the first operational V2 to hit its target.

Later, in Holland, he and his unit were acutely embarrassed when the Dutch Underground in the northern (occupied) provinces firmly warned him that it would discontinue helping the Allies if the severe RAF bombing of the V2 launch sites near The Hague continued. Pat cannot now recall whether this followed the attacks of early 1945 or the disastrous raid by medium bombers of the 2nd Tactical Air Force on Saturday, 3 March which provoked an extremely strong protest from the Dutch Government representatives in London. Details of the raid, which killed some 800 civilians and caused some 100,000 to be evacuated from their homes, had been brought to Brussels by the secretary of the KLM airline company who successfully crossed the lines and reported 'The temper of the civilian population has become violently anti-Allied as a result of this bombardment'.

The RAF raids were the result of urgings by Herbert Morrison, Minister for Home Security, that more should be done to stop V2 launchings by using the RAF heavy bombers in mass attacks on the Dutch launch sites. This advice was firmly rejected by the prime minister and the Chiefs of Staff because of the inevitable slaughter of Dutch civilians.

Exactly what went wrong on March 3 is uncertain. None of the 70 tons of bombs fell within 500 yards of the wooded Haagsche Bosch where the rockets were being made ready for launching. Officially, the blame was put on the briefing officers, with the RAF telling the prime minister: 'Investigations are not yet complete but it seems likely that the responsibility for what occurred will be traced to one or more officers who were responsible for briefing on this occasion... This lapse is at present the subject of a Service enquiry which may result in Court-Martial proceedings'.

In effect, it was soon recognised that the V2 was virtually immune from any form of counter-attack, other than physical occupation of the launch sites. However, it proved possible to reduce the scale of the attack by some disruption of production in Germany and by attacks on the lines of communication. The possibility of SOE sabotage or Commando raids was mooted but abandoned due to the high level of German security surrounding the V2 operations.

Although, in the event, the Nazis gained little military advantage from the rockets that had taken them so long to bring into operation and that had consumed so many valuable resources, the long-range rockets opened up vast new possibilities in the conduct of military operations: 'In future the possession of superiority in long distance rocket artillery may well count for nearly as much as superiority in naval or air power' (from Public Records Office file at Kew).

None of the public records at Kew in PREM 3/111 relating to the Big Ben rocket mention what may well have been in the minds of some, namely, that such missiles might, before long, be capable of carrying nuclear warheads, each powerful enough to wipe out whole cities.

On the credit side, it has to be recognised that the wartime development of long-range missiles and liquid-fuelled rockets led directly into the Space Age and all that this has entailed.

I have another reason to be interested in Steve's work at Crowborough, namely the minor role played in the operation by my father. He was sent as part of the team to ensure that supplies of materials – especially wireless gear – was made available to the engineers in the shortest possible time. Before he left Whaddon for Crowborough, he was called in to see Gambier-Parry, and the importance of his role explained to him. He told father that if anything – *anything at all* – were to hamper the work from the supply side, then father should contact him personally and immediately.

After the war, Steve continued with the unit in the newly formed DWS. He worked at Century House (the post war replacement for Broadway), with the Technical Group for a while, and then for the remainder of his service, with HMGCC at Hanslope. There he was in charge of scientific research and development, dealing with the very sophisticated wireless and listening devices used in the Cold War.

When he retired in 1981, he told me, he was Superintendent of HMGCC at Hanslope and had a staff close to two hundred who were mostly based there. That Steve was a brilliant engineer had been widely recognised from his early days at Whaddon Hall, and his post war career in the Firm confirmed that view.

He moved to Looe in Cornwall and involved himself in numerous activities including voluntary work for the deaf and disabled, photography, rambling and a host of other interests. For most of his life he wrote a most detailed diary including recording his work on the V2. His whole life is also recorded in the most meticulous family photo albums. A truly remarkable man.

Sadly, Steve Dorman died in November 1998, but by then he already knew that I had started work on the book – charting the wartime history of Section VIII – he had so encouraged me to write.

Chapter 33

Jack Whitley's Story – Handling Our Agents' Traffic

Jack Whitley was born on 16 November, 1916 at Agbrigg, just outside Wakefield in the West Riding of Yorkshire. His mother died in 1918 from the influenza epidemic that killed more people than World War I, so he was brought up by his maternal grand-parents. He enjoyed what he describes as 'a most happy young life'. He left school at 14, the usual leaving age at the time, but found it difficult to find work as unemployment was rife in the early 1930s. Fortune smiled on him when he was befriended by leading lights in the NSPCC and they found him a post in one of the local offices.

In 1936, he joined Jackson's Stores, a Yorkshire furnishing company based in Leeds, with branches in a number of towns. He was doing well, with good prospects, but on the outbreak of World War II, he left his job in the furniture business and volunteered, before his call-up time, to join what was then the Royal Corps of Signals.

Jack's interest in Morse and radio had been fostered by his future father-in-law's notes taken during World War I. He was posted to 59th Division Signals in Liverpool for a spell of 'square-bashing', then to Rhos-on-Sea, Colwyn Bay for some sea air. In March 1940, the Army remembered the recruits at Rhos-on-Sea were supposed to be signalmen, so they were posted to Hull where, housed in dusty Londsborough Street Barracks, they attended Hull Technical College for three months, to be taught Morse code by GPO telegraphists.

The drill routine of the unit was supposedly so good that the Mayor of Hull complimented them, during a visit between air raids, as the best contingent of men he had seen on Church Parade. Jack's future wife said the same! From Hull, Jack and the other men were split into groups. Some of his new friends ended up in the Western Desert or Eritrea, while just a few of them found themselves posted to a holding unit, the 4th Line of Communications, Royal Corps of Signals, at St. Johns Wood in London.

Eventually Jack was posted out of London, taking the usual route via Bletchley station to Whaddon. He and his fellow soldiers were met at the entrance to Whaddon Hall by a swashbuckling gentleman by the name of Captain Emary. Jack was now in SSU1 and wondering what it was all about.

From then on, Jack experienced the most varied amount of radio work imaginable – along with some army life. His little group was billeted on the good people of nearby Stony Stratford. He recalls the happy family life of 'the Parkinsons' at number 7, Coronation Road, Stony Stratford.

Jack's unit was called upon at times to do guard duty at Windy Ridge, for reasons best known to others. Early on, the only weapon was a truncheon, but later he was issued with a Ross rifle which was so badly warped that it could possibly have shot round corners, for he never once hit the target during his musketry course! By now, he had been moved to Gee's farm ('Gees'), at Little Horwood, where for a time he and his fellows had to sleep in cow-stalls. Gees had a teaching wireless room in which Jack improved his Morse code in the grimmest farmyard surroundings.

Jack just home on leave but the rifle had to go with him on the journey

One day, to the unit's amazement, a fleet of magnificent Packard motor cars arrived, all fitted out with transmitters and radios, mostly Hallicrafters. The Major I/C told them that some of them were to operate from the Packards and that they would be going out into the countryside. They did more of their training work on the move, contacting HQ and other stations.

Afterwards, Jack realised that if an invasion had occurred, the SIS communications system would certainly have been mobile and effective. Their contacts would have been the various Army commands, including the RAF and the Admiralty. Windy Ridge (before the 8th Army period) worked the War Office (44), Admiralty (63), RAF (87), and Southern, Western, Northern and Scottish Commands. The numbers of the stations were the codes used, to and from Windy Ridge, instead of the actual name of the receiving station.

Jack's great pal during this period was Clifford Haigh, an ex-reporter on the Leeds Mercury newspaper, who, after serving at the British station in Stockholm for the rest of the war, became editor of the Times. Cliff was best man at Jack's wedding in Kensington in August, 1941 and they kept up their friendship in London and Folkestone, until Cliff died in 1999, aged over 90.

There was a special machine at the top end of the Operations Room in Whaddon Hall which churned out tape and was usually dealt with by the charge-hand of the watch. It handled only incoming Stockholm traffic which had become so important, because only Stockholm was left to be able to monitor our agents' work in northern Europe, especially the growing resistance movements in neighbouring Norway and Denmark.

Jack and Cliff spent periods together at Windy Ridge and also in London, mainly in the basement of the War Office (44) and in a dug-out in St James Park (RAF 87), as operators contacting HQ (Windy Ridge), and other commands. Our transmitters for RAF 87 were on Duck Island in the Park. One morning, Jack recalls going out of the station very early to take the air and found himself face-to-face with none other than Sir Anthony Eden and escort; pleasantries were exchanged. Unfortunately, during his time in the Whitehall area, Jack never encountered the great man, Winston Churchill. Jack and Cliff used their off-duty periods to attend a wireless college in Brixton and eventually gained their PMG (Postmaster-General) Certificates in wireless communications. During this period, the SSU team slept in the Quartermaster-General's House in Northumberland Avenue which was connected to the War Office.

The bombing was heavy. Jack particularly remembers going out in Whitehall after the night of the fire-bombs – it seemed like the end of the world.

At the end of 1942, the unit returned to Whaddon. Cliff was then sent to Stockholm and Jack had a period at Main Line station, handling embassy and legation work as well as some Ultra traffic. His charge-hand of the watch was 'Dinger' Bell. At this time, Jack was transferred out of the army and into MI6 Section VIII. He was then issued with a new Part 1 Army paybook (AB64 Part 1), which read 'This man is on Special Duties with permission to wear civilian clothes' and was signed by Lt. Col. Rooker. As he was now 'Not paid army funds' but paid by Section VIII, his AB64 Part 2 paybook was taken away from him.

The wireless sets at the Hall were the remarkable HROs; no more Hallicrafters, and only a few AR88s. Jack was billeted out again, this time with a railwayman and his wife, the Copperwheats, of Osborne Street, Bletchley who became good friends. Jack's first wife, Margaret, joined Bletchley Park and worked, he believes, in Hut 3 with Major Turner, Captain Price-Jones, Peggy Hooton, Jill Menheniet and others.

Main Line contacts at Whaddon Hall included Moscow, New York, Delhi, Australia, Teheran, Beirut and Stockholm. Jack remembers working New York throughout the night and early morning on ZZZ traffic (very urgent). He was told this was unusual, as the station invariably faded after midnight at that time of year, but he used his knowledge of frequency-changing when needed, which 'Dinger' Bell did his best to provide.

When Germany invaded Russia considerable changes occurred at the Hall. Around that time, Jack met Lord Sandhurst, who put him to work in a room in the loft of the Hall, to do some 'intensive listening' on an HRO. Shortly afterwards, he went to Nash station at the top of a nearby hill. Nash only handled SIS agents' traffic and was not concerned with Ultra or Foreign Office traffic.

Jack believes that one of those involved in setting up this station was Major Jan Ware who was billeted near him in his next billet with the Frosts in Old Stratford. Captain Harry Tricker was also involved, but he was sometimes at the nearby Weald station. Jack did a few watches at Weald, along with Dick Neaves and others, working mostly to French agents, but his main work was at Nash with a motley crew on a three-watch system. Jack's watch included a Mr. Krone, who had an amazing ability to tune his HRO and transfer his results to the typewriter and sometimes to Braille – yes the man was blind!

Another member of the team was an RAF Sergeant known as Smithy, Alan Manson, a very popular Section VIII charge-hand, a Royal Observer Corps man called Waite, a hefty Scottish lad called Jock Dowie with a fist that you would have thought hardly delicate enough for the Marconi keys, who seemed to like the only AR 88 in the wireless shack. There was another civilian called Aggi or Agey and Monty (whose surname escapes him) who became the charge-hand and enjoyed sucking his pipe. The whole watch were excellent operators and became good friends.

The sets were of course, HROs, except for Jock's AR88, making about eight sets altogether, which would also be the number of men on each watch. The most unusual thing about the Nash station was that it had a 'battery man' called Kenny. Somebody, at sometime, had decided the sets would provide a clearer signal if they were battery-powered, to limit noise and interference for the delicate work done at this station.

The agents they worked to were P9s in Norway, P8s in Holland, P7s in Belgium and France, also the occasional Mihailovich man in Yugoslavia. These agents had little time or inclination to advertise their presence by placing an aerial in a prominent position, merely to provide a strong signal. So Nash traffic from our agents usually consisted of a very weak signal with attendant QRM (noise). There was one exception, however, a French agent codenamed Bertie who, the watch said jokingly, must have slung his antenna on the Eiffel Tower!

Our own aerial array at Nash stood out more like Daventry on a sunny day. All the work was in code, but one day Jack received the following message *en clair*: 'Your beautiful set I haf found on the shore, wud u like I to use it 4 u.' All were utterly staggered and no immediate reply was made, but no doubt Whaddon would have coded a reply for a later sked (schedule) to this Dutchman to find out if he was genuine.

Nash did a great deal of work with P9 Norway contacts, as did the back-up station at Forfar. There were agents with names like Upsilon, Mu, Lerken, Pi and many others. Cliff Haigh, now living in Stockholm, met some of these marvellous agents when they needed support but he was always very conscious of surveillance by secret agents from the German Embassy.

The operators at Nash had no knowledge of the contents of the coded messages they had to send at prearranged times or of the incoming traffic from agents. The work became constant at the rate of six or more contacts an hour during the eight-hour watches. The only time they had any inkling of what they were doing involved the P9 area when it was threatened by the *Bismarck*, *Tirpitz* and other German battle cruisers.

Of course, every contact received maximum attention at all times, since the operators were very aware of the dangers being encountered by the agents with whom they were in contact. If they had a quiet period, Kenny, the battery man, would be busy and, amongst other things, would enthral them with yarns of his dancing prowess. He was a Victor Sylvester fan and Bletchley Park was one of his venues – or should that be happy hunting grounds?

The generator and battery rooms at Nash wireless station

With the defeat of the once-invincible Wehrmacht on the Russian front and in North Africa and Italy, and the allies' rapid advance in Europe, the Nash team was split up. Jack, along with some of the army personnel still in SCU, joined a group of stations, called the Liberation Group, which moved with the advance into Germany.

He travelled out to Paris and Brussels alone, with a covering note signed by Brigadier Gambier-Parry. It asked anybody who stopped Jack not to question him, but to contact Gambier-Parry personally, if necessary. Captain Harry Tricker and Lieutenant Jimmy James were the officers-in-charge of the team. They crossed the Rhine at Dortmund and went over a heap of rubble which was once the city of Essen. As they arrived at Suchtein, 'Monty' negotiated the peace.

In spite of the cessation of hostilities, the work continued from our units HQ in Bad Salzuflen, with several outer stations working in co-operation with Army Intelligence. The job of the unit was to rout out the Nazis. Jack took over stations at Bad Rehburg and Lubecke (not Lubeck), home of the War Office Liaison Group. The colonel-in-charge wanted to ask for Jack's transfer to his section, in view of his wireless work in the mobile van, and also in setting up a telephone switchboard – said to have been confiscated from Telefunken! Needless to say, he declined gracefully, as he felt Section VIII had been exceptionally good to him. He therefore served out his time with SCU and was demobbed in February, 1946.He reverted back to the rank of Signalman, as his higher rank of sergeant was only war substantive, as was the case with many of the nominal ranks held by those in Section VIII.

I will let Jack finish in his own words:

'I have recently been back to Nash with my son and this visit brought back a flood of memories. The old shack had gone, as had all the aerials. A metal firm now occupies some of the area. The owner remembers the demolition of our 'happy hunting lodge'. Kenny's old battery shed still stands!'

Jack returned home to work for the NSPCC and has devoted his life to that most deserving of charities.

Chapter 34

Pat Hawker – His Many Roles in the Secret Wireless War

Pat Hawker served in a number of SCUs during his wartime service with Section VIII and is therefore a real authority on our units operations. He has helped me immensely in reviewing my work and with contributions. This is his own story.

Pat was born in the West Country seaside resort of Minehead, Somerset in 1922 and soon was sharing the special fascination of wireless broadcasting for youngsters growing up in the 1920s and 1930s. The ability to pluck music and speech out of the ether, with simple equipment that could be built at home, sparked for him a sense of wonder that somehow has never gone away. There was the progression from crystal sets to simple valve sets powered by 2V accumulators (rechargeable at the local radio shop or garage), expensive 120V HT batteries, and soon, for some, the thrill of listening to distant broadcast stations on the newly blossoming 'short waves.'

He recalls that it must have been 1930-31 when he was about eight or nine years old that his eldest brother made a crystal set on which he attempted to receive the low-power BBC Cardiff transmitter across the Bristol Channel. This was not a success and Pat's brother bought, through the columns of Exchange & Mart, sufficient Black Cat cigarette coupons to obtain a two-valve 'KB Kitten' battery receiver. This brought the family some BBC programmes but Pat's own interest was stirred by the building of the new 'Twin Regional' high-power BBC transmitter at Washford Cross, just six miles from his home. This opened on 28 May 1933 on 309 and 261 metres – although quite a few of the Regional programmes were in Welsh. The strong signals soon encouraged Pat and some of his friends to build their own crystal sets for which even bedsprings provided a satisfactory aerial.

By 1933-34 he was buying for 3d (1.25p) the weekly Amateur Wireless (later absorbed into F. J. Camm's Practical Wireless), the monthly Wireless Magazine and experimenting with his first valve. This cost him 5s 6d (27.5p)- a considerable sum for schoolboys in those days. Financed, in his case, by an early morning summer round ringing the door bells of 'bed and breakfast' houses to sell fresh bread rolls (ld each or 7 for 6d) for a commission of 1d per dozen.

Then in 1935, he discovered the fascination of the 'short waves' after building a simple two-valve receiver with regeneration ('reaction') with slow-motion tuning capacitor and a coil wound on a toilet-roll former after baking it in an oven to dry out any moisture. Saw cuts were made across the terminal-type Bakelite valve holders. Nothing had to be soldered, but care was needed to ensure that 'reaction' was smooth with an absence of 'plopping'.

It must have been that autumn (1935) when he heard his first amateur wireless station: Norwegian LAIG speaking in English on what proved to be the 20-metre band. Later came a clutch of English amateurs on 40 metres. This soon led to a yearn to participate in this interesting hobby. Determined to master the technology and procedures Pat, in company with a school friend, Charles Bryant, the necessary documents including those of their long-suffering fathers were assembled, applications made, and in 1936 Pat at the age of 14, became '2BUH' and Charles' 2BXZ'. That year both Charles and he joined the Radio Society of Great Britain. It also saw Pat's first appearance in print: an article in the school magazine describing the thrills of listening to short wave broadcasts.

The AA licence was intended as a stepping stone to the full licence. This gave him two years to learn Morse code to the licence requirement of 12 words per minute. He admits he found this an irksome and difficult task, and gave up more than once. Then Charles Bryant found a former, aged, Royal Navy wireless telegraphist who for a modest fee gave them both some training sessions. He may once have been a good operator but Pat later commented that if they could copy his appalling 'fist' then they should be able to copy anyone.

But he adds that he has never regretted the time spent, and believes that a Morse requirement is a unique way of encouraging a lasting commitment to the hobby. He took his Morse Test (at the local Post Office) at the end of September 1938, at the height of the Munich crisis, and received his radiating licence G3VA that October at the then minimum age of 16 years. For some months he enjoyed operating his home-built equipment although studying for his Higher Schools Certificate; after taking this at Bristol he left school.

September 1st 1939 saw the official closing of all amateur-wireless transmissions in the UK, with the removal of Pat's transmitter by Post Office officials the next day, the eve of Neville Chamberlain's announcement that Britain was at war with Germany. For Pat, by then aged 17, this put all his own plans into abeyance. His father helped him become an articled clerk with a local firm of Chartered Accountants. He concentrated on improving his Morse, copying a lot of the Press Messages on Rugby Radio, GBR and the German Transocean agency transmissions from Nauen, etc. He also bought the final disc of a set of 78rpm Morse training records issued by Columbia, gradually improving his ability to copy at speeds up to 25wpm.

Then, in early Spring 1940, a mysterious letter arrived out of the blue from a Lord Sandhurst asking whether he would be prepared to do some voluntary work on behalf of the war effort. [He learned later that the RSGB had provided 'Sandy' with a list of their licensed members and these were then approached subject to a 'nothing known against trace' by the local police and the Security Service (MI5).] Enclosed was an extract from the Official Secrets Act that he would need to sign and return to Box 385 Howick Place, London SW I before any further information could be disclosed. So began his connection with the Radio Security Service. He found that Charles Bryant, the only other amateur in Minehead, had been similarly approached. For the next 18 months he worked in his spare time as a Voluntary Interceptor (VI) copying 'wanted' transmissions, initially on the family's Philco radiogram (to which he fitted a beat frequency oscillator). Later, on a Hallicrafters S20R 'Sky Champion' receiver loaned by RSS which soon established a South West Regional Office in Exeter under Captain D H Norton. In October 1940, RSS changed its mailing address to 'Box 25, Barnet.'

For an 18-year-old, it was thrilling to feel part of what appeared to be some sort of secret service, though little was disclosed about the transmissions he copied with their strange prefixes and some amateur style 'chat' and were obviously not normal Service procedure. It was some years before he learned that the VIs were tapping into the covert wireless networks and agents of the German Abwehr, with their messages ending up at Bletchley Park. In early 1941, a few meetings of the scattered south-west VIs were held at the Regional Office at 27 Dix's Field, Exeter. Pat was amused to note that what appeared to be a scrambler telephone was under the table and guessed [correctly] that the meeting was being monitored at Barnet.

In the early summer of 1941, he received a letter from RSS inviting him to become a full-time interceptor, as a special enlistment, in a new military unit (SCU3). With the letter came conditions of service, including attractive rates of pay, UK service only and accommodation in brick-built huts, that made it all sound like home from home. A document that later was chewed over endlessly by many 'barrack-room-lawyers,' when reality did not quite live up to promises!

Pat accepted the offer but then it was a matter of several months waiting while continuing as a VI and failing to pass another medical when 'called up' for a National Service medical at Taunton.

Finally, in early November, 1941, he was told to return the Hallicrafters S20R receiver to Exeter, to destroy all the paperwork relating to his VI work and to report to Barnet. He duly reported on Friday, November 7th, to (Major) Lord Sandhurst, went to Ravenscroft Park where he passed a 25wpm Morse test. He was told he would be working in the Discrimination section at Box 25 and allotted a civilian billet in Byng Road, Barnet, along with three other former VIs – all pre-war radio amateurs – who had reported to Barnet earlier that day.

A few days later (12 November, 1941), he was formally attested by Lord Sandhurst as Signalman No 2600077 for 'special duties'. A number of the standard questions on the Attestation Form had their answers stricken through. These included such questions as 'Are you willing to serve outside the United Kingdom,' and 'Have you received a notice paper stating the liabilities you are incurring by enlisting and do you understand and are you willing to accept them?'

Pat was issued with an Army Book (AB) 64 Part 1, with a slip of paper signed by Capt. MacIntosh for Lt. Col. SCU3: 'This soldier has been enlisted for special duty, on termination of which he is entitled to a free discharge under K.R.'s 1940, Para 390, XVIII (c). This man is on special duty and has permission to wear civilian clothes.' Later, this slip was removed from personnel at Hanslope Park, although by then Pat had moved on, and still retains the slip!

In October 1942 his AB64 was amended on the page 'Record of employment as an army tradesman' to read: 'Op (Special) Group B, Class III, Classified & Mustered', The purpose of this amendment remains obscure, since the page was intended 'For men in receipt of tradesmen's rates of pay only'. And the only time that Pat received Army Pay was in 1946 during his demobilisation leave!

He was thus the seventy seventh recruit to this special '26' series of Royal Corps of Signals army numbers, but in reality issued only to the Special Communication units controlled by Section VIII of MI6/SIS (though this was never disclosed to the newcomers).

For the first few months, Pat took advantage of the permission to wear civilian clothes when going on leave or visiting London or Northampton on his days-off, but soon found that the uniform was warmer and that young men wearing uniform were given preferential treatment!

The four ex-VI recruits that week – Jimmy Adams, Peter Gourlay, Ron Delahunt and Pat (all four pre-war radio amateurs – were intended to work in 'Discrimination' (traffic analysis) run at Arkley View, Barnet by Major Kenneth Morton Evans (a regular officer and keen wireless amateur) who had been seconded to MI5/RSS that year. Pat soon found he had possibly made a tactical mistake in being the only one of the four who had at the first attempt passed the A-grade (then 25wpm) Morse test. After a few days at Arkley View in 'Discrim', it was decided that he should be posted to Hanslope Park as an intercept operator, leaving his B-grade colleagues to see out the war in the comparative comfort of Barnet – though he admits that in the outcome, the posting led to a more interesting war!

Many years later, in the 1980s, he was to write: *'Browsing through the 'Appointments' section of Wireless World my eyes alighted on a large illustration of cottages set around a strangely familiar church steeple. An 'invitation by HMGCC for graduate-status engineers and scientists to come and work in the 'high-tech countryside' of Hanslope Park, Buckinghamshire.'*

'My mind went back to a depressing evening in November 1941 when, still in my 'teens, I accepted an 'invitation' to this country estate – to find myself working in a hastily converted granary. Nor do I recall, as the advert. puts it,

that the Park was "a mere stone's throw away from this delightful rural village"'with memories of the long footslog back to the Park from the four pubs, the one tea-shop, and later the excellent WVS canteen run by the good ladies of the village.

'Times change. Nobody then suggested that it would be particularly helpful' if I described 'the type of working environment most suited by my career plans.' Rather I recall a highly irate adjutant (Captain Ash) telling me in no uncertain terms that it was not my job to think!

'The village served mainly as a dormitory for those employed in the railway and printing works of nearby Wolverton. The vicar made the "News of the World" for his alleged activities on visits to wicked London. Hanslope Park, itself, had been the scene of a notable pre-World War I murder (when, in 1912, the owner was shot by his gamekeeper just outside the gates of the Park)

'Nevertheless, "The Farmyard" as it was sometimes called, was not without distinction. Among those who worked for a time there was the brilliant Alan Mathison Turing, pioneer of digital computing mathematics, and advanced cryptography, though he clashed with the local constabulary by riding his bicycle to work wearing a gas- mask (a sensible precaution in view of his hay fever, a problem with rural workplaces. Engineering was under Dick Keen whose book 'Wireless Direction Finding' was for long the classic text in this field.

'The intercept activities were at first in charge of Captain Prickett, an easy-going character who once told Reg Cole and myself never to ask permission to do something since he might have to refuse: "Do it and ask afterwards" he said, though I did hear that he was later in difficulties for putting his precepts into practice.'

Pat came to the Park with Jack Kelsall, a former maritime wireless officer, and two Welsh amateur operators, Stan Thomas and Les Garley. Problems arose immediately. The first of the four brick built accommodation huts was already occupied by the earlier operators and engineers, so room was made for the newcomers in Hut 2 occupied by the General Duties soldiers including some old lags who not only represented a security risk but were a culture shock for the former VIs who still considered themselves at least 'semi-civilians'. The following morning Jack Kelsall protested at the arrangement and the GD men moved out of Hut 2.

A storm in a teacup but one that underlined that the Adjutant, Captain Ash, was taking seriously the task he had been given of turning the special enlistments into something akin to 'real soldiers'. This became all too clear over the following months as further batches of operators arrive at the Park from Barnet, including not only amateurs but also former sea-going and Post Office telegraphists, etc. RSS, like Whaddon, was clearly seeking out experienced or newly trained Morse operators from wherever they could find them, enlisting not only former VIs but also those Post Office telegraphists who had worked for RSS under the original Post Office contract – terminated when control of RSS passed in Spring 1941 to Richard Gambier-Parry. Pay parades, occasional training sessions under a regular army sergeant, continued in the Park for months.

Soon there developed low morale among the Hanslope operators, though all were anxious to carry out diligently the duties for which they had volunteered. It all culminated in Ash leading off-duty operators on a 'route march' through the local villages about March 1942. Due on watch [at The Lodge] immediately after their return, the operators made their feeling known to Barnet by complaining on their log sheets that they were too tired to listen effectively. The result was that Ash's military ardour was damped down although, at least during Pat's time at Hanslope never entirely eliminated. It sometimes seemed that a few of the early RSS officers who apparently were on regular army pay and had to pay mess bills resented the special enlistment pay of the SCU3 volunteers. Compared to those called up for compulsory National Service, the special enlistment "toy soldiers" had little real reason to complain, but many of them felt that compared to the promises made in the original 'conditions of service' document, Hanslope Park left much to be desired – and were not conducive with the skilled task of intercepting weak signals from an enemy increasing learning new ways of making their task more difficult.

In the first few weeks, some of the operators made their feelings known to F J ('Dud') Charman who as a prominent radio amateur and Council Member of the RSGB was one of the earliest RSS VI Group Leaders. He was a professional engineer with EMI and had been seconded temporarily to Hanslope Park to design the pioneering 'Aerial distribution broadband amplifiers' for the new station. His design resulted in low-noise, highly linear broadband amplifiers at the leading-edge of the then technology. They became a standard not only for Hanslope Park but also for many other intercept stations – and indeed remained in use for many decades. His design stemmed from the work he had carried out in the late 1930s in developing a television distribution system for Radiolympia. He achieved remarkable broadband performance by using a medium-power transmitting valve (type 807) for amplifying the multiple incoming signals without incurring the problem of intermodulation, each amplifier feeding some eight receivers.

Charman was in touch with Lord Sandhurst who at the end of December 1941 had handed over his large body of VIs to Major Sabine at Barnet and had been promoted Lt. Colonel in charge of Section VIIIP at Whaddon Hall, running the clandestine communications with Western European agents in occupied countries.

'Sandy' (also known to VIs as 'Dogsbody') was no longer in a position directly to influence RSS policy; indeed both he and particularly his wife had little affection for [Colonel) Ted Maltby who had been made Controller, RSS by Gambier-Parry. Unlike most of the original Section VIII senior personnel, Maltby had not come from Philco (GB) but had been chief salesman in a leading London hi-fi and recording firm well used to ingratiating himself with his customers and superiors. Sandy did what he could and the situation at Hanslope began to improve. In June 1942, in a letter to Dud Charman, Sandy wrote from SCU 1 at Whaddon:

'Glad you think that, on the whole, they are happier at the Farmyard than they were in the winter; they certainly ought to be and, as you say, will be very much more so when 'Dud's Masterpiece' is working. I am afraid I cannot agree that it would have been better for them not to have moved to Hanslope; it would have been quite impossible for them to have done what they have done, and are doing, in their own homes. The difference between the volume of work since they took over and that produced by our friends the Post Office is just nobody's business. The Military side of it, particularly the over-militarisation, is being rapidly and firmly dealt with.'

Between the end of November 1941 when the small station in the granary, equipped with some half-dozen HRO communications receivers, was closed. In May 1942, a new temporary intercept station was set up in 'The Lodge' about half a mile away at Bullington End, on the road towards Wolverton. The early Hanslope operators were accommodated there as well as working in The Lodge, which for most seemed a welcome relief from the pseudo-military atmosphere ruling in the Park. [The 'Lodge' is now the Hatton Court Hotel].

Hanslope was very different from the villages of Pat's native West Somerset. It served mainly as a dormitory for those employed in the railway and printing works of nearby Wolverton (now part of Milton Keynes). It was served by the LMS Castlethorpe railway station, destined to disappear post war in the Beeching cuts.

The early operators found the Lodge much more relaxed than the Park and were left largely to run the place themselves. Many lasting friendships were made – Pat shared a room with Watson (Bill) Peat. Johnny Bowers and Des Downing (a Northern Irish medical student who left a few months later to become a commissioned officer for RSS in India). All four were amateur enthusiasts in their teens and had been VIs. At the Lodge, Pat intercepted Group 5 ('Patrick') services that used different procedures to the main Abwehr services. He believes that these transmissions may have been those of the Hungarian intelligence services in the Balkans, closely linked with but independent of German Military Intelligence.

Several of the Hanslope operators took to expressing their disquiet in doggerel verse. For example, the excellent Scottish intercept operator George Proctor penned with some feeling:

When first I came to Hanslope
And saw its lovely huts
I said the Army's lousy
Why did I join? I'm nuts.

But now I am confounded
For it is plain to see
If I think the Army's lousy
It thinks the same of me.

The new purpose-built 32-bank intercept station, with each bank equipped with two (or in two cases with three) HRO communications receivers was opened in May 1942 by which time the number of intercept operators and engineers – mainly ex-VIs but also former Post Office staff and ships' wireless-officers, etc. – had risen to well over a hundred. The new station was equipped with a formidable set of directional aerials (rhombics and vee beams) covering all points of the compass, feeding to multiple receivers though the distribution amplifiers designed by Dud Charman. The aerials were erected, under the supervision of Robin Addie, by a detachment of the NCC (Non-Combatant Corps) under the vividly expressive Sergeant 'Digger' Buick, an Australian with an extensive 'non- Parliamentary' vocabulary who had been one of the many VIs recruited by Charman at EMI. The NCC squad, mostly highly-intelligent conscientious objectors who had been clearing bomb damage in London, seemed to find the onerous task of erecting the high wooden poles a more attractive proposition, spurred on as they were by Buick's shouted obscenities. Later they erected many aerials for the Whaddon SCUI stations as well as the RSS station at Forfar.

Pat worked at the station until about April 1943. At first concentrating on the Abwehr Centre 5 (2/500) services in the Balkans, a busy and skilful network spreading from Sofia down to Salonika, Athens and the Greek Islands, etc. Of this traffic, it has been written by [Cmdr.] Ewen Montague: 'A number of Abwehr and other stations in the Aegean and Greek Islands we were sending excellent and informative reports.... for a very long time they constituted virtually our only information from these areas.'

Later, Pat agreed to take charge of the 'General Search' section of four banks, totalling ten receivers. For this he rose to the giddy height of Lance Corporal. Most weeks he travelled down to Barnet to be briefed in Discrim on new developments by Jimmy Adams who had joined SCU3 on the same day as himself.

Life at Hanslope gradually became more pleasant as the facilities improved, with a medical hut, a large mess hut and a NAFFI. Some of the operators, including Pat, brought their bicycles to the Park, bringing the back-projection cinemas at Wolverton and Newport Pagnell in easy reach; similarly, the WVS at Stony Stratford – much frequented by WAAF teleprinter operators from Bletchley Park, etc. Although in some respects the Park still retained its attempt to imitate a genuine army camp, this became far less onerous. Invitations to dances at the Bletchley Park WREN outstations at Gayhurst and Wavendon were welcomed. Within the station, there were flushing toilets, although the Park still depended on cesspits.

The station kept continuous 24-watch in three eight hour shifts. The station became something of a show place for VIPs, with Col. Maltby clearly in his element in showing them round, at least until one of the visiting 'brass' wrote DUST on a massive disc recorder.

The practical importance of intercepting and decyphering the flood of messages transmitted by the large number of Abwehr networks throughout Europe is well explained by (Commander] Ewen Montague of Naval Intelligence Section 17M in his book 'Beyond Top Secret U' published by Peter Davies Ltd in 1977. There is however, no mention of the Radio Security Service, SCU3, or the VIs as being the agency which was actually responsible for receiving the 'Special Intelligence' he used 'in defeating the German Secret Service, the Abwehr, at its own game and using it not to inform but to mislead, and sometimes vitally mislead, the German General Staff' – to quote from the foreword by Hugh Trevor-Roper [Lord Dacre].

Then, about March or April 1943, Lord Sandhurst, who as we have seen, had left RSS at the end of 1941 for SCU1 at Whaddon Hall to run the Section VIII(P) clandestine wireless links with occupied Western Europe, came to the Park to seek out more operators for his Weald control station. About 20 of the Hanslope operators including Pat's friends Johnny Bowers, 'Bill' Peat and Reg Cole volunteered to transfer to SCU1. Pat, as a Group Leader, had more difficulty in transferring, but succeeded a week or so later.

So ended his three years of working for RSS, first as a spare-time VI at home and then some 18 months as an SCU3 intercept operator at Hanslope Park. Interception of the Abwehr traffic was a highly skilled but often rather tedious job, with little information given to the operators on the cypher traffic they were taking or the actual locations of the stations. Pat recalls that he spent many off-duty hours working out from the single Great Circle bearing given to operators to enable them to select the most appropriate rhombic aerial just where the main Abwehr network centres and sub-centres were located. It was not until some RSS files were placed in the Public Record Office in the 1990s that he found that virtually all his calculated locations were correct – no easy task since there were no cross-bearings; the only other information to go by were the frequencies used, giving a very rough indication of the distance between the stations.

Pat still feels that relatively little attention has been paid by the media to the secret listeners of RSS and the value of their work to British Intelligence, counter-intelligence and deception. Throughout the war they remained a separate organisation from the 'Y' service; expanded from a tiny number in September 1939 to a peak of almost 3000 including over 1300 operators, 80 engineers and 470 administrative personnel plus 125 civilian clerks and some 1200 or more VIs throughout the UK. It had intercept stations in the UK, Middle East, Gibraltar and North America. As a 'Most Secret Source' it penetrated to the heart of Abwehr.

MI5 records, released recently to the Public Record Office, show that this 'Most Secret Source' warned them in advance of German plans to infiltrate spies into the UK enabling them to be picked up on arrival and then 'turned' to become part of the 'Double Cross' operation.

It played a vital part in deception operations including 'Operation Mincemeat' (the floating ashore on the Spanish coast of a body carrying false documents suggesting that the Allies planned to assault Greece rather than Sicily). Even more important was its role in 'Fortitude' (the deception plan for the Normandy campaign), confirming that the German High Command was being successfully misled by the radio link between London and Spain carrying the traffic of the double-agent Garbo.

The Garbo link, run by peacetime radio amateurs seconded to MI5 from RSS, was also of immense importance to BP since the Abwehr so trusted their 'agent' that they sent him their latest and most complex agent cyphers and signal plans.

Pat spent a year at the Weald station, run by Captain Harry Tricker, with, as his charge hand, CQMS Jimmy James, a Canadian who had previously been a 'Main Line' operator at Belgrade. Weald with some ten operating positions formed the control station for MI6, BCRA and some M19 agents, mainly in France P1 and P5 (the Polish-French group) but with some Belgium and at least one in Denmark. For a time during 1943, Weald also communicated with the clandestine vessels [Slocum's Navy – see chapter 38] sailing from Cornwall or the Scilly islands to Brittany to rendezvous with the French CND intelligence group led by 'Remy.' This involved 'continuous watch' during their voyages – always a tedious duty, but made rather more interesting by the use at Weald of the RAF Syko cypher with its small abacus-type 'machine.' There was a short daily contact with the Beagle weather group in Belgium led by Albert Toussaint that remained on-air until the final liberation of Belgium in September 1944.

Although the station remained open throughout the 24 hours, unlike Hanslope there were only two 'watches,' with the recognition that half the operators could take a four-hour rest in the adjoining hut but during the long evening-overnight watch – it must be admitted that a bench served as a 'bed', no straw mattresses, and commonly shared, never washed blankets that were decidedly smelly. However there was a complete absence of military 'bull,' with some of the operators remaining civilians wearing Royal Observer Corps uniforms – a privilege that Gambier-Parry had secured in November 1941, not only for Section VIII operators but applied also to the RSS VIs.

If arrival at Hanslope Park had been a culture shock, Pat found Weald something of a technological shock with only crude, semi-vertical wires suspended from relatively low cantilever wires as aerials. It was not until 1944 that 'Digger' and his crew of NCC aerial erectors arrived to erect high masts and improved aerials. The station depended upon batteries rather than mains-electricity until after Pat had left. At night, lighting came from a few hurricane lamps.

The transmitters at nearby Calverton were a mixed bag of American (REL) 750-watt units and various British and Whaddon 100-watt and 30-watt transmitters – of which only the REL units seemed to be equipped with fast-acting keying relays. The early 'signal plans' were crude and would have given little difficulty to skilled German interceptors; these were substantially improved soon after the SCU3 operators reached the station. But the overall impression was that Section VIII was giving less priority and resources to the agents than RSS was able to give to interception of the Abwehr and other enemy transmissions.

Most of the agents were equipped with simple and rather crude low-power transmitter-receivers such as the 'Paraset' (Mark VII) built by SCUI in the workshops at Whaddon and Little Horwood. Some of the agents were good operators but for some it was a struggle to copy their cypher messages. Agents using the MkVII needed 'safecrackers fingers' to tune the two-valve regenerative receivers which covered 3 to 8 MHz in a single band. It is much to their credit that they succeeded in transmitting their messages at all under the stress and hazards that they must have been suffering,

At Weald, there was the thrill of copying their faint signals on the poor aerials. But there was also the trauma of knowing, when a-station suddenly went off the air, that the operator could have fallen into the hands of the German direction finding teams, or was engaged in a shoot out in a desperate attempt to escape. Or he (or she), may have been 'blown' by the 'V-men' penetration agents that the Funk Abwehr used, often with devastating effect.

With the approach of the Normandy campaign, Pat, with several of his amateur-radio colleagues and other of the younger operators from Weald and Nash, were marked down to join a new Section VIII unit – SCU9 with Major Tricker as CO. This was to form a mobile unit in connection with World War II's largest single Allied Intelligence operation. Christened 'Sussex' it involved parachuting 50 two-man wireless-equipped teams of French secret agents in a wide sweep from Brittany to the Belgian border.

Their function was to report German troop movements and other intelligence independently of existing Intelligence or Resistance groups. Sussex was jointly run by MI6/OSS/BCRA. Responsibility for the radio links was divided between Section VIIIP (Brissex) and OSS (Ossex). Most of the French agents were recruited in French North Africa by one of de Gaulle's most experienced agents, Colonel 'Remy' (Gilbert Renault). They trained at St Albans, with those intended as wireless operators, receiving further training at the Section VIII wireless-school behind Harrod's in Knightsbridge [23 Hans Place] and during practice exercises in the UK.

The first agents were dispatched in April 1944 (signal plan WAYFARER). The 25 Ossex teams were to work to the OSS Special Intelligence wireless station 'Victor' at Hurley, Berkshire but were equipped with Whaddon MkVII agent-sets and with the Whaddon ASCENSION R/T equipment for working to aircraft. To provide liaison between 21st Army Group and Brissex (and other secret intelligence operations), a No 2 Intelligence

(Underground) Section was set up, providing a cypher section and administration of both Intelligence Corps and Signals personnel 2I(U) Sect. The total staff numbered about 40 compared with 20 (with 12 operators) in SCU9.

Bad weather and the loss of one of the two Mulberry floating harbours slowed down and delayed the many Allied units waiting to go over to Normandy. SCU9 managed to get a small forward party (one small signals vehicle, CQMS Gerrish and Les Holyord and Jock Dowie, both VIIIP operators from Nash) over in June. SCU9 was equipped at Whaddon with two signals vehicles, a main vehicle with some six operating positions fitted with HRO receivers and MkIII transmitters and one higher power MkX transmitter, that with an 813 output valve, could provide some 150 watts output. Electric power could be supplied by a 2kW generator or 150-watt petrol electric generators both made by the American firm Onan, as well as British 'Tiny Tim' petrol-electric battery chargers. The vehicles were not intended for operating on the move, but with guyed wooden poles supporting relatively low aerials.

But the main SCU9 party spent some six weeks hanging about; partly at Weald, partly at St Albans (at the Sussex agents training school TS7), partly at Leigh-on-Sea where they joined up with 21(U)Sect, and then five days on the American LST 958, mostly anchored off Southend. Then, at last, through the Straits of Dover (with the French coast still under German control) to the Mulberry at Arromanches.

Even in Normandy, Pat recalls a series of moves. First the local population at St Gabriel objected to their using a convent as a mess hall (it had been similarly used by the Germans), then the unit was moved from near Port-en-Bessin when the Americans claimed this was in their zone of operations. 21(U)/SCU9 came to rest making camp in the grounds of a rural chateaux near Juaye, where the main disturbances were the almost daily thunderstorms. SCU9 communicated with Weald and the operators filled in time listening for the Brissex agents, although these were using Weald as their control station. The Allied forces were still bogged down with the British and Canadians halted outside Caen. It remained virtually a static situation until mid-August when the Americans made their break-out in Brittany.

Pat recalls this brought about a change in the role of the VIIIP Signals Section, with operators attached to individual Intelligence officers or to forward units etc. The first was his friend the late Watson ('Bill') Peat, who accompanied Colonel Henderson of 'Sussex' to Rennes and newly liberated areas of Brittany.

Then, on Monday, August 21, the British press and broadcasts carried stories that the Americans were fast approaching Paris where, it was claimed, a revolt had broken out. German bulletins admitted that 'irresponsible elements in Paris have taken up arms' and the nightly curfew had been extended. It became clear later that the prime mover was not organised French Resistance, but the Paris police who, with exceptions, had previously done little to support the underground fighters but now were anxious to jump on the Allied bandwagon. By Wednesday, Radio France (Algiers) even announced that Paris was 'again a free city.'

Acting on the belief that Paris had been or about to be liberated, the SCU9 signals vehicle nicknamed 'Eskimo Nell' was manned by CQMS Gerry Gerrish and Bill Peat and were told to accompany a small 2I(U) detachment in making their way to Paris. This vehicle was a Guy 15cwt wireless van fitted out by Mobile Construction at Whaddon Hall and it carried a MkIII 30-watt transmitter and HRO receiver. However, the week passed without any replies to Juaye's increasingly urgent calls after a personal priority message came from London, via Weald, for General de Gaulle! It soon became clear that the news that Paris was free had been premature and that the Germans were close to suppressing the Paris revolt. The Americans with General Le Clerc's French armoured column at its head were held up outside Paris until a secret armistice was signed late on Friday.

Come Saturday, August 26, 1944 and there was still no wireless contact with 'Eskimo Nell.' The de Gaulle message was causing acute embarrassment not only in Juaye but also in London and Whaddon. The decision was taken to send two jeeps into Paris, one carrying an Intelligence officer to deliver the message by hand; the other with Pat to find out what had happened to Gerrish and Peat.

Pat recalls how the two-Jeep convoy left Juaye early on the morning of Sunday 27th August, making a long detour south of the Falaise gap where the Germans were still holding out. They passed through the destroyed Normandy villages of Flers, Villers and Conde. Then fast along the poplar-lined avenues of a sunny countryside and through Alencon, Chartres, Rambouillet, virtually unmarked by war, until, they came to the heavily bombed railyards outside Paris.

While the first wild frenzy of Liberation had partly subsided, the welcome everywhere was overwhelmingly warm, with the British uniforms in the two jeeps hailed almost as liberators. By late afternoon, they arrived at the Hotel de Ville to deliver that overdue message for de Gaulle.

The missing operators had reached Paris late on the Saturday, joined in the celebrations, and on the Sunday installed their radio in a house in the fashionable XVIth arrondissement, finally making contact with Juaye while the new party were already en route to Paris. With the arrival in the next few days of Major Tricker with John Bowers, another SCU9 operator, Gerrish (deemed responsible for not attempting to make radio contact until reaching Paris) was ordered back to Normandy (he later served a spell in the glasshouse for running a major black-market operation in Brussels).

More happily, Bill Peat was soon heading for Holland, arriving in Eindhoven as the Arnhem airborne landings were taking place. After some months in Eindhoven running a very busy link with Weald, assisted by Stewart Francis, he then accompanied an Intelligence officer into Germany. He was later commissioned and stationed at the SCU11/12 base at Calcutta in India. Pat remained in touch with him until his death in May 2001.

Pat says his own arrival in Paris and subsequent stay there for the next six weeks was a memorable experience. It seemed the war in Europe was rapidly moving to its conclusion, although he recalls being woken up early in the morning of September 8th by a distant double-thump explosion. Later it transpired that this was the first V2 rocket to be launched in earnest by the Germans, the only one to be targeted at Paris, some twelve hours before the first V2 aimed at the UK landed in Chiswick, London.

Pat Hawker in SCU9 wearing a much-prized leather jerkin

Pat often took the opportunity to look around central Paris. One occasion, accompanied by Johnny Bowers, was on Tuesday, August 29. With the Metro still not running, this involved a long walk to the Trocadero, the Etoile and then down the Champs Elysees. The wide boulevards, the stately squares and the neat parks were virtually untouched by the earlier fighting except for the pock marks of small-arms fire. Compared with London, the fashionable gift shops along the Boulevard Hausseman, the modists of the Rue de Rivoli and the Rue de la Paix were all soon displaying luxury goods virtually unknown in wartime Britain. Only the food shops appeared empty. Returning up the Camps Elysees, they were surprised to see an enormous parade of American troops marching, perhaps twenty or more abreast, military vehicles interspersed by a half-dozen French military bands, coming from the Etoile, clearly conceived as a Victory Parade.

Absent from the parade were any British or French troops and, when reports of the parade began to appear in the British press, a row blew up with criticism of American political insensitivity. This was to have the result that the parade was soon declared by the British authorities never to have happened! They insisted that there had been no Victory parade, just a few American troops passing through Paris on their way to the Front! If this had been true, Pat and Johnny had witnessed the most curious advance into battle of World War II. But with the passing of years, photographs of that parade have been published, an endorsement of the phrase 'to lie like a communiqué'. But that afternoon some Frenchmen, noticing the absence of British uniforms from the parade, insisted that Johnny Bowers and Pat should adjourn with them to a nearby café. Vive l'entente cordiale!

On 10th September, another attempt was made to assassinate de Gaulle. While speaking at the Trocadero he was fired at by an unknown marksman from the Eiffel Tower. That afternoon, finding the tower re-opened for the first time since the Liberation of Paris, and undaunted by the 'non marche pas' of the lifts, Pat was painstakingly climbing the narrow spiral iron staircase towards the first platform when a young Frenchman came rushing down. Pat politely squeezed aside, little guessing then that this must have been the would-be assassin.

In those early days of the Liberation, this was still a Paris without buses, without the Metro, with electricity switched on only fitfully. Bicycles were in vogue, including bicycle-taxis. Civilian cars, many taken over by the FFI (French Forces of the Interior), trailed 'gaz de bois' wood-burning stoves that helped to relieve petrol shortages in France during the occupation. SCU9 depended on Onan petrol-electric generators.

When the Metro finally reopened on some routes on September 11, the SCU9 operators found themselves the centre of a small riot when an attempt was made to buy tickets. For four years the 'Grey Mice' had not used tickets. Parisians were determined that the Allied troops should fare no worse! Although there were at first very few British servicemen in Paris by the end of September, Paris was turning into a leave centre, mainly for the Americans. Entertainment shows were organised at the large Olympia Theatre at which American and French stars performed, including Fred Astaire and the Quintette of the Hot Club of France with Django Reinhardt. The Resistance film of the Liberation of Paris – shots from which still turn up on television programmes – was showing at the large Metropolitan cinema.

For a time, wireless traffic for the newly re-established British Embassy in the Rue St Honoré was handled by operators from SCU9, with gear stripped out of its larger 6-position wireless vehicle. Then Section VIII sent over two 'Main line' operators from Whaddon to form an Embassy station (sadly losing Major Jack Saunders, who was in charge of 'Main Line', when his Lysander vanished over the Channel on his return to Britain).

All good things come to an end. On October 10, the Paris section of SCU9 moved up to join the main unit by now established in Brussels, with outstations in Eindhoven and Liege.

Three weeks later, Pat was again on the move, this time as personal operator for an Intelligence officer heading for Nijmegen where there existed a carrier-telephone system at the local electricity-generating station in touch with Dutch resistance across the Waal and Lower Rhine. Pat was given a double transposition 'poem' cypher (LMT cypher) a break with the normal MI6 rule separating wireless operating from any coding operations other than Syko. For a day or two, Pat joined Watson Peat and Stewart Francis at their station at Eindhoven, extremely busy sending Dutch Intelligence reaching their centre at the Abbe Museum.

Moving on to Nijmegan, then virtually the limit of the Allied advance, Pat found himself mixed up with IS9 (WEA), the M19 escape/evader unit. IS9 was virtually another private army' under Major Airey Neave (the first British officer to escape from Colditz, later as an MP, he was killed by an Irish terrorist car bomb at the

The Belgian 'Meteorological Service' for the RAF – a group of young Belgians who provided a daily service of weather reports and other intelligence through several years of occupation. Despite the efforts of the German direction-finding teams the service was never closed down although not all the group survived. It worked to our station at Weald.

Houses of Parliament) and Captain Hugh Fraser, a former SAS officer – later also an MP. IS9 forward section, with its batch of tearaways, Belgian SAS and Dutch resistors, was soon planning Pegasus, a second attempt to bring back more Arnhem evaders still hidden by the Dutch across the rivers (the earlier Pegasus 1 had rescued some 136 airborne troops).

Unfortunately, Pegasus 2 was an abysmal failure with the evaders and their Dutch guides ambushed by the Germans, some 36 killed and almost all the others captured. Pat moved with his 21(U) officer to Helmond where the Guards Armoured Division was based until suddenly ordered south by Montgomery during the Battle of the Bulge.

Pat returned to Brussels for Christmas, but while there he was asked by Major Tricker whether, instead of returning to Helmond, he would go to Eindhoven on loan to the Netherlands Intelligence Department (Bureau Inlichtingen). The Dutch badly needed two skilled wireless operators for the two Eindhoven control stations of its clandestine radio network with the still occupied northern provinces.

At Eindhoven, as a Local Sergeant, he worked for the remainder of the war in Europe as chief operator at the Abbe Museum (the BI headquarters), the only Englishman in a network set up and controlled by the Dutch. (The other SCU9 operator, Sgt. Bert Lawler. Was similarly loaned to BI for a second group of station controlled from a separately located control station). Copies of much of this traffic were forwarded to Weald by the Eindhoven forward 21(U)/SCU9 'Liberation Group' link.

The network had been set up by Jan Thijssen of the Council of Resistance (RvV) and the Orde Dienst (OD) but later incorporated into the main Dutch Forces of the Interior under Prince Bernhard. The radio group included a number of Dutch pre-war amateurs as well as some professional wireless-telegraphists.

Pat still recalls, with tremendous respect, the work of Jack Verhagen, a former Dutch marine radio officer, who sent him a stream of cypher messages at average speeds (timed) of over 25 five-letter cypher groups per minute from a secret station set up by Jan Zadbergen, PAOZY, in the nurses bathroom of the St Elisabeth hospital at Alkmaar, north of Amsterdam. This station, carrying traffic from the Amsterdam HQ of the Dutch Resistance Forces, was one of the very few that survived. Most of the other operators and associates, including Jan Thijssen, were arrested and executed by the Germans.

In the North, the Dutch civilians experience famine conditions with some 15,000 starving to death in the dreadful 'Hunger-Winter' of 1944/45. There were also many civilian casualties from the RAF bombing of the V2 launch sites. This nearly resulted in the Dutch Underground withdrawing the cooperation that resulted in a stream of Intelligence reaching Eindhoven, not only by the covert wireless links but also by secret telephone systems and Dutch line-crossers.

A few weeks after the end of the war in Europe, Pat rejoined the main SCU9 party in the British Zone of Germany, moving soon to Bad Salzuflen, but staying there only a few weeks before spending some weeks back in Brussels as a relief operator and later leaving to set up an out-station in Bad Godesberg and later another near Bonn. Six months in the Rhineland gave Pat an insight into the complexities (and often absurdities) of most secret intelligence operations. A life that more closely resembled Graham Greene's 'Our Man in Havana' than James Bond, or even the more realistic creations of John Le Carre.

The Rhineland gave him the unusual experience of once decoding a one-time-pad message that read 'Hawker is not repeat not to have access to code books.' When Pat showed this to his officer, he laughed and told Pat to carry on as before – he had no fancy of doing the time-consuming coding himself! BP and Whaddon were apparently shocked at the idea that some of the outstation operators were using SIS codebooks and one-time pads, especially as they had received no formal training at BP!

Pat returned to Bletchley at the end of January 1946, working at the Hall and then later again at Hanslope Park as a 'main line' operator for what was, in 1947, to become the Diplomatic Wireless Service. He had however, decided not to stay with the organisation and was duly discharged in the autumn of 1946, returning to his home in West Somerset.

Like other Special Enlistments, he was discharged under Kings Regulations (KRs), 1940 Para 390/XVIII/a 'His service being no longer required for the purpose for which he enlisted' – the KR more normally used for getting rid of undesirables! But at least his glowing 'Testimonial' included 'Was specially recommended for his good work in Holland during operations.'

But after six years in the closed and secret world of Signals Intelligence and then Secret Intelligence Pat found it difficult to resume an uneventful civilian life as an articled clerk. He could not settle down to study accountancy and in September 1947 he departed for London as an assistant to John Clarricoats, the General Secretary and Editor of the Radio Society of Great Britain, lured by the prospect of combining radio with editorial work. For three years he made Amateur Radio a profession as well as a hobby.

In early 1951, he joined the Technical Books Department of George Newnes Ltd and was soon editing and compiling the many volumes of 'Radio & Television Service', 'Radio & Television Engineers-Reference Book' and many other titles, building the foundation of a later career as a technical journalist and engineering information officer with the Independent Broadcasting Authority. In retirement Pat continues to be a highly respected writer on radio and television technology.

Chapter 35

Bob King, a VI and Member of RSS

Bob King is a well known authority on VIs and was a member of RSS which came under SCU3 – part of Gambier-Parry's organisation. This is his story.

Bob King became interested in the mysteries of wireless at an early age. By the time he was 14, he was making short-wave receivers from old battery broadcast receivers. Discovering so many Morse signals which he could not understand, he taught himself to read Morse code and was thus able to listen to radio amateurs throughout the world.

Being too young to obtain a transmitting licence, Bob fixed up a pair of wires across a field to a similarly interested friend some distance away. Using buzzers and Morse keys made from brass strips, they sent Morse code to each other. During a thunderstorm, a vivid display of lightning discharge took place across the key contacts!

Bob was only 16 in 1940 when Harry Wadley, a local radio ham and former naval officer already working as a voluntary interceptor (VI) for the RSS who knew of Bob's ability, asked him if he would like to use his skills on war work. Bob, who was already in the Air Training Corps, allowed his name to be submitted to Box 25. This was followed by a police check-up and a visit from Captain Hall, a Royal Corps of Signals Regional Officer. After being sworn to secrecy and signing the declaration under the Official Secrets Act, his spare time was soon fully occupied intercepting hundreds of messages from what was, to him, an unknown source.

Under the conditions of total war, with the likelihood of enemy invasion still in people's minds, Bob's family accepted that secrecy was paramount and asked no questions. He managed to afford to buy second-hand the cheapest receiver in the Hallicrafter range, which he modified with the addition of a regenerative radio frequency amplifier. With careful handling, it gave excellent results.

In due course, Box 25 contacted Bob to suggest that he work full-time there before his call-up papers arrived. Giving up all thoughts of possibly shooting down ME 109s with his Spitfire's Browning guns, he presented himself in the snowy January of 1942 at the now revealed address, Arkley View, near Barnet. After being tested on his Morse speed and operating knowledge, he was transferred to the Royal Corps of Signals and instead of being sent off to distant parts as an operator, was assigned to the General Search department at Arkley, where he remained for the duration of the war. Most of his colleagues were licensed radio amateurs so he could continue to study in preparation for the technical examination that he required to obtain his own licensed call-sign, G3ASE. He did this as soon as permitted, in late 1946.

Upon demobilisation, Bob took various jobs in the wireless industry, then transferred to local government and administration, eventually training as a teacher at Trent Park, Middlesex, now part of Middlesex University. This delightful mansion, set in extensive grounds, was originally the home of Sir Phillip Sassoon. It was used

Name H.KING (1) Region HN
Address **R.S.S. Log Sheet** Group RO
File No. V/HN/353 Sub-Group OXFORD
BERTIE

(1) Date and Time G.M.T.	(2) Call as signalled	(3) Message and Remarks	(4) Frequency kc/s	(5) Wave Length M.	(6) Type of Signal	(7) Strength
9.12.41. 1700	CZE	QSA0 PSE CALL K SRI QSA0 QRX NEXT NW73 GBVA	5400		CW	3
1800	E	NIL IIK ND 22ND 349RM (QRM BLOTTED E OUT)	5400		CW	3
1800	GGG	QTC (LISTENED TILL 1815 BUT ND)	6200		CW	2
1815	WER URK	QTC CT 935/71 = (VY HEAVY ATMOSPHERICS)	4880		CW	3
		AOHLC VUBAK -NK42 MHPXN LTUPK				
		XFGEL WOFYH RDZFZ RZAXP DFJZB				
		TYHNN OPLDU NECIW FSJGS QVYIG				
		DVDQX WBDMJ MHRWW NKLFW UFF-O				
		THZGQ OOVKW CZEHQ MWEKE HZ-ZM				
		ZQDD MIWP VXAOU KQYYW MESKE CNCXX				
	DUF	RPT W35 PSG 457	5190		CW	2
	URK	... CT T FX	4900			
		QFK H APSY- OYIIR LCHUP FBJJL				
		YJUHO RIPBG WLYRY MMTUR OMIUQ				
		MFDQY EOVEG DQEZB IUVCU ADWNY				

WATCH PLEASE
NOTED THANKS
STILL WANTED
U.K. COVERED THANKS
after 1600

An actual RSS Log Sheet completed in December 1941 by Bob King showing the German Abwehr group being monitored – 'Bertie' based near Berlin. All German groups were given names starting with the same letter as the base. For example – 'Violet' – for Vienna, 'Willie' for Wiesbaden, and 'Bertie' for Berlin.

during the war to house senior captured German officers for interrogation. His room still had the evidence of bars at the windows and it has been related that information was obtained by 'bugging' the rooms.

During Bob's period at Trent Park Teacher Training College, he installed his transmitter and receiver in the attic and made opportunistic use of the large flat roof for the aerial systems. From here, hundreds of radio contacts were made all over the world at a time when such communication was not usually experienced by the general public. After spells of teaching in schools in London and Ipswich, Bob became Head of Science in a large comprehensive school in Cambridgeshire, a post he held for 27 years.

For four years, Bob was a part-time gliding instructor for the ATC at Martlesham Heath, near Ipswich, and has built two of his own homes. His other activities include amateur radio, making models including steam traction engines, bell-ringing, cycling, computing, voluntary hospital work and gardening.

Bob King today, an enthusiastic 'Ham' and organiser of the annual meetings of old colleagues of SCU/RSS, held at Bletchley Park.

Bob King is an enthusiastic amateur radio operator and still keeps in touch with several of SCU wartime operators through the Ham radio network. He is a much admired member of our dwindling band of ex-SCU/RSS colleagues and gives freely of his time and energy to ensure a true record is made of our unit's work. Each year, with kind permission of the Trust, he organises a reunion for ex-members of the various Special Communications Units at Bletchley Park.

Chapter 36

'My God They're Shooting at us' – and Other Stories

John Lloyd with 2nd SAS in France

John was an authority on our wartime activities and a member of SCU8. He had joined up at just over 17 years of age, under the 'Y' (Youth) training scheme, and entered the Royal Corps of Signals with an army number 2492886. Under the scheme, he went on courses to various civilian schools, including Marconi and Ericsson for factory experience, at the same time as receiving Corps training. He later passed his Post Master General 'Special Class' Certificate in wireless telegraphy.

In mid-1944, whilst at the Royal Signals camp at Catterick, John was approached by Section VIII on account of his skills in Morse and wireless. He then went for an interview with Don Lee at 54 Broadway, after which he was recruited into Section VIII with a new army number 2602372 – signifying that he was not paid out of army funds. He expected to be trained to join a Jedburgh team – combined intelligence units parachuted into France – to provide intelligence prior to, and during, the D-Day invasion period, as well as joining with the various resistance units. These were two man teams drawn from US or UK forces; were known as 'Jeds' and each team had a wireless operator attached to it. However, the speed of advance across France meant that 'Jeds' were no longer required, so John was withdrawn from the scheme, and instead, posted to SCU8.

He went over to France just before Christmas 1944 and joined an SCU/SLU attached to the 2nd SAS, part of the SAS Brigade under Brigadier R. W. McLeod. Their SLU was based in Guy 15-cwt vans fitted out at Whaddon by the Mobile Construction team although the unit was very mobile using numerous Jeeps. It was also very well armed and supported by elite troops.

This particular SCU/SLU had already been in France since D Day+3. Its role was quite different to other SLUs from Whaddon, as it was in close touch with mobile intelligence units such as Phantom, and independent troops of the SAS. Normally, our units would be based at an army group or army headquarters, well behind the lines. Quite apart from any other consideration, it was essential to preserve the integrity of the Ultra secret that might be jeopardised – with catastrophic results – if overrun by enemy troops.

The SAS Brigade was made part of Lieutenant-General Frederick ('Boy') Browning's British 1st Airborne Corps. Much to the irritation of the SAS members of the newly formed corps, they had to give up their distinctive sand-coloured berets and exchange them for the airborne maroon berets. John arrived in time to be caught up with the German counter-attack through the Ardennes, known as the Battle of the Bulge. He quickly found himself working as an operator with the unit in close proximity to the actual fighting. The 2nd SAS went in close support of Patton's Third Army during the battle. John had a high opinion of Patton, who seemed to remain in control when panic appeared to be the order of the day. The German attack was eventually contained and the 101st Airborne Division relieved at Bastogne.

Later, the 2nd SAS, complete with its SCU/SLU, was involved in crossing the Rhine. Some of our SCU/SLUs crossed by boat and some – including John Lloyd – flew over. It makes John's the only airborne SCU/SLU!

As soon as the unit had crossed, a pre-arranged signal was sent to Gambier-Parry at Whaddon – *'The cork from the bottle has gone Pop'* – so that 'Pop' was one of the very first to know that the Allies had finally made that momentous crossing into Germany. On one occasion, John's companion thought the figures across the fields were our troops, until he sang out 'My God, they're shooting at us!' and everyone dived for cover. They later became caught up in several fire fights and although issued with .38 revolvers, John took to carrying a Lee Enfield rifle at all times for greater protection.

John was subsequently injured and returned to Whaddon and then to Hanslope where he spent time in the workshops on DF vans for the Middle East, and later went on to Creslow. He left the unit in 1949.

During his retirement, John spent a considerable amount of his time working on features of the cryptology trail of Bletchley Park.

Wilf Lilburn – always a key member of Section VIII

Wilf Lilburn's name appears continuously throughout this book because he played such a crucial part in the affairs of the unit from its earliest days, and even after the war, when it gained a new lease of life as the Diplomatic Wireless Service.

Wilf was born on 20 August, 1910 in Stanley, County Durham. He was at school with Spuggy Newton, and together they attended the North-Eastern School of Wireless Telegraphy from September 1927 to April 1928. Wilf went on to obtain a Government Certificate of Proficiency in Wireless Telegraphy. He immediately became a wireless operator in the Merchant Navy but left the service in November 1929 after serving on three ships. His 'Continuous Certificate of Discharge' shows that each captain signed him off under the headings 'Conduct' and 'Ability' as 'Very good'. You will have seen his picture alongside Spuggy Newton at Durham Railway Station in 1933, further proof of their close relationship. Spuggy was being seen off by his friends to take up his new job at Philco. It is fair to assume it was he who later obtained a job there for his friend.

Wilf Lilburn photographed in Glasgow in 1934 where he was working for Philco – prior to joining Section VIII in 1938.

Wilf specialised in the development of short wave and especially its application in two-way wireless in motor cars. He later became Philco's Service Manager for Glasgow and was largely responsible for equipping the city's police cars with wireless. It is not so well-known that his expertise in the field led him to be chosen to install a wireless telephone in a Rolls Royce belonging to HRH Edward, Prince of Wales. He was first sent to Mulliners (one of the Rolls Royce specialist body-builders), to learn how the seats were designed, and the upholstery applied, as the Prince demanded that no wiring should be visible if possible. Having built the set back at Philco to the required specification, he then carried out the installation work at Mulliner's factory.

Once again, the 'Philco factor' came into play and so by late 1938, Wilf was to be found working, alongside his life-long friend Spuggy Newton, for Gambier-Parry and MI6.

Wilf then began wireless installations with John

Darwin, and he is mentioned in the extracts from Darwin's diary (see Chapter 21). He installed wireless gear on the MV *Cecile*, a large and luxurious yacht loaned to 'C' for the duration. It was owned by Mansfield Markham, a wealthy businessman and pre-war film director.

I first met Wilf at our home at Caterham in 1939, as he stayed with us whilst working on Funny Neuk at nearby Woldingham, as I explain in Chapter 38. He was constantly being called away and I know one of his trips was into neutral Holland where MI6 had run into serious difficulties, including the loss of one of our agent's sets. This is usually known as the Venlo affair, and is mentioned in more detail in chapter 14.

For some reason, Wilf is shown as being on the 1940 Electoral Roll for the house known as Funny Neuk in Woldingham, together with 'C' – Admiral Sinclair – although Sinclair had died in late 1939. When I joined Whaddon in 1942, Wilf was working in the R&D hut that opened directly onto the metal workshop where I had started, so I saw him frequently. I think he was then particularly involved with 'Ascension', the unit's air-to-ground wireless system on which I was to work at Mobile Construction, later in the war.

Wilf was much in demand as an engineer and moved constantly. I have his passports for the war years with customs stamps and visas for many countries, although he also travelled on Courier's Passports and, no doubt, sometimes as a King's Messenger. I lost touch with his movements but he spent time in Lagos, Nigeria, in North Africa at the 8th Army HQ and in Cairo.

Later, he appeared in Ceylon and then India, which is where I next met him again in 1945, and subsequently in Singapore, in 1946. During this time, he was largely involved in wireless station design. After the war, he worked at Hanslope and travelled widely for the DWS. He later married Joyce Hill (see chapter 24).

David Bremner – from sending an SOS at sea, to teaching Morse to SIS agents

Dave was born on 26 June, 1902 in Forfar, Angus. His story has come to me from his sons James and Neil. Wanting to go to sea, he fancied the uniform of a wireless operator with its smart cap and gold braid round the sleeve. At the age of 15, he enrolled as a student at the Wireless School in Dundee but was clearly under age for the course. It is believed he was only accepted due to the severe shortage of wireless operators at the time. He received his First Class Certificate before his sixteenth birthday. He left Forfar in December 1917 and reported to Marconi House in London, where he was immediately ordered to join a ship, the *Airedale*, at Middlesborough which set sail on 27 December, 1917.

David Bremner when at Hans Place Knightsbridge

Dave's first voyage was to Gibraltar and Huelva in Spain. From there, the ship returned to South Wales to discharge its cargo and load up with coal for the Italian Naval fleet lying in Taranto harbour. This trip was the last young David saw of the war for a while because on 10 April, 1918, the ship was torpedoed in the Straits of Messina. He reported to the bridge for orders and was told to send out the distress call 'SOS'. Fortunately, the ship was only five miles offshore and the crew were able to run it aground on a sandy beach.

23 Hans Place just behind Harrods in Knightsbridge. Used as an SIS school to train agents in Morse and the use of our agent's wireless sets.

The crew were taken ashore in a little place called Reggio and put up in an hotel. Somebody at the Admiralty decided the ship could and should be repaired, so Dave and the crew sat around for seven months waiting for the work to be done. Before it sailed again, however, he was sent to Genoa to become the wireless officer of the Magdala, replacing its own operator who had a severe case of influenza. The ship sailed the next day for Baltimore with a cargo of wheat, and it was whilst under way between Genoa and Gibraltar, that he received the famous Armistice message. 'Armistice will be signed this morning at 11 am. Thereafter all submarines on the surface will be treated as peaceful.' It was on this voyage he heard his first wireless telephone message. He was astonished to hear a distinctly American voice breaking in amongst the Morse signals, giving instructions to someone.

Dave's Merchant Navy career took him to many parts of the world but he decided he had had enough of the sea and left Marconi in 1923. Shortly afterwards, he and his brother Jim set up as wireless engineers and electrical contractors in their home town of Forfar. The firm of Bremner Brothers was very successful. David married in 1932.

Dave's son James, who was only six when World War II broke out, realises that his father had become what he now knows to be a Voluntary Interceptor. He had a wireless receiver at a desk in a corner of the living room and used to sit for ages wearing headphones, writing on a pad covered in squares. This, as we now know, was the standard RSS pad. About this time, his father started giving Morse lessons to a number of young men seated round the dining-room table. Until recently, he thought his father was teaching them faster Morse prior to their service in the forces but realises they were probably being trained as Voluntary Interceptors.

David was later called away to London and the family subsequently saw little of him. It is likely he was enlisted at Arkley as a full-time operator on interception. He moved to Hanslope where his skill at Morse instruction must have been welcome. What is certain is that from Hanslope, Lord Sandhurst transferred him to one of the SIS safe houses in London – 23 Hans Place, just behind Harrods in Knightsbridge. There he taught SIS trainee agents how to use our wireless equipment and rudimentary Morse. The agents' sets at that time would have been the Whaddon-made MkV, the MkVII suitcase sets and the MkIII transmitter.

Those who know the houses in the Knightsbridge area will be aware that they are very tall and one of Dave's first jobs was to resite the aerials on the roof. He had suffered with a heart condition for many years and sadly he died suddenly at Hans Place on 25 July, 1943.

A most fulsome letter of condolence from Lord Sandurst to his widow is shown opposite and is couched in most sympathetic terms. It should be noted it *appears* to come from him at HM Government Communications Centre (HMGCC), 54 Broadway, SW1. HMGCC was the frequently used cover for MI6 Section VIII, and Broadway was the headquarters of MI6. However, the postmark on the hand written envelope is Bletchley because Lord Sandhurst was stationed at nearby Whaddon!

Other Morse instructors at 23 Hans Place, were Bert Gillies and Bill Wort. The latter, I remember from my

H.M. GOVERNMENT
COMMUNICATIONS CENTRE,
54, BROADWAY, S.W. 1.

OUR REF. ....

YOUR REF. ....

27th July, 1943.

Dear Mrs Bremner,

It is with the greatest sympathy that I am writing to express my sincere grief at the tragic loss you have suffered.

As you know your husband and I have been associated for some years and have been working closely together for the last two. He has not only proved himself as a friend but also as the most willing, capable and reliable of Assistants.

As an instructor in one of my Schools he has done a really great job of work for the nation, and his loss is going to be severely felt, not only on the instructional side but also by his fellow instructors to whom his friendship meant a very great deal.

For myself, I have lost one in whom I had explicit faith and the nation has lost a very great and enthusiasticly loyal patriot.

It is not possible to even imagine what your loss must mean, but it will be a comfort to you to know that he had striven and succeeded, far above the average, to help win this war, and has given his children a magnificent example to follow.

Believe me,
Yours very sincerely, Sandhurst.

A truly sympathetic letter dated 27th July 1943 from Lord Sandhurst to Mrs. Bremner just touching on her husband's role as "..an instructor in one of my schools..." He was in fact, training SIS agents in Morse, and the handling of our wireless sets prior to them being transported into Europe. The letter appears to come from Broadway in SW London, whereas it was written at Whaddon Hall, and posted in Bletchley.

The envelope containing the letter sent to Mrs. Bremner – postmarked 'Bletchley.'

last months of service in Singapore; he is pictured in a group photograph at our wireless station at Adam Park, included in my own story, Chapter 38.

John Riley – export or die

John's involvement as an interceptor was due mainly to his background. From boyhood he had always had a keen interest in wireless, starting from the time he met a neighbour who in those days would have been classed as a well-informed amateur wireless enthusiast. It was this older man who taught him how to build the early type of crystal receiver and at the age of fourteen he built and sold several sets to others in order to raise pocket money. This was at the time when the first broadcasting stations were opened in England, Station 2LO in London at Savoy Hill and later Station 2ZY in Manchester.

John was further helped when a businessman in textiles, a refugee from the Russian revolution, came to live in John's remote village. Here was a man who could afford one of the more expensive two-valve receivers and double-ganged headphone plugs enabling several people to listen at one time. Loudspeakers had not been invented, they came much later.

John Riley after being recruited into Section VIII as 2602121 Signalman Riley

Valve receivers were greatly superior to crystals, which were not always sensitive to weak signals and the sound tended to fade as the crystals decayed. All the same, they were regarded as modern wonders and there was always some improvement in the design of the tuning arrangements, from fixed coils to variometers, from fixed to variable condensers, and so on. John progressed through these stages and eventually built his own three-valve receiver which unfortunately had to be powered by dry battery (for high tension), and wet accumulator (for low tension) since there was no electrical power supply in his village! This interest continued and he regularly experimented helped by articles in the then popular weeklies *Wireless World* and *Amateur Wireless.*

Later in life, during the few years before the start of World War II, he was in charge of exports at a company producing artificial silk (known today as rayon). One of the economic requirements, in wartime particularly, is to earn foreign currency with which to pay for essential imports, especially food.

It is difficult now to recall what an effect the submarine war had on life in the land. I am not talking here about dramatic events like the loss of ships that made the headlines, nor the tragedy of the many ships lost while carrying vital war materials – petrol, munitions, food, wool, tanks and aeroplanes, for example. With the desperate need for these materials, room could not be found on ships for such luxuries as bananas, oranges and a whole range of other goods we now regard as necessities.

All the materials coming into the country had to be paid for, and it was essential that we maintained our exports, as far as possible, to earn foreign currency. However, ships carrying these exports encountered the same dangers from enemy attack as the vital imports we so desperately required. Since the Ministry of Labour considered exports so vital in wartime, John was classified as being in a reserved occupation, and not subject to call up for military service. He was fortunate to be able to stimulate his firm's export business with orders from the 'free' or 'neutral' overseas countries and particularly with the hard-currency dollar countries. When two of his junior colleagues had been called up, and in each case lost their lives, John became unsettled and tried to find a means to release himself from his reserved status. The only opening seemed to be in the merchant navy in which all officers were 'reserved'.

Normally, merchant ships only carried one wireless officer but under the new conditions of warfare, every merchant ship was expected to maintain a continuous watch and this meant carrying three wireless officers, working in shifts. At fixed times every hour, it had become an accepted convention for normal wireless communication at sea to be suspended. Wireless operators would then listen on what was called the 'distress band' for any SOS calls. All ships recognised this humanitarian precaution as an unwrittten rule. The war at sea eventually demanded almost continuous monitoring on the distress band.

Since merchant navy officers were classified as civilians engaged on essential duties, it occurred to John that here was his opportunity. He discovered that to deal with the acute shortage of wireless officers, a wireless school had opened in Manchester. This was staffed by two experienced officers, one who taught the Morse code and the other who dealt with the technicalities of transmitting and receiving apparatus. John joined this class without telling his employers – who would almost certainly not approved – and for the next six months he studied in the evening and at weekends and eventually qualified. The class of about thirty men were trained in the basement of a large office block in the city centre, sometimes whilst German bombers were attacking overhead.

The next step was to find employment. A visit to Liverpool revealed that wireless officers were, in fact, not employed by the shipping companies but by agencies. There were three such agencies, Marconi, Siemens and International. Having discovered that due to the shortage one could almost make a personal choice, John accepted a post with Siemens and was told he would sail to Buenos Aires within fourteen days as a civilian, and return on another ship as second wireless officer. His choice was governed by a desire to work on a cargo ship rather than an oil tanker. In earlier years, he had gained a working knowledge of Spanish as he had considered taking a post in textiles in South America.

John was given a number of clothing coupons and purchased his merchant navy uniforms, including tropical kit. However, just a week before he was due to sail, his employers intervened through the Ministry of Labour. He was told that his reserved occupation in charge of exports was just as important as that of a wireless operator and he would not be permitted to sail.

John was just beginning to settle down once more in the export department of his firm, feeling that he would have to accept the Ministry of labour ruling, when he received a telephone call from a certain army captain. He was invited to attend a meeting at an office in Preston, Lancashire. During the conversation he was told his name had appeared on a list of 'unemployed wireless officers' and that he might be interested in a proposition involving wireless.

He accepted the post, which he understood was in something called RSS, and was told that he would later be enrolled as a member of the Royal Observer Corps – as a cover. Subsequently, he was visited by one of the local 'hams' (amateur wireless operators), who was then acting as Group Leader in charge of about ten other hams in an extended local area. Having been sworn to secrecy, he was shown what his tasks entailed. Briefly, the work was to listen on certain short wave wireless bands for 'suspect' enemy transmissions. In some cases, these were known, and were already being monitored. The service had extended its function and was also monitoring other 'suspects' in enemy territory. John knew that he was now a VI – a voluntary interceptor.

The headquarters of RSS was at Barnet. But, as the organisation came under the control of MI6, the main intercepting station was built at Hanslope, and the unit then designated as SCU3. There was a great demand for skilled staff to man the new station and the obvious area of recruitment was from the ranks of the most proficient VIs.
Thus it was that John was asked to go to Arkley where he underwent tests to demonstrate his ability to read Morse at 25 groups a minute. He was then taken out of his reserved status and recruited as a Special Enlistment into the Royal Corps of Signals. His Army number, 2602121, was selected from those reserved for NPAF enlistments. It was remarkably close to mine – 2602902. From then on, he worked at Hanslope on interception, along the lines already described in earlier chapters.

John's own record of his wartime role is illustrated with sketches, including an RSS intercept bank. At the end of the war, in company with so many from our units, he found it difficult not to be able to talk about his war work and its importance. Like the other interceptors at Hanslope, his sole reward was the note from Ted Maltby at 'PO Box 25'. I hope that in telling the story of these highly skilled operators, some belated honour will be paid to their devoted and vital work in intercepting German wireless traffic at the highest level.

Bernard Gildersleve – with 9th US TAC in Europe and SIS in Germany
In 1942, Bernard was living in Dulwich and was employed at a bank in Park Lane. He was only sixteen when he and some other teenagers volunteered for the Home Guard at a unit based with an army anti-aircraft battery on Dulwich Common. They were trained as members of a plotting team and were on duty during air raids on London.

When he was nearly eighteen, he had a medical examination in preparation for his military service. Afterwards, he was asked whether he had any preference as to which service he would choose to enter. Having explained his Home Guard experience on an anti-aircraft battery, and with some knowledge of gunnery and gun-laying, he assumed he would be suitable for the Royal Artillery. That was duly entered as his choice. He was therefore greatly surprised when his actual call-up notice arrived, to find it ordered him to report on 24 March 1944, to 'The Royal Corps of Signals, SCU1 at Whaddon, Bletchley, Bucks'.

When Bernard arrived at Bletchley railway station, it appeared that a number of others had had the same orders. They were gathered together and driven, not to Whaddon, but to Gees at Little Horwood, the home of SCU7. There they were kitted out with uniforms, then moved to Ashbys in Stony Stratford for their six

Bernard Gildersleve with a Whaddon made 'Coffin' set.

weeks' basic training. Bernard believes he was in the first draft to use Ashbys and found the same instructors as described in Martin Shaw's story. At the end of their training, they were moved to SCU7 back at Gees for intensive instruction to make them into wireless operators. He reports going on day-schemes in the Packard wireless cars as part of the training, all under the 'kindly guidance' of Squadron Leader Matthews, RAF and Lieutenant Murray of the Royal Signals.

Bernard duly passed his B3 test and was promptly transferred to SCU8 with the huge distinction of swapping his forage cap for a beret which, at that time, was the 'trade mark' of a member of SCU8 – destined for duty overseas. He was allocated to a team going to a 'forward station' under CQMS 'Lofty' Day. There were three other operators in the team, Brown, Downes and Willis, in addition to two drivers.

The mobile unit drove their two vehicles, a converted Dodge ambulance and a US army 15 cwt truck, to Bushey Park, near Hampton Court, and met up with the American contingent of the Ninth US Tactical Air Command (TAC). From there they went to Southampton and boarded a tank landing-craft in which they had to wait – being tossed about by the seas – for eight days! Finally, they crossed the Channel and arrived on Utah Beach on 7 September, 1944. By this time, the beach was quiet.

After spending a few nights en route at various places, the unit finally reached Charleroi in Belgium where it camped under canvas in a field on a hillside overlooking the town. There, the men formed SCU/SLU 8 – attached to the US Ninth TAC. They opened up their station, which they called 'Mermaid' in the Dodge ambulance and settled down to their schedules. They logged each message as it came in, and then took them to the RAF cypher team who worked in a large tent about thirty yards away. The SCU team slept in another large tent nearby, leaving the duty wireless operator to work the eight-hour shift on his own.

The RAF cypher sergeants worked under the leadership of a junior RAF officer. The US Army officer commanding the whole SLU was a Captain Hoopes. There was another more junior US officer whose name escapes Bernard but his nickname does not. He was known as 'Dead Loss' since he was always at a dead loss to know what to do!

All the messages were in five-figure groups which were decyphered using the one-time pad system. They were then taken and shown to Major General Elwood R. ('Pete') Quesada before being incinerated. During all this time, the Ninth TAC was providing air support for the US Army as they advanced through Belgium so they had to move forward, which they did in due course, to Verviers.

At Verviers, they were billeted in a first-floor flat above a garage, directly opposite the Palais de Justice (courthouse) which had been occupied by the Ninth TAC HQ. They then transferred all the wireless gear from the ambulance to the flat where they set up a station in comparative comfort. On one occasion, however, they were subjected to a dive-bombing attack by a whole squadron of planes who dropped their bombs one after the other. It was later rumoured that the planes were our own Lockheed Lightnings from the Ninth US TAC, seemingly intent on bombing their own HQ! It was said they had mistaken the town for Aachen only twenty miles away but the truth of the matter was never discovered – or perhaps and more likely – it was hushed up!

In December 1944, von Runstedt launched his Ardennes offensive which became known as the 'Battle of the Bulge' and some German parachutists dropped behind the Allied lines, necessitating an evacuation of the wireless station as far back as Liège. Most of the Ninth TAC went right back to Charleroi. At Liège, they spent a most uncomfortable Christmas under frequent air attack. When the German offensive had been repulsed, they returned to Verviers and as the armies moved into Germany, the SCU/SLU followed, first to Brühl, near Bonn, then to Marburg, and finally to Weimar, where they celebrated VE Day.

After the war in Europe, the team was sent to Versailles and then to the UK via the Dieppe-Newhaven ferry. Back at Gees, they were transferred into SCU1 and billeted under canvas at Nash camp, ostensibly for training for the Far East where the war with Japan was still in progress. After a few weeks 'toughening up', a few were flown to Bad Salzuflen where they were designated SCU9. This was a spa town and by this time there was a NAAFI and a Church Army presence leading to the feeling they were being catered for both physically and spiritually.

The unit was described as 'No. 1 P & EU' (Planning and Evaluation) but this was really a cover-up for the highly secret work involved; they neither planned nor evaluated. From their base, the operators were sent to one-man out-stations to provide wireless communication for small secret units consisting of a few army officers and some civilians, probably from MI6. The first of these out-stations was at Plon in Schleswig-Holstein.

Later, Bernard went as a relief operator to a number of these out-stations, most of which were in comfortable homes, with German catering staff, so they were well fed. There were visits were to Benthe near Hanover, to Düsseldorf and to Berlin. Bernard recently sent me a picture of himself with one of Whaddon's so-called 'coffin' sets being used at the Düsseldorf out-station. The coffin set was made at SCU1's Little Horwood workshop and consists of an HRO receiver with its set of coils, a MkIII transmitter and a power pack. The whole set was contained in a wooden case made by Len Warner our very talented cockney carpenter.

The Berlin station was called Stowaway, described to the uninitiated as 'No. 4 Economics Assessments Detachment'. There, Bernard found himself employed for other purposes than just operating a wireless set, which added interest to his last days with the SCUs. As his demobilisation date drew near, he returned to Bad Salzuflen where he spent time improving his Morse speed and was upgraded to B2.

Bernard returned to the UK in 1947 and, after being demobbed, rejoined his old job in a private bank, but this time in Fleet Street instead of Park Lane. Working in London, he discovered that an SCU unit had been formed in the Territorial Army, so he subsequently rejoined the SCU in the City, just off Farringdon Street, as a member of its TA. Bernard now lives in retirement in Dorset.

Charles Tracey (right) in Columbo 1944 with Derek 'Jock' Leslie who was later at Kunming with Tom Kennerley, then in Calcutta and Singapore with me.

Charles Tracey – from Highland Light Infantry to Royal Corps of Signals

Charles was called up in January, 1940 and posted to the Highland Light Infantry with an army number 3319873. In those days, it was possible to claim a move into your brother's regiment, and as his brother Peter was in the Royal Signals, he applied to be transferred. With the thousands being called up and the hugely complex army administration system, he was astonished to find someone had actually read his request and approved the move. Thus he found himself at Little Horwood with his brother, but with absolutely no wireless experience as an engineer or as an operator.

Charles spent many months – he says it seemed like years – in just sending and receiving Morse signals. In those early days, before the construction of the Nissen huts, the wireless training was in the stables of Gees. Later, he and the other trainees spent time out in the Packard wireless vehicles sending dummy messages back and forward to each other's cars.

After passing his B2 test in 1942, Charles was posted to Windy Ridge as an operator, in what was then known as 'A Group'. There were about twenty others in the teams and they operated in shifts. At that time, Windy Ridge worked stations in the Middle East, mostly handling Ultra traffic.

Charles' brother Peter was posted out – as a civilian – to Tangier in 1942, then on to the Embassy in Madrid, where he was later joined by James Tully and Tony Morrell. All of these men had been trained at Little Horwood and now found themselves acting as civilian wireless operators for MI6 in Madrid.

Charles was posted to Delhi (then SCU11), in December 1943 with about twenty others, including operators and drivers. They left from the Clyde in a large convoy and arrived in Bombay three weeks later, after zigzagging across the Atlantic, to avoid German submarines. From Delhi, he was sent down to Kandy in Ceylon, where the unit was part of ISLD (Inter Services Liaison Department), the cover name for MI6. The staff at Kandy included 'Colly' Colbourne, as well as Wilf Lilburn, who had designed and installed the wireless station. The unit was headed by Wing Commander 'Jock' Adamson.

To begin with there were two operators, Charles and Derek 'Jock' Leslie, who went on to Kunming with Tom Kennerley. I met Jock – who was quite a character – later in Calcutta and then in Singapore. In Kandy, the operators worked to London via Delhi on HROs and MkXIV transmitters. Occasionally, one of them went to our station in Columbo, where they helped with the traffic to Calcutta.

Max Houghting (seated) at the ISLD office in Naples, October 1944 – with colleague

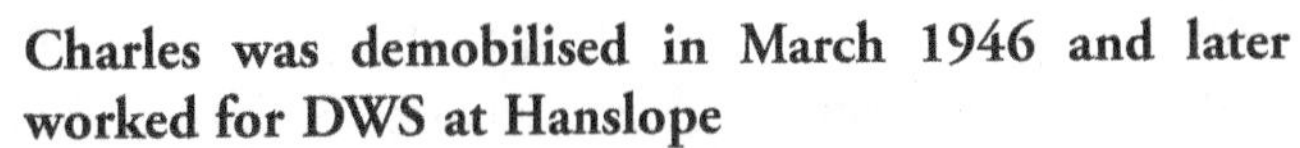
Charles was demobilised in March 1946 and later worked for DWS at Hanslope

Max Houghting – from RAF to Section VIII

Max was trained as a RAF wireless operator and on 17 November, 1940 he left Liverpool on the *SS Strathnaver* for the Middle East, via Freetown and Durban, arriving at Tewfik on 28 December. Within a week, he was in the desert, taking part in Wavell's push that eventually took the army all the way to Benghazi. He remained in the general area of Libya and Egypt, depending somewhat on the movements of General Rommel and his Afrika Corps, until November 1942, when he was posted to Malta.

On arrival in Malta, Max was told he was to be seconded to a 'special unit'. He reported to a large house called Villa Anna, on Birkirkara Hill, St. Julians. It was a large unit named GSI (J) – General Staff Intelligence Department 'J'. There he was told he would work under a Royal Signals Sergeant and be contacting two stations in Yugoslavia.

Max recalls that one of the stations was called 'Bullseye'. The stations to which he worked were run by the Tito and Mihailovich factions. The receivers were Hallicrafters and the transmitters were housed in another part of the building. Communications were patchy so all coded traffic was sent twice. The operators also did interception work when these same stations were working to Cairo.

In February 1943, there was a reorganisation. Max and another RAF operator 'Midge' Middleton were taken under the wing of Bill Furze from Whaddon. He told them they were now members of *'Section VIII of the Foreign Office'.* In fact, they remained in the RAF but, like many others recruited into the unit, they had their RAF pay 'made up' from Section VIII funds. They were then joined by 'Gravo' Graves, a senior operator from Istanbul, and a wireless engineer from Whaddon – Sergeant Griffiths. A station was opened up in the Governor's Palace at Attard using a MkIII transmitter and an HRO receiver.

They had considerable traffic for Whaddon and Cairo and were also in contact with two agents in Sicily. However, the main agent was a Frenchman codenamed 'Pappillon' operating in Tunisia. Sometimes 'Pappillon' would go off the air in the middle of a 'sked' (scheduled transmission), which indicated the Germans were cutting the power to isolate and locate. Then it was a matter of listening until he came up again – perhaps several days later. 'Pappillon' was brought out by submarine which gave Max and the others an opportunity to learn the conditions under which he was working. The British authorities wanted him to go back and he did so, after being decorated by Lord Gort, who was Governor of Malta at the time.

Max and his fellow operators had been housed in comfortable accommodation in Sliema but later transferred to a large house near the Palace. The team then consisted of operators and four cryptographers Ian Sadler, Ben Bolt, Bill King and Fox.

Len Stone and Alan Parkinson who had been at Nash, joined them as operators. The invasion of Sicily started and soon afterwards, Max, 'Midge', 'Gravo' and Sergeant Griffiths flew to Syracuse where they set up a base in a large block of flats overlooking the harbour. The station was erected in a laundry room on the top floor where they started working to Whaddon and Malta on an HRO and a MkIII. Bombing of the harbour took place from time to time so the site was not very comfortable, indeed, Max tells me *'it was hell!'*

The unit, now officially named ISLD (Inter Services Liaison Department), expanded and, after the invasion of the Italian mainland, it was moved on to Bari. Here the CO, Major Henderson, was replaced by Major John Bruce-Lockhart who was flown in from the ISLD HQ in Cairo.

ISLD was especially concerned with working to agents in Yugoslavia and northern Italy. A larger transmission station was built, containing an American 200 watt Collins transmitter instead of the 30 watt MkIIIs, which considerably improved performance. More operators – 'Ginger' Allen and Harry Brett – joined them and Captain Edgar Harrison in from Cairo became Signals Officer. Later, two young operators – Dennis Herbert and Dudley Bradford – whose fathers held senior positions back at Whaddon, joined the growing team.

Shortly afterwards, 'Pappillon' turned up, resplendent in a captain's uniform. He told Max he was going round the Italian prisoner-of-war camps to see if he could find any Italian wireless operators willing to be flown into northern Italy, as additional agents for ISLD. Unfortunately, the first such agent dropped by parachute promptly went over to the Germans so it is likely other volunteers were dropped and lost, before the double-cross was realised.

On 23 December, 1943, there was a bigger-than-normal air raid on Bari harbour and about sixteen ships were sunk, including two ammunition carriers. Although the ISLD station was a half a mile from the harbour, the blast was so huge, it blew in the shutters, and knocked three AR88s off their benches – and they are a very considerable weight!

Following the allied armies advance, Max moved up to Naples but after four years service abroad he was posted back to Whaddon. He remembers being driven to the airport in one of the unit's Packards, which were still useful, even though no longer employed as wireless vehicles. After some leave, he arrived at Whaddon in October 1944 and started work at Main Line station in the field in front of the Hall. He recalls that one of the watch supervisors there was Claude Herbert, and the other 'Dinger' Bell. The operator cubicles had HRO receivers.

After a period at Whaddon, Max was returned to the RAF and posted to an Air Ministry Unit based in an Oxford Street store in London. There, he helped sort German documents relating to RAF aircraft that had crashed in occupied Europe. He found they had kept most meticulous records of the event, including details of the squadron and names of the crew. Additionally, they had to assemble the considerable quantity of personal effects of the deceased crew that had been returned. It included love letters and money found on the airmen. This was carefully batched and returned to the airman's family.

He left the RAF in 1946, and after a short time with in the police force, Max started work as a wireless engineer in civil aviation at Heathrow.

Chapter 37

With Rommel, the Other Side of Enigma by Siegfried Maruhn

This book is about the wartime operations of MI6 Section VIII, much of which involved the interception and dissemination to the allies of Ultra traffic. I have been fortunate in finding colleagues who were connected with that aspect of our work, and several chapters are about their handling of Ultra traffic.

There were obviously two sides to this wartime wireless traffic, however, the sender and the receiver – in our case the 'Y' service. So far, the book has been about our 'reception' of Enigma and the use we made of it – rather than how it was sent in the first place. I was therefore very lucky to meet Siegfried Maruhn who worked with Enigma cypher machines whilst serving in the German army and get his story.

Siegfried was born on 13 April, 1923 in Tilsit, then East Prussia, now a Russian town called Sovietsk in the Kaliningrad Oblast. His family subsequently moved to Frankfurt-on-Main where he finished school in 1940. He volunteered for the army as his father was already an army officer so he felt compelled to join that branch of the service. His father was killed on 9 April, 1945, in Thuringia in combat with American troops, just before the war in Europe ended.

Siegfried soon became disillusioned with army life but had no option but to carry on. In 1941, he became a signals operator and was shipped out to Africa in September of the same year. Shortly after arrival, he fell ill and was sent back to Italy, only to be reassigned to the Afrika Korps in 1942. He arrived there just in time to live through the battles of Alam Alfa and El Alamein and take part in the 2000-mile German retreat to Tunisia.

Siegfried was wounded during a fighter-bomber attack on 9 April, 1943 and shipped back on a hospital boat to Europe on 29 April. He was lucky never to be sent to the Russian front but was returned to duty at the Western front in October, 1944 only to be captured on November 19 in Metz by the FFI (Forces Françaises de l'Interieur) and handed over to the U.S. Third Army the next day. By then he was a sergeant in the signals section. Here is his story, in his own words.

"I first became acquainted with the Enigma machine, when I joined the Tenth Panzerarmeenachrichtenregiment (Panzer Army Signal Regiment) in Africa. This was in June 1942. The regiment was stationed near the front lines at El Alamein. I was then 19 years old.

I had been in the Signal Corps since 1941 and had served in Africa before, but as I had been with lower echelon staff officers I had not worked on an Enigma machine before. Now our job was to maintain connections between Rommel's headquarters and the German and Italian Army Corps and Divisions under his command. At this level, Enigmas were used on a regular basis because they were considered safer than lower grade encryption methods.

Siegfried (standing centre), training to use German wireless equipment at a wireless school at Wiesbaden in 1941

At first, our signal troop, comprising a sergeant (Wachtmeister) and eight lower ranks, was stationed at the headquarters itself. Shortly before the battle of Alam Alfa we were assigned to the headquarters of the Italian Corpo d'Armata Ventesimo (XX Tank Corps), which comprised the Italian divisions Ariete, Trieste and Livorno. Our headquarters were somewhat behind the front lines, away from the Mediterranean. We reported to the German liaison officer, a colonel, who was with the Italian Corps commander. Our role was to exchange messages with Rommel's headquarters on one side and with the Italian divisions further back. These divisions also had German liaison officers and signal troops belonging to our regiment.

Our traffic was exclusively in German, interpreters taking over when messages had to be given to Italian staff members. The food we were given consisted of Italian army rations of very poor quality. The Italian army of that time still had the traditional class system fully in force.

I had a very short introduction to the Enigma machine. Actually, it is not very difficult to use. The trickiest part is setting the initial conditions of the machine. One had to follow the instructions laid down in a master plan, issued monthly. There were five wheels (Walzen) from which to chose. Three were inserted in the prescribed order. On each wheel a rotary ring had to be turned to a given position. Then a number of cables had to be switched on at the front of the machine to connect pairs of letters. These initial conditions had to be established on both sides, sender and receiver. After all, no message can be safely delivered if sender and receiver have not completely synchronised their machines. The initial setting of the wheels was part of the message itself.

From then on, all you had to do was to hit the typewriter keys of the Enigma and take down the letters lighting up on the display. Usually, we worked in teams of two, one soldier typing, the other jotting down the results. The encoded text thus acquired would then be transmitted as Morse code in groups of five, each message preceded by the Morse code 'ka' (-·- ·-), closing individual messages with an 'ar' (·- ·-·) and ending the transmission with 'sk' (··· -·-). We named this last code 'Schluss Kamerad', meaning 'end, comrade'.

Receiving and decoding a message was similarly simple. The signalman on duty took down the letters of the message transmitted in Morse code and then, usually with the help of another soldier, decyphered the message on the Enigma, reversing the encryption process. He typed the letters of the encrypted message and received plain text. This was taken down on forms and forwarded to the addressee.

Very important or secret messages were marked 'Durch Offizier entschlüsseln', which meant that only an officer was authorised to handle the Enigma message from this point on. At least that was the theory. Often, the officer-in-charge was lazy or careless enough to let a lower rank do the encrypting or decrypting work. Usually they were better at it and faster as well.

So for most of us, the Enigma machine was just a black box. You put something in at one end and received something else at the other. Input and output were connected in some mysterious way. The Greek name Enigma seemed to express this mystery perfectly. I recall that I knew at least the elementary workings of the machine. For example, I realised that on the first level the individual letters of the message were converted by the Steckerverbindungen of the plug-board in the traditional 1:1 way.

Then the letter just acquired was sent to the first wheel and changed again, in way that depended on the internal wiring of the wheel. This process, I could see, was repeated twice, further variations being produced by the wheels turning at a prescribed rate. I did not know at the time that there was a reflector at the end, sending the signals back through all three wheels before it finally reached the display.

Still, we had the impression that this was a very intricate process, making it very difficult for the enemy to break the code. We were told and assumed that our messages were completely safe and could not be decyphered, at least not soon enough to make the knowledge worthwhile, which, after all, is the prime objective of any code.

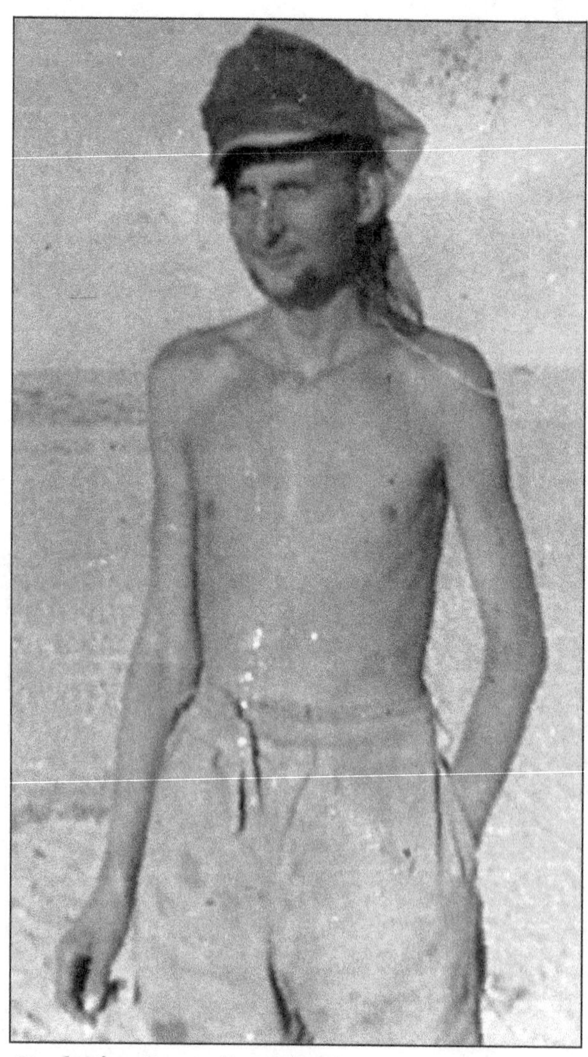

Siegfried at Rommel's Afrika Corps HQ, in the Western Desert.

The only real danger, we were told, was that a machine, its wheels and the current tables of settings would fall into enemy hands. We therefore had strict orders to destroy the machine and the tables first of all if capture seemed imminent. When such a loss was reported or speculated on, the settings were changed prematurely. I remember at least one such incident during my tour in Africa.

It seems that our defeat at Alam Alfa has been helped by the British being able to decrypt the Enigma messages in time. I wonder whether a similar feat was responsible for the sinking of the Italian troop carriers Oceania and Nettunia on 18 September, 1941. The ships were on their way from Taranto to Tripoli. I was on the Oceania and spent some time in the rather cold waters of the Mediterranean off Malta before an Italian destroyer picked me up and delivered me and most of my comrades – albeit without weapons – to Libya. At the time we speculated that our Italian comrades-in-arms were responsible for the ostensible breach of secrecy.

During the battle of Alam Alfa, our signals troop was trapped between two British barriers of landmines. We had crossed two of them in our Kfz 17 (wireless signal car) but could not go any further because the Italian tank force we were travelling with was bogged down front of us. It was a very uncomfortable position because the British opened fire on us at dawn and their aircraft dropped bombs. We were lucky to be able to make our way back to our former positions.

During the battle of El Alamein, the Italian Army Corps that I was with was almost completely annihilated. A few remaining tanks, together with vehicles under repair at the time of the battle, were used to form the Kampfgruppe Ariete (Ariete Combat Group), a modest outfit consisting of no more than 30 tanks. Our signal troop stayed with them all the way from El Alamein to Tunis. We were quite busy since, of course, there were no telephone lines so all communications were by wireless. But when, at a temporary rest stop near Buq Buq, still in Egypt, we heard on our wireless that the Americans had landed in Africa we knew pretty well that the war, on this continent at least, was nearing an end.

I, for my part, was not permitted this early and easy way out. On the road from Sfax to Sousse in Tunisia, I was severely wounded by an American P38 Lightning fighter-bomber and sent back to Europe on 29 April, just a few days before the Italian and German forces capitulated in Africa.

After a three-month hospital stay I was sent back to my base unit in Coblenz, serving for the next year as Funklehrer (Radio teacher), helping young recruits master Morse code (sending and receiving) and instructing them in encrypting and decrypting, including work on the Enigma. Later, in 1944, I was attached to a Feste Heeresfunkstelle (stationary army signal unit) at Coblenz, which handled traffic between senior staff officers. This material was encrypted exclusively by Enigma and most of it had to be processed by officers.

When, in the autumn of 1944, the German front in France collapsed, I was declared fit enough to return to frontline duty. I was sent to Metz, heading the signal troop of an artillery battalion, consisting of myself and one other soldier. Because there were plenty of telephone communication links in the fortress of Metz, we had nothing to do. At this lowest level, there were no Enigmas.

When the end came on 19 November, 1944, we put a hand-grenade into our wireless transmitter, burned our code material and looked for someone to surrender to. Our battalion staff had left during the night without telling us, but we met up with them a few days later – as fellow POWs.

This is my somewhat unspectacular story of working with the Enigma. Much later, I learned of the Allies' success in breaking the code and it seems to me a spectacular achievement, and certainly worth regarding as one of the most important events of World War II.

I believe this is the first time that the work of a German Enigma operator on active service has been told as part of the overall Ultra story.

As a POW, Siegfried spent four months in American camps in France, but was then shipped to America at the end of March 1945 where he worked on farms in Mississippi and in Georgia, but spent most of his time as an interpreter. Years later, Siegfried was to serve as a captain in the Bundeswehr, the present German armed forces, though only as a reservist.

On his repatriation in the autumn of 1946, he decided to become a journalist, working first for an American newspaper and from 1952 for the Essen-based *Westdeutsche Allgemeine Zeitung*, the largest regional paper in Germany. He was its editor-in-chief from 1970 to 1988, when he went to Washington, serving as White House correspondent for that newspaper and several others. He retired in 1992, working only occasionally as a journalist and author. Siegfried is married with four sons and seven grandchildren.

Chapter 38

Geoffrey Pidgeon – His Own Story

This is the story of how my family became involved with Section VIII and of my own minor role in these great events, starting with what has to be regarded as a very odd beginning.

My father's introduction through 54 Broadway

When the war started, we were living at Caterham in Surrey. My father was in the theatre business, working at Cecil Roy's, a ticket agency in London. However, all the theatres closed immediately on the outbreak of war for fear of bombing so after many years with the firm, he found himself without a job. He had been a volunteer ARP (Air Raid Precaution) Warden for the local area and, fortunately for us, he was later offered a full time job as a warden, based in the specially-built concrete Warden's Post at the end of Greenhill Avenue where we lived. I do not know how much he earned but I think it was about three pounds a week.

The rapid expansion of the local aerodromes, Croydon, Kenley and Biggin Hill involved the employment of large numbers of builders, most of whom were billeted in the nearest towns and villages. We had our fair share in Caterham, with Kenley only a mile or two away. I do not recall whether this was a voluntary arrangement but mother was not too keen to have unknown building workers in her home. As the war progressed, there were rumours that Canadian troops would be brought into the area and billeted in local houses and this eventually happened. This prospect did not please my mother either and being a shrewd woman, and determined to have some control over who was billeted upon us, mother arranged with Mrs. Luck, our local newsagent to put a carefully worded advert in her shop window offering accommodation, subject to certain conditions. I suspect this also constituted a handy additional source of income but it meant that at least mother could 'vet' the applicants. If they proved suitable, she could then claim that the house was fully occupied. Billeting officers were moving into the town and there were no ifs or buts about the process of billeting, if there was space in your home you had people billeted on you.

I have no doubt that Mrs. Luck was in cahoots with mother in selecting suitable people to send to us. On the very first day of the advert, two men came to the door and introduced themselves as Bob Chennells and Wilf Lilburn. They told mother that they were working at a wireless station which later turned out to be Funny Neuk at nearby Woldingham. Mother liked them instantly, they liked the room she offered, so they became our first (and only) lodgers. That is how my family first became involved with Section VIII.

Our lounge had been emptied – apart from the piano – and two single beds installed with some cupboards and a wardrobe. Bob Chennells was an extremely engaging and outgoing person, willing to spend time with us, my brothers and me. Wilf Lilburn, though more reserved, was amiable with a good sense of humour. Both were very polite, an important attribute in our household, and we quickly took to them. On reflection, I think my brothers and I were most impressed that both Chennells and Lilburn had motor cars of their own, at a time when car ownership was by no means as common as it is today.

Bob and Wilf then asked if they could put up a wireless aerial along the length of the garden; most homes had one in those days anyway. One day, I chanced to look in his room and found Wilf Lilburn with an ordinary looking wireless set, in the wooden case of the day. I could see the back of it which proved to contain a telephone handset. Wilf was speaking to someone *on the telephone* and I quickly withdrew, very impressed yet puzzled.

One Sunday morning, Bob Chennells had to go to the wireless station and asked if I would like to go along with him for the ride. He owned a Morris 8 four-seater coupé, and we drove up through woods to the gates of the station where I had to remain under the stern gaze of the guard, who peered out from his hut in the trees. I am sure I could find my way there today. This later became an important wireless station for the Czech Intelligence Service.

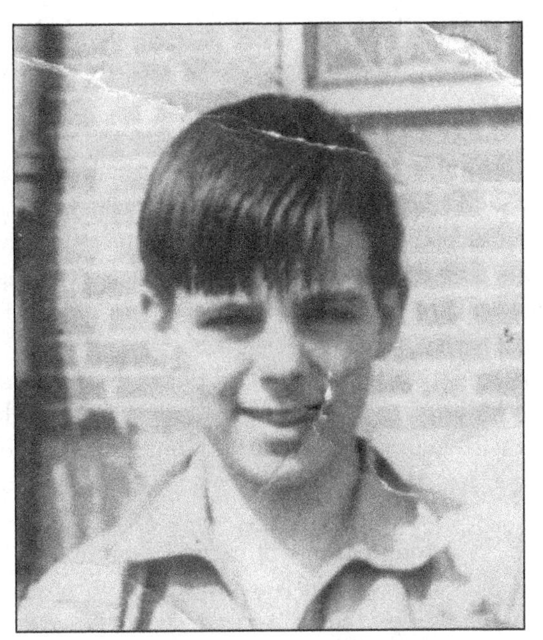

Myself at Caterham, around the time I first met Wilf Lilburn and Bob Chennells as they started work at the new wireless station called Funny Neuk.

Father had started his ARP training back in 1938 and at the same time, mother began training as a Red Cross Nurse, based at the local Red Cross Headquarters at 'Roseneath', near Kenley railway station. As a member of the Caterham School Boy Scout Troop, I had become an ARP messenger – I thought it was all very exciting.

The end of the 'phoney war' in 1940 saw the Blitzkrieg, the defeat of the Low Countries and France, then the retreat via Dunkirk. At that time, mother became a full time Red Cross nurse. Although the salary was very small, it must have been helpful to the family income at that time. My younger brother and I were at Caterham School and there were expenses to be met every term. After she qualified, mother was attached to a mobile operating theatre that had been built into a single-decker bus based at the Kenley HQ. She was on duty in it when the first heavy bombing of Croydon aerodrome took place, during the Battle of Britain.

Wilf Lilburn went on a dangerous mission to Holland before Hitler invaded. He actually *told* Mother he was going to Holland and I remember wondering at the time how he would get there. He returned about a week later, and shortly afterwards he left us. Bob Chennells stayed on a little after that, but the timing is now hazy. I can only now guess at the dates, but I think these two were with us from late summer 1939 until early 1940 but were coming and going all the time.

In late May or early June 1940, father was telephoned by Bob Chennells who asked if he would consider taking a job 'connected with wireless'. Father had built several of his own wireless sets, from his earliest 'cat's whisker' crystal set, to a full-blown wireless from a kit, in the early 1930s. I remember that wireless set, had a horn speaker just like the one in the HMV gramophone advertisement. Perhaps it was an older model and he picked it up cheap. However, it worked and he was justly proud of his skill.

Father was asked to go to an address in Broadway, Westminster for an interview. That address, 54, Broadway, is now widely known to have been the wartime HQ of MI6. At the interview, he was asked to describe various wireless components, a valve, a resistor, a condenser, etc, and their functions, without being told why.

He was asked for references, and I know he mentioned the names of a number of important clients from his theatre ticket agency days, including royalty, but I do not know if they were followed up. He used to personally look after members of the royal family if they wanted to book seats through his firm in South Kensington, and would attend the show at the same time to ensure all went well – especially if he had recommended it. He had his favourites, including the elderly daughters of Queen Victoria living at Kensington Palace, the Queen of Spain whenever she visited London, and others. However, although these references might have been impressive, I am certain it was his Chennells and Lilburn connection that actually got him the job.

Father was called back a few days later and told he would be employed at a place 'in the Midlands' in 'a position connected with wireless' and was then given a railway ticket for Bletchley in Buckinghamshire, 'where you will be met'.

On arrival at Bletchley, he was driven in a Packard saloon straight to Whaddon Hall, in the village of Whaddon, about five miles west of Bletchley. He was then interviewed again and his name appears amongst a group of some twenty men who were being recruited into the unit around that time. Some were transferred from army units but most had been 'sought out' as having special skills and some of them, like father, were 'not paid army funds'. The primary source of the funds was SIS – MI6.

My father's starting salary (and I have the details signed by Captain 'Charlie' Crocker) was £260 per annum which rose to £350 during 1944 and a little more later. All this was, of course, like all our salaries, tax-free. He was billeted in nearby Stony Stratford with the Crowe family (the combination of the two names caused some hilarity) at what was then Number 6, Calverton Road.

Father was soon put into uniform and given an AB64 Pay Book. He was made a sergeant in the Royal Corps of Signals and issued with an army uniform and equipment that included a revolver. A brave soldier then took him to a clump of trees near to the Main Line wireless station at Whaddon Hall, and on one tree was a target. He was told to fire six rounds at it with his revolver. Firing towards a clump of trees without proper supervision is extremely dangerous but that is exactly what happened. This was the sum total of father's military training in the five years he was in the unit!

Father was put in charge of the wireless stores, then based in just a small part of the stables of the Hall. His boss, with a somewhat wider brief, was Ewart Holden. The stores grew in importance in the ensuing years. Indeed, the biggest single structure erected at Whaddon was the stores extension, between the stables and the drive.

Considering how many thousands came eventually to be employed at Bletchley Park, Whaddon, Hanslope and the other outlying stations, it is a miracle that the secret of the work being carried on there did not leak out. I am sure that part of the reason was that many of those who were recruited came through personal recommendation from others known to them who had already been accepted into the fold, making for a close-knit community. Many of the staff at Bletchley Park came from the universities where one could vouch for another. So far as Whaddon was concerned, many came from Philco and other wireless manufacturers, others were vouched for in different ways but in the early days, a personal introduction was almost essential.

When father started at Whaddon in 1940, the family remained at our home in Caterham, Surrey. The town lies roughly between the airfields of Croydon, Kenley and Biggin Hill so that, during the Battle of Britain in July and August of that year, there was much activity in the skies above us. Our house was slightly damaged during German bombing raids, whilst the battle went on overhead. Following a particularly heavy raid, my parents decided to move near to Whaddon, where father had already settled in.

We soon found a home at 95 High Street, Stony Stratford. It was above and behind a branch of Canvin, the local butchers and later became the premises of the Trustees Savings Bank. Under the prevailing circumstances, we considered ourselves very fortunate to find this quite acceptable home in view of the great pressure on accommodation locally as a result of so many people being evacuated to this safer region,
away from London.

Later, there was the great influx of people caused by the expansion of Bletchley Park and the various SCU units in the area, and this made finding a home of one's own in the area quite impossible.

1942 and I join SCU1

With the pressure on school places in the comparative peace of Buckinghamshire, I had to wait nearly six months to find a place at nearby Wolverton Grammar School. I was there until June 1942. In that year, I had made some model battleships for an exhibition and my father took them to Whaddon to show to friends. It appears that Lieutenant-Commander Percy Cooper RNVR, who was in charge of the wireless workshops, was one who saw them being shown off by a proud father. This coincided with a decision to expand the workshops and I was asked to go to Whaddon Hall where I was interviewed by Commander Cooper. He was in charge of all the wireless workshops for SCU1 under Bob Hornby, including R&D and manufacturing.

Commander Cooper thought my models showed a mechanical aptitude and asked if I would like to work for the unit. My 'scholastic career' was clearly going nowhere so I gladly accepted the offer. He showed me into the workshops where men were assembling wireless sets but I chose to go into the metal workshops, making the chassis, rather than into the wireless rooms. Wiring sets seemed to me, at first sight, to be more like knitting. So, I started work at Whaddon in July, 1942.

My first 'boss' was Captain 'Charlie' West of the Royal Corps of Signals but he was one of the many 'specially enlisted' men who were NPAF, ('Not Paid Army Funds'). Like many others, he had come from Philco. I was placed under the supervision of 'Dickie' Bell, a civilian who was a highly skilled metalworker. This section dealt mostly with the construction of the chassis and metal components for the unit's own numerous designs of agent's wireless sets, including the MkIII transmitter, the famous MkV transceiver in an attaché case.

When I joined in 1942, the staff in our 'hut' consisted of a cross-section of civilians, 'real army' personnel and 'special enlistments'. They included Bert Norman; 'Frank' Franklin (a tool maker who was living in retirement at Nash when the war started); 'Polly' Perkins who described himself as a 'sheet metal basher', John Harding, 'Nobby' Clark, 'Dickie' Bell and myself.

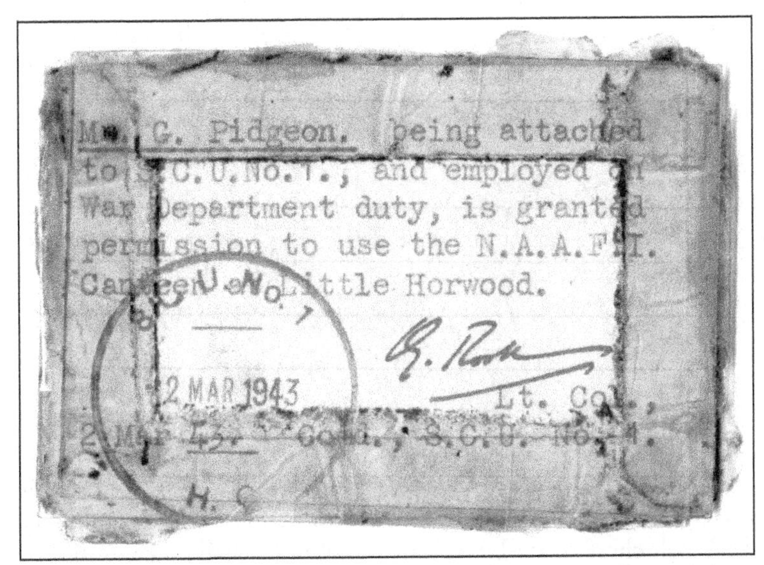

Mr. G. Pidgeon, being attached to S.C.U.No.1., and employed on War Department duty, is granted permission to use the N.A.A.F.I. Canteen Little Horwood.

12 MAR 1943

G. Rooker Lt. Col.,
2 Mar 43. Comd., S.C.U. No. 1.

March 1943 and still a civilian in Section VIII, this pass enabled me to use the Little Horwood NAAFI where one went for a tea break. It says '...attached to SCU No. 1 and employed on War Department duty.....' and it is signed by George Rooker brought in by Gambier-Parry to instill some measure of 'real' army to the unit. I had another when working at Whaddon.

I was not the youngest employee in the Whaddon workshops. We had a forge, handling metal sections for the larger chassis, under the direction of Joe Hill. He had a young local boy as an assistant who I believe was only 14. Otherwise, I suspect most of the men were somewhat older, having had established careers in the various skills required in the organisation.

My initial training was not very exacting and I was naturally given the more mundane tasks. I think my first job was punching out ventilation holes in the back panels of transmitters. It felt as if I had punched millions of holes in my first stint which actually lasted just three or four days. It bored me to tears and it was not the first time that I considered that 'knitting' wireless sets might have been a better career!

After a discussion with Bob Hornby, Charlie West arranged for me to have a one-day a week release to Wolverton Technical College. There I studied maths, physics and mechanical drawing, and I also attended on one or two nights a week. I worked hard during the day release and started to do so in the evening classes, especially in mechanical drawing which I particularly enjoyed. Unfortunately – or perhaps fortunately as it turned out later – Jane, my wife-to-be, was also at the college studying typing and shorthand so our intended evening classes were somewhat neglected!

Unknown to me at the time, three chaps of my age were being taken on as 'apprentice' operators. They were Dennis and Maurice Herbert, sons of Claude Herbert and Dudley Bradford, son of Charlie Bradford. They were taught Morse and wireless procedures from scratch and, in time, became important members of the unit. So it appears the decision to employ a youngster like me as a 'trainee' engineer was not an isolated occurrence. They had taken on these three trainee operators at the same time, all four of us sons of well-established personnel.

I made progress in the workshop and was soon given more responsible tasks. I was shown how to use drills, the milling machine, and a lathe, on which I made knobs for our own Morse keys out of plastic rod and a plastic material known as RM70. The lathe was manufactured by The South Bend Tool Company and, like so much else of our equipment at Whaddon, came from the USA.

Soon, I was allowed to cut sheet metal to size on the guillotine and make chassis for our various models. I then punched the holes for valve sockets, spacers and grommets. Best of all, in those early days, I enjoyed making the fronts of the sets out of early forms of plastic sheeting on which I then engraved such information as 'On-Off' and the like. Examples of many these sets can be seen in the Wireless Museum in Hut 1 at Bletchley Park, and in the tower of the mansion, where they have reproduced the Section VIII wireless station known as Station X. The MkIII there shows the type we made in my early days, complete with wooden cabinet, whereas other sets are in Hut 1.

Production of agent's sets was stepped up in late 1942 and early 1943. Clearly, there was no more space to expand at Whaddon so a plant was set up at one end of the vehicle workshops at Little Horwood Manor. 'Gees' had a large workshop built for the repair of farm equipment but, in 1942, most of it was being used as a workshop for our unit's army vehicles, which were under the command of Major Freddie Pettifer. About half of this building was taken over as a wireless factory and then put very strictly out of bounds to all others at Little Horwood.

The transport section at Little Horwood included a large number of despatch riders who were used to carry the messages received from the wireless stations to Bletchley Park and then back again for onward transmission. The unit's extensive fleet of Packard motor cars was based there, as were army lorries of all types and the coaches used to ferry people from their comfortable billets in the surrounding villages to our various places of work.

I was told I would be working there when it opened in early 1943, and was delighted that I was now to become a 'fully fledged' member of the metal shop team. Before moving there, Charlie West told me to collect a full kit of tools from the stores and gave me an official order which I proudly showed my father who issued them to me personally. Whilst I had earlier been given files, rules, a set square and other tools, a 'full kit' was very extensive and included a splendid tool-maker's oak chest made by Moore and Wright. It had a number of drawers each full of tools, and I now had my own micrometer, so I felt I had arrived!

Bert Norman became head of the new metal shop and we continued making wireless chassis and metal components with 'Polly' Perkins, 'Nobby' Clarke, myself and one or two others who later joined us. I now used the new capstan lathe, and we had a large power guillotine, making cutting sheet metal much simpler and faster.

The whole production unit was managed by Lieutenant Hugh Castleman, again nominally of the Royal Corps of Signals, but actually in Section VIII. The wireless assembly side was initially under Johnny Lester. Later, Len Dewick, who owned a wireless shop in Stony Stratford High Street, joined us and took over that section. I think that was mainly because he was 'getting on' in years, was well over six feet tall, and had a real air of authority. The stores, at one end of the workshops, issued the parts for wireless production and was a sub-division of father's wireless store back at Whaddon, with one of his team in charge.

The wireless assembly room contained about four rows of linoleum-covered benches with about six men to each table. I recall just a few of their names – Bob Lester, Ken Bromley, Joe East, Syd Hutchins, 'Mac' McGilvery, and the ubiquitous Jock Denham as general factotum.

I remember Jock Denham well. He was considerably above average age, probably in his mid-forties, and had grown up in the Gorbals, the slums of Glasgow. He was a large and amiable man, willing to impart his views on life generally, and women in particular, to the young men around him.

On one occasion, a batch of about fifty MkVII sets were completed for urgent despatch, when someone spotted that they did not have their serial number on the power packs. The number was engraved on a small plate about 25 mm (1 inch) long and Jock Denham volunteered to work late to complete the work. Unfortunately for all concerned, Jock decided to fix the serial number on top of the multi-voltage transformers and used a very long drill. The result was that he drilled through the coils and ruined every single one. Fortunately, the disaster was discovered before the sets were sent out to agents in occupied Europe, but the whole workshop then had to work flat out to replace them.

By the end of November 1943, I was at the minimum army recruitment age of seventeen and a half and my father, and (more importantly) Lieutenant Commander Cooper, decided that I should join the Royal Corps of Signals. My joining the unit was posted in 'Part 2 Orders of No. 1 Special Communications Unit' a copy of the document can be seen in Chapter 9. It shows that I was 'Not Paid Army Funds' although like my father and others at Whaddon, I was given a Part 1 Army Pay Book which showed one's service record, training, last will and testament, next of kin, etc. However, we were not issued with a 'Part 2 Pay Book' which is the actual record of a soldier's pay.

On one or two occasions, when I travelled abroad with army personnel, the time came for the weekly pay parade. I was naturally forbidden to explain that I was not paid by the army, and the resulting puzzlement on the face of a sergeant major, demanding that we stand in line to receive our pay, was a delight to behold. We usually bluffed our way out of these sticky situations but I am very glad they were rare occurrences.

Having read the stories of the rigorous military training of Wilf Neal, Martin Shaw and others at Ashbys and Little Horwood, I am astonished to learn how 'militarised' the unit – outside of my privileged inner-circle – really was. It shows up too in the complaints by the early VIs at Hanslope, expressed by Pat Hawker, who unlike Wilf and Martin, was at least a 'Special Enlistment'. I am sure Gambier-Parry did not approve of an overly 'military regime' for his men and the early members of the unit were largely protected from it. However, the demand for growth of the units – particularly the need for operators to travel abroad in the forthcoming invasion of Europe – meant that they should at least have a basic knowledge of soldiering, if only for their own protection. Some of the 'spit and polish' was applied by those who actually enjoyed the pretence of being soldiers – even if they were Section VIII or 'toy soldiers' as some called us. Much to my father's dismay, I was not to be one of them.

I was taken by father to be 'sworn in' over in the 'military' part of Little Horwood across the road from the workshops, leading to the village. When that was over, we went to collect my uniform and all the paraphernalia of army gear, including mess tin, knife, fork and spoon. With all this packed around me we were driven home to Stony Stratford in a Packard saloon. My indoctrination into 'army life' now appears so very different to the initial experiences of Wilf Neal, Martin Shaw, and many others joining SCUs. The only other time I went into the 'military' camp was two days later, to have the required inoculations.

Now ostensibly a Signalman in the Royal Corps of Signals, my work continued at Gees' until nearly the end of 1943. Then Dennis Smith, one of the unit's most skilled wireless engineers, was instructed to expand the 'Mobile Construction' team to carry out its increasing tasks. It was based in a new hut erected near the exit lodge at Whaddon Hall under the high wall of the kitchen garden. The curve of the drive gave plenty of parking space for the vehicles on which they worked and the team's own transport.

He was free to choose his team and it eventually consisted of people with a wide selection of skills. Jock King, Tony Wheeler, Charlie Dunkley, a local first-class joiner, whose name I sadly forget, Wallace Harrison (brother of Edgar, see chapter 31), Norman Stanton, Jock Denham and myself. Jock Denham was included because, in spite of his faults, he was a good 'all-rounder' and as such, useful to a team with a need of varied skills.

Dennis, having been in R&D with Wilf Lilburn and Alf Willis, was now responsible for fitting different kinds of wireless equipment into aircraft, ships, MTBs and wireless vehicles. It was the most glamourous work I could think of – infinitely better than being involved in manufacturing wireless chassis – one hundred at a time.

Incidentally, I should mention that security at Whaddon was in the hands of the Corps of Military Police. They were on guard on the entrance gate to the Hall grounds and at the exit gate close to Mobile Construction's workshop. As with so many things at Whaddon, they were unusual in that they had blue tops to their peaked caps instead of the red usually associated with a Military Policeman. That is because they were 'VPs', the Vulnerable Points Wing, selected to guard secret establishments.

The great joy of being in Mobile Construction was the sheer variety of work undertaken. We could easily be working on aircraft at Tempsford one day, and in the next day or two be off to a south-western Channel port to work on an MTB. The imminent invasion of Europe involved the unit in constant work on SLUs, from the time I started there until late 1944, but we all shared the visits to airfields and ports.

I will now explain the work in the different spheres of Mobile Constructions activity.

Aircraft

MI6 used aircraft based at Tempsford with Whaddon-designed Ascension air-to-ground short-wave wireless equipment fitted for contacting agents on the ground, and locating drop zones for them. The first planes used were Douglas Havocs (A-20), known as the 'Boston' when in use with the RAF at Tempsford with 161 Squadron. Flight Lieutenant Maurice Whinney (of Section VIII), flew these on operations in 1942, carrying the Ascension equipment for contacting French fishing vessels off the Brittany coast.

In effect, it was like using a wireless-telephone with a French-speaking passenger in the plane who would be able to talk to patriots or agents on the ground without the need for them to be skilled in Morse. There was also the benefit that the plane could be 100 miles away, and by using very high frequency, it was almost impossible for the Germans to locate the agent on the ground. Later the Bostons were replaced by Hudsons, although I believe Bostons of the Free French continued to operate from Hartford Bridge.

The RAF's 161 Squadron had a varied role but was mostly concerned with SOE, as opposed to SIS agents and operations. Its Lysanders could land in a very short distance but incredibly the much larger Lockheed Ventura of 138 Squadron were also landed in occupied France to put down agents and supplies.

We would travel to Tempsford daily from Whaddon, usually by Packard saloon which, albeit camouflaged, was still a rather ostentatious vehicle for a secret unit. It caused quite a stir with the sentries at the entrance gates who must have thought at an Air Marshal, at the very least, was arriving in such a splendid motor car.

At one time, we fitted wireless gear into a Halifax but I am not sure what the gear was exactly, nor why we did it. Otherwise, all the work I was involved with at Tempsford was on Hudsons and Venturas. On the airfield were four-engine bombers, like the Stirling and Halifax, that had been retired from Bomber Command, as the superior Lancasters took over operations. These were used to supply freedom fighters with arms and supplies as well as for dropping agents by parachute into occupied territory.

After installing our equipment, we went on flights in the planes to perform wireless tests of all kinds. In spite of the importance of our work, we always took a lunch break at a restaurant called 'The Buttery' in nearby Sandy. Travelling there and back in Packard saloons made me feel how lucky I was to be leading such a civilised existence at the height of the war.

We also carried out work at Hartford Bridge (now called Blackbushe) near Camberley in Surrey, which was an aerodrome in the Tactical Air Force. There we performed the same installations and tests on a squadron of Mitchells, as we did on the planes at Tempsford. So far as I am aware, their role just consisted of contacting agents on the ground, via Ascension and sometimes with supply drops. One squadron was Free French but another seemed to consist of crew from a number of Commonwealth countries.

The furthest we ventured at the time was out over the coast. Even at that stage, there could have been enemy aircraft about, and I am glad that beside testing the wireless equipment we had fitted, they also tested the guns!

While on the subject of testing guns, I must record a story about my father. He had expressed a wish to Dennis Smith to have a flight as he had never previously travelled by plane. One day, at short notice, Dennis had to go from Whaddon down to Hartford Bridge to check a plane's wireless gear and offered father the opportunity to go with him. They arrived late with the plane – engines running – absolutely ready for take off. My father was unceremoniously bundled into the rear of the plane; perhaps I should explain that the Mitchell's bomb bay completely divides the plane into two halves, there is no way from front to rear.

He crouched in the plane's fuselage just under the rear gun-turret and in just a few moments they took off. Shortly afterwards, when the rush and excitement died down, father realised he had scrambled into the plane without an oxygen mask, flying gear or parachute. These facts, combined with the increasing cold, the unexpected vibration and noise, rapidly made him feel very uneasy. Then a few minutes later, just to add to his misery, the rear gunner tested his machine-guns. A stream of hot cartridge cases spewed out all over him and I believe that this was one of father's worst ever experiences. He never flew again!

As I have indicated, we lived well as we had an open expense account. When working at Hartford Bridge we did not use the RAF mess but ate out in a nearby smart restaurant called The Ely on the road to Camberley. A table occupied by a Royal Signals sergeant and a couple of scruffy signalmen, sometimes raised an eyebrow or two amongst the senior RAF officers who often used the place for entertaining high-ranking colleagues. In the evening, we always returned to the nearby comfortable hotel, then called The White Lion, at the bottom of the hill leading to Hartley Wintney.

One evening after work, we went straight into the hotel bar and Dennis Smith called for three pints of beer from the landlady, Mrs. Temple-Cook. He put down his papers and a small metal box on the bar. After talking for a while about the day, he lifted the lid on the box and we were absolutely astonished to hear a repeat of the cheerful greeting from Mrs. Temple-Cook, Dennis's ordering the beer and our chat! It was the first time we had heard our own voices and they came from the latest version of the RCA wire-recorder Dennis was going to test for use with Ascension.

The only other wartime aerodrome I flew from was Little Horwood. This was only a few miles from the 'Gees' farm and our base there. Little Horwood was used for emergency landings but it was actually an OTU (Operational Training Unit) and had a few Wellington bombers based there when I worked there. In Chapter 17, I recounted my foolish escapade of driving a Packard coupé along its runways but I did not explain our reason for being there. Our unit had an Avro Anson aircraft which was, in effect, a flying test-bed for the products of the R&D workshops at Whaddon. The Avro Ansons were intended for training airmen to be bomber crew and, in common with all training aircraft, they were painted bright yellow, making them very conspicuous. As a result they were kept well away from the coast.

I went on several short flights in the Anson with Dennis when he was testing new equipment. The pilot was usually Maurice Whinney. My role – as with most installations when working directly for him – was to 'solder this', 'couple that' or 'drill those'. My job might not have won the war but I am confident that Dennis had a competent assistant beside him as he performed the wireless work at which he was so proficient.

Vehicles

I have explained how, in 1940, the Unit had purchased every single new Packard car in the country from the Packard showroom and fitted them with receivers and transmitters. Some of these were dispersed across the country. A few were later sent to North Africa but proved unsatisfactory off-road abroad. Our 'A' Detachment in Cairo stripped the wireless gear from the Packards they had taken with them and fitted them into more suitable vehicles.

In the chapter on Mobile Construction, I described the team's work on the various SLUs which continued from late 1943 right up to D-Day, in fact one or two were not even finished by that time. That was not a problem since they were intended for units in the follow-up forces. The SLUs were a major project for our smallish team and occupied a great deal of our time but there were still visits to airfields and ports, almost always with Dennis Smith.

Taken at a rally at Bletchley Park in 2001 when I was delighted to see a Dodge Ambulance – exactly as they looked when delivered to us at Mobile Construction – for fitting with our wireless equipment. Used by SCU/SLUs in the US military areas in the invasion of Europe.

MTBs, MGBs and MFVs – in Slocum's Navy

The NID (Naval Intelligence Division) and SIS had joined forces to provide a 'ferry' service for agents and the delivery of mail and arms, across the seas to the coast of occupied Europe. This was controlled by Captain Frank Slocum RN who had been recruited into SIS by Admiral Sinclair in the mid-1930s. They used MTBs and MGBs, but MFVs (Motor Fishing Vessels) were in common use as well. The latter were trawlers and other fishing vessels – some specially built – manned by Royal Navy personnel disguised as fishermen. Ships in Slocum's Navy, as it became known, were used on the coast of France, in the North Sea, on the Norwegian coast, and in the Mediterranean. Their duties might also include interception, weather reports, and their special wireless gear was fitted by personnel from Section VIII.

The Flemish fishing vessels who contributed the 'Flemish fishing code' – which was almost unintelligible to anyone else – were particularly important for weather forecasting.

I came late into this aspect of the work, which had been going on since the days of the SS Cecile way back in 1939. In 1944 I worked with the team on installations to MTBs at Dartmouth and at Brixham. Even after D-Day, much of the coast of France was in German hands for a considerable time, and Norway longer still.

Although Dennis was running Mobile Construction, he was still very much R&D orientated. For example, he designed gear that was intended to make it easier to locate an agent arriving off-shore. To test it, he drove down to Dartmouth in a Packard leaving me to follow on with a Guy 15 cwt wireless van which we could use as a workshop and with some of the experiments. I packed a wide variety of tools and then he gave me my instructions. I was driven there by one of the pool Royal Signals drivers from Little Horwood.

Those 'of a certain age' will recall how all directional and town signs were removed in 1940 due to the invasion scare and were not replaced until after the war. There were no signposts and if there were an advertisement hoarding, any reference to the location would have been obliterated. All this made it very difficult to move about, coupled with the fact that if you asked the way, most people were reluctant to help.

On arrival at Dartmouth, I was to telephone HMS Westwood Ho – an old River Dart pleasure boat – which was used as the base ship for the MTBs on which we were working. It lay in the harbour between Kingswear and Dartmouth. The contact name given to me – if all else failed – was that of our Major Cox, or if he was not available, I was to ask for the First Lieutenant on board and tell him I was from Section VIII.

The journey went very well and I arrived in time for a late Sunday lunch on board. Since it was served with a half-mug of navy issue rum, I slept the rest of the afternoon and most of the night!

We installed the wireless gear on an MTB and around midday we drove off with the Packard and Guy van to Slapton Sands. This was part of the Hamms area of south Devon where a number of villages had been cleared of civilians to provide a training ground for US forces. Slapton Sands looked very much like the beaches in Normandy they were to assault on D-Day. The area was still closed to civilians and we had to pass through a tight security guard before driving on. We passed through desolate villages which was quite eerie. I recall one large house with the windows blown out but with the net curtains still in place, flapping in the stiff breeze from the sea. It seemed like a lot of handkerchiefs waving at us as we passed.

A party of naval officers and one or two from the army had already assembled on the sands, including Major Cox. Meanwhile, the MTB had arrived from Dartmouth. It came close inshore where Dennis boarded it by dinghy, after telling me that when they signalled, I had to speak into the microphone of the wireless-telephone he had installed in the van. The MTB moved offshore some two hundred yards when I saw a lamp flash and I started to recite the ABC. I did that two or three times as the vessel picked up speed and then raced away. I then tried counting for a change but soon became tired of that so thought a little light relief was in order. Therefore, I presented the assembled company with this little ditty:

Mary had a little lamb
She also had a bear
I have often seen her little lamb
But never seen her bare!

I am not quite sure what it is like being out at sea in a force 9 gale but it must be something similar to the way in which a senior Naval officer (covered in gold braid) descended upon me demanding that I cease immediately, and return to the previous recital. The others there seemed highly amused but I was careful to stick to the script in any future work of that kind.

The 'Westward Ho' was a River Dart pleasure steamer moored in Dartmouth harbour, and the base for a number of craft operating in Slocum's Navy, including MTBs, and MGBs (Motor Gun Boats). Most were fitted out with Section VIII wireless equipment, in addition to their standard naval wireless gear. The MGB shown here is 502, which was based at Dartmouth, and used on cross-channel activities.

On another trip, we went to Brixham. From there, late one evening, we went out right out into the channel in an MTB to perform some tests with an 'agent' in a rubber dingy. He was put overboard and we pulled away with the idea, I believe, of tracking him by wireless. I went down and sat beside the navy W/T operator who was not involved, and was again was offered rum. I thought rum was a daily ration but they always seemed to have plenty for visitors. He took down a message en-clair to the effect that a floating mine had been seen earlier, on such-and-such a bearing. I went with him up to the bridge to where Dennis was working. The skipper plotted the position and shouted that it was exactly where we were slowly drifting whilst listening for the 'agent's' signal!

The whole crew went to the sides of the boat peering into the dark to see if we could spot the mine, whilst Dennis tried to coax the 'agent' back without telling him of the problem he – and we –were facing. After a while, we hauled him aboard and moved slowly back to Brixham, thankful we had not disturbed the mine, or met a German E-Boat whilst we were so vulnerable.

The MFU

I first encountered the MFU at Tough's Boatyard in Teddington, in the autumn of 1944, when Dennis Smith and I were returning to Whaddon from a trip to Brixham. We had made a diversion there for him to see the progress being made on this, the most secret of the unit's projects.

We went into a large shed and he said I would be working on 'that' shortly. He had not mentioned it earlier, or prepared me for the strange vessel that I now saw in front of me. The shape came as a complete surprise. It looked like a big steel cigar, tapered at each end, not like a torpedo but perhaps more like a submarine without a conning-tower. We met (or rather Dennis met) the owner, Douglas Tough, and discussed certain details with them whilst I clambered all over it. I was very puzzled but in those days one didn't ask too many questions.

Tough's were not told what the vessel was intended for, only that its code name was MFU, which they thought stood for Motor Fuelling Unit. The cylinders had been made elsewhere and delivered to them to be made to look like fishing boats. They also had to make the sealed covers and install the engines. These were Perkins P6 60 HP diesels with a fuel capacity that allowed them to cruise at 8 knots for up to two days. Pressure trials took place in Sunbury docks on the Thames so that if the vessel did not come to the surface it could be raised with a nearby crane.

If the MFU was not a submarine and could not move under water, what was it for? I learned much later that the intention was to tow it near to a coast occupied by the enemy, where a team of agents would assemble the canoes and their gear, ready to go ashore. They would then press a switch, causing the vessel to sink after a short, pre-determined interval. It had free flooding chambers fore and aft to enable it to go down to the sea-bed and up again on cue. It could be programmed to rise to the surface at specific times at night when the team would return to the craft and use its wireless for short periods. The wireless was high frequency and so could easily be used to communicate with aircraft fitted with Ascension. Its brief transmissions would not be detectable and shortly after broadcasting, it would be gone again. It was a floating base for the team, capable of moving on the surface if needed, albeit at relatively slow speed.

For some time, I had been aware of a new wireless receiver/transmitter being developed, by Dennis, in his own office inside our unit's hut. He had made it up on a baseboard festooned with components pointing in all directions – a complete 'lash up'. When he was satisfied with its performance, I was told to make a steel chassis for it, to a certain maximum size, with holes for valve sockets, screws, and grommets. When that was completed, the valves, coils, condensers, etc. were put in place and the whole thing tested. My next task was to make a front for the set from sheet plastic, complete with holes for knobs, screws and dials. On this, I engraved the various 'legends' such as 'power on/off', 'transmit', etc.

I was in my element and made two sets of each chassis and front. The second, once approved, went to Bert Stacy, our draughtsman, who made drawings from it. That was quite the reverse of normal practice. To the best of my recollection, the whole thing was about twice the length of the MkIII and perhaps overall, looked like two of them joined together. I regard this as my best achievement – something I did completely on my own – utilising skills I acquired under such good instructors. Dennis was pleased and I believe it erased one or two of my earlier 'black marks' such as the ditty at Slapton Sands!

Little did I know that this new set was to be used on the MFUs or that, in December 1944, Dennis would ask me to pack my bag for a trip away from home over Christmas. We drove down to Teignmouth and stayed in the Undercliff Hotel at nearby Shaldon. From there, we drove over daily to work on the MFU which had changed somewhat since I had seen it earlier, back in the autumn.

It now looked more like a boat and the 'cigar' shape was lost in the partial decking, sides and cockpit that had been added at Tough's boatyard. It was smothered with men carrying out every kind of job – wiring, welding and painting. They were also checking store space in the front compartment. This area was for Folbot canoes,

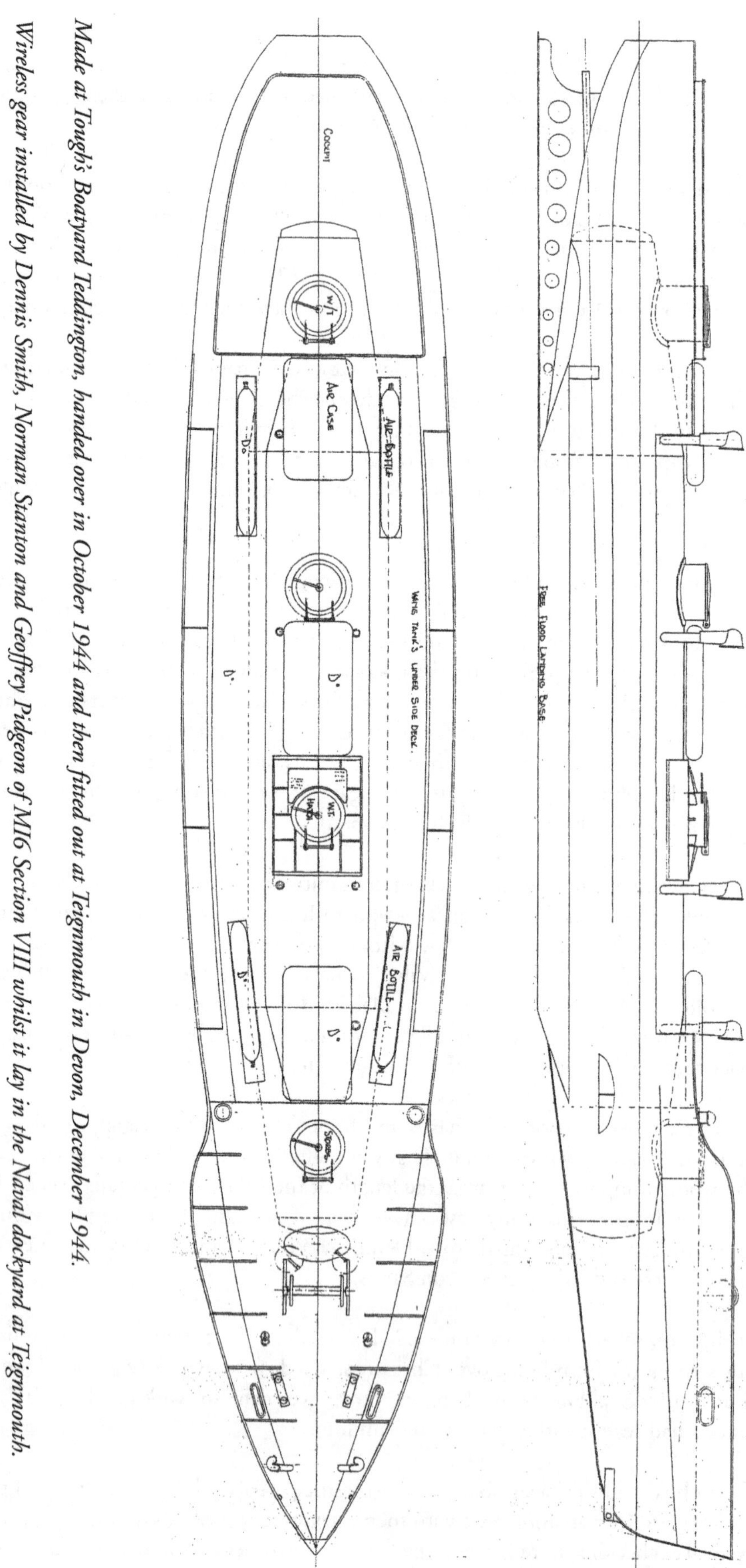

Made at Tough's Boatyard Teddington, handed over in October 1944 and then fitted out at Teignmouth in Devon, December 1944.

Wireless gear installed by Dennis Smith, Norman Stanton and Geoffrey Pidgeon of MI6 Section VIII whilst it lay in the Naval dockyard at Teignmouth.

weapons and provisions. The middle compartment was the control room with its switches, radio charts, compass and electrical controls. Our radio was to be fitted in the front end of this space from left to right – or rather port to starboard! The rear compartment contained the engine, air compressor, and so on.

Dennis Smith, who is 6'3", had great difficulty moving around inside the wireless and control area, so Norman Stanton and I did the final wiring and connections, under instructions from Dennis leaning through the hatch. I think we spent about ten days there, not because we had that much to do, but because of the sheer number of different trades trying to complete their work at the same time in confined spaces.

Bearing in mind the great secrecy surrounding the project, my memory of it is quite good but I have no recollection of us performing any tests, other than Dennis confirming that his set was working, as anticipated. There would have been no question of us going out to sea on it, or being involved in the 'sink' and 'recover' process but our job may have ended with the successful operation of the wireless equipment.

The wartime boatyard records at Tough's have recently been found, showing the commencement and completion of every vessel launched and I have been given a copy. It appears from these that only one MFU was fully completed and the rest scrapped. Both Robert Tough, the son of Douglas Tough, and I, know that this is not accurate. One had certainly been tested at Teignmouth before we arrived, and had not come up from quite deep water off a cliff feature, called 'The Ness'.

When I arrived at ISLD in New Delhi in mid-1945, I was interviewed by Major Robin Addie and Captain John 'Tommy' Tucker on my future role there, and they asked me about the MFU. It appeared they were connected with the project in the earlier days and I gather it had also been intended for use in the waters of south-east Asia.

I remembered that the time-switches that made it sink and come up later were made by Venner & Co. of Kingston. 'Tommy' Tucker told us that the wrong choice of switch was certainly the reason for the failure. Perhaps that remark was something to do with the fact that he was Tucker of the family firm of Tucker

The MFU in the Thames, alongside an air-sea rescue launch at Tough Brothers. Its tubular shape has been fitted with sides, bow and stern. It could then be fitted out to resemble a fishing boat.

Switches? However, in recent discussions with Robert Tough he tells me that after the MFU refloated, it would have been vital for the returning crew to get on board quickly since the valves would have to be shut off before the compressor stopped working. Perhaps that was the reason for the failure during trials at Teignmouth, and not the Venner switches.

Perhaps the MFU could have been turned into a success but before the problems could be dealt with, the war in Europe came to an end, and shortly afterwards so did the war in the Far East against Japan.

While I was at Tough's in 1944, Robert Tough was an apprentice to his father, Douglas Tough, and of my own age. I visited him again in 1996, when I decided to write about the unit. He had no official drawings as they had all been taken away but he well remembered the strange contract. Certainly it was odd for a shipyard making MTBs, motor gunboats, and the like. He kindly made a sketch for me and then later found two photographs of the craft taken before it was handed over to the Admiralty – probably the NID. Obviously, the project was highly classified but the pictures were taken of an Air-Sea rescue launch that was being repaired and luckily, the MFU just happened to be alongside. I visited Robert again in 2003, and was delighted to find that he had discovered one of the original working drawings of the craft scaled at ½" to 1ft. He gave me a copy which is reproduced here and is quite remarkably similar to his sketch from memory.

VE Day

On 8 May, 1945, VE (Victory in Europe) Day, I worked during the day and later went to the Red Lion, a pub in Mill Lane, Stony Stratford, for a celebration drink with close friends. My father had earlier suggested that the family go to the Cock Hotel in the High Street for the evening, with some of our mutual friends from Whaddon. I willingly agreed and later met them there. Along with my mother and father, the crowd included the Watson family who, with the Peacheys, owned the Cock Hotel, Sid Wickens, whose father ran an outfitters a few doors north of the 'Cock' and Ewart Holden and his wife, Tessa, from our unit, who lived along the High Street. My older brother, Ron, later joined us with some of his Home Guard friends. Dennis Smith and others from Whaddon were there, with a sprinkling of Bletchley Park staff.

The saloon bar was fairly full when we arrived at about 8.00 p.m. but it soon became totally packed out. No doubt there were many 'locals' there but I should add that it was then such a friendly town, there was no feeling of our being 'outsiders' in any sense of the word. There were also a few local servicemen on leave with their families in the saloon, and in the public bar beyond. Normally beer and spirits were limited – if not actually rationed – but no such restriction was imposed that night!

Much toasting went on and Churchill's name was frequently mentioned. We were excited and elated, perhaps as much by the day's unfolding events, as by the alcohol consumed. There was singing and great merriment so that the time flew by. It was somewhat after 11.00 p.m. when we realised that we were on licensed premises well after closing time. Somehow that fact seemed, in a strange way, to emphasise the importance of the day!

A week or two earlier, we had worked on clearing our gear from a couple of Mitchells at the airfield at Hartford Bridge. A Canadian pilot gave me a Verey Pistol and a large satchel of cartridges, with words to the effect that they would help light up the place when the war ended. I only remembered them about midnight and collected the satchel from the shed where I had concealed these items from my mother.

I went outside the main entrance to the 'Cock' and fired the first cartridge up the High Street over the Bull Hotel sign at an angle of 45 degrees, towards the Plough public house, to the huge delight of the crowds pouring out of the many pubs that then existed along the High Street. I fired the rest off in the same general direction. Only afterwards did I think of the damage my over-exuberant behaviour might have caused. Fortunately, there was no damage and perhaps I helped to round off the evening's festivities. Soon afterwards, we wandered home, tired but exhilarated by the day's momentous events. The next day, I met a friendly but senior member of the Whaddon staff, who liked to drink at the Plough. He told me I should have stayed up later because someone put on a firework show. I did not tell him it was me!

Bound for the Far East

With the fighting in Europe over, the emphasis moved to the continuing struggle in the war against Japan. By this time, the Japanese army was being pushed back from the high points of their earlier conquests. They were retreating through the Pacific islands pursued by massive US forces, in China by the National and Communist forces, and in Burma by the now famous 14th Army.

General MacAthur had moved his HQ from Australia to Manila in the reconquered Philippines and Group-Captain Winterbotham wanted more SLUs to receive Ultra. He wanted them to set up in the forward cities of reoccupied countries, as the advance towards Japan itself continued. Naturally, the first wish of service personnel in Europe was to be demobilised and return home to civilian life. Certainly, the older service personnel had no desire to go to the Far East and this general approach existed in the SCUs as well. Quite rightly, our 'lords and masters' sought out the younger members of the organisation to transfer to the war in the East. At just 19, and the youngest, I was one of them. I was interviewed by Spuggy Newton and it was decided that I should be trained in various other skills. The first was a course in maintenance of our wireless sets, for which I was sent over to the wireless repair shop under the eagle eye of Jack Buckley. He was a good tutor but with a bitter tongue, so I kept a lower-than-usual profile, getting on with the job of learning to repair faulty HROs and the like.

There were three others with me but they were wireless engineers, whereas I was a basically a fitter with wireless knowledge. However, they in turn had to learn new tricks since the idea was that each of us should go to one of these new SLUs, help set them up and maintain the equipment, whether it be the wireless sets, electrics, batteries or aerials. The three others with me were Brian Birch, Fred Stapley and 'Fergie' Ferguson whom I recall, had the wonderful Christian names of Andrew, Carnie, Lackie – no wonder we simply called him 'Fergie!'

Later we were sent to a tented camp just outside the nearby village of Nash, which was set up in the same field as our wireless station. We were issued with blankets and allocated to bell-tents, six to a tent. It would have been most uncomfortable for anyone, but especially for those of us who had been living in comfortable places like Tattenhoe or – in my case – at home.

Fortunately, there was no formal military training but most days we had to listen to lectures in a marquee from various 'old sweats' as they were called. These were older soldiers with experience of service abroad and life in general. One in particular had a very rosy glow – indicating that he had considerable experience with the bottle – as well as the problems of mosquitoes. After a couple of days of this, I spoke to 'someone' and arranged (or 'wangled' in army parlance), a pass to 'sleep out' and so returned to my comfortable bed at home each night. If there were no lectures scheduled, we four went our various ways to continue our training – me to the tender mercies of Jack Buckley at Whaddon.

One day, we were issued with solar topees or pith helmets, a form of headgear more in keeping with the pre-war Indian army. We thought of them as being long since redundant. Nevertheless, they caused much amusement for a day or two but I doubt if a single one was actually taken abroad, some I know were tossed into nearby hedges!

We were in camp for about two weeks before we were assembled for a parade before Brigadier Gambier-Parry. I suppose there were some sixty of us in the draft to go to India and beyond. Of these, only four of us were from Section VIII, so when the time came for the parade, the real soldiers were lined up in three ranks on the grass, and we were put some twenty yards away to the left. The army contingent included drivers, clerks and operators. Gambier-Parry arrived in his Packard limousine with a small party of officers, including a Captain

Ford from Little Horwood camp. Rumour had it that Ford had been a uniformed commissionaire outside a cinema before joining the army at the outbreak of war. I had not heard of him before but he was very military in his approach to the assembled company, and plainly obsequious to Gambier-Parry.

They walked amongst the ranks to our right with Pop being his usual friendly self. Here was this wonderful man, immaculate in his uniform, but talking in the most informal way to his men. He spoke to many of them, then he moved over towards our small party of four, who were rather less military in their bearing. A few paces away we heard him say 'I suppose this is the Section VIII lot?' which, I suppose, really showed what a scruffy crew we really were, and how different to the rest of the parade in military appearance. The others with him agreed as they moved towards us as, when without orders, we shuffled to attention.

He spoke to Brian Birch for a moment, asking about his work, and then he stood in front of me. Someone whispered 'This is young Pidgeon, sir' and the great man said 'Aha, young Pidge!' He then asked if my father approved of my going overseas and what he thought about it. He said it just as if father's opinion truly mattered to him. He then asked what work I had been doing, and I told him I had been part of the team building the SLUs, and had worked on aircraft and MTBs. I recall all this came out in a jumble since I was in awe of the man. He smiled, nodded and mumbled something like 'Well done!' and moved on to Fred Stapley and Fergie Ferguson. You may wonder how I recall this brief meeting in such detail but to understand that, you would have had to know, or have worked for, Richard Gambier-Parry.

He then stood back and said to us collectively, 'Have any of you ever fired a rifle?' To which we all replied 'No', although in my case that was not correct, having earlier been an Army Cadet. He turned to Ford and told him we could not be sent into a war zone unable to use a rifle 'so see they do so at once'. Ford went off in a car to nearby Little Horwood and was back in twenty minutes complete with four Lee Enfield rifles. We were driven to a sand pit off the London Road at Stony Stratford and each of us fired a clip of five bullets – not at a target, as none existed – but simply into the sand at the other end of the pit. We then drove back to Nash and on the bonnet of the Packards in which we had travelled, Ford signed our AB64 pay books 'Fired musketry course' and the date. That was the full extent of my preparation before going off to win the war in the Far East, indeed the sum total of my entire army training!

'... urgently required in New Delhi...'

A few days later, we attended a meeting at Whaddon where we were told we would be taken to the British Overseas Airways Corporation (BOAC) terminal at Victoria in London and from there to an airport near Bournemouth to fly to India. We were due to leave the following day so I went home to say goodbye to the family and friends. We were each given a document marked 'To whom it may concern' saying we were 'travelling on special duty' and were '...to report to I.S.L.D. (SCU.2 "B" Group) G.H.Q. New Delhi'. It went on to say **'They are urgently required in New Delhi, and every facility should be given them to expedite their arrival'.**

We were driven to Victoria and our luggage checked in. We then spent a few hours wandering around the West End, and as the only one who had lived in London, I became our guide for the day.

We returned to Victoria in the late afternoon and waited in the Art Deco lounge before being taken to a train waiting at the private platform alongside the building. We were shown into a reserved First Class compartment and the waiter asked us if we would like drinks before we started. I clearly remember ordering a scotch and dry ginger which seemed appropriate in such a splendid setting, though I had never had one before.

The train then drew away and eventually we arrived at Bournemouth station. There a small coach took us to Sandacres Hotel where we were shown to our rooms. We were told we had an early call so went straight into dinner. I was roused at 5.15 am with a cup of tea, then after breakfast we were driven to the airfield. A Dakota was waiting for us but it was unlike any I had seen before, since those had been camouflaged and used for troop

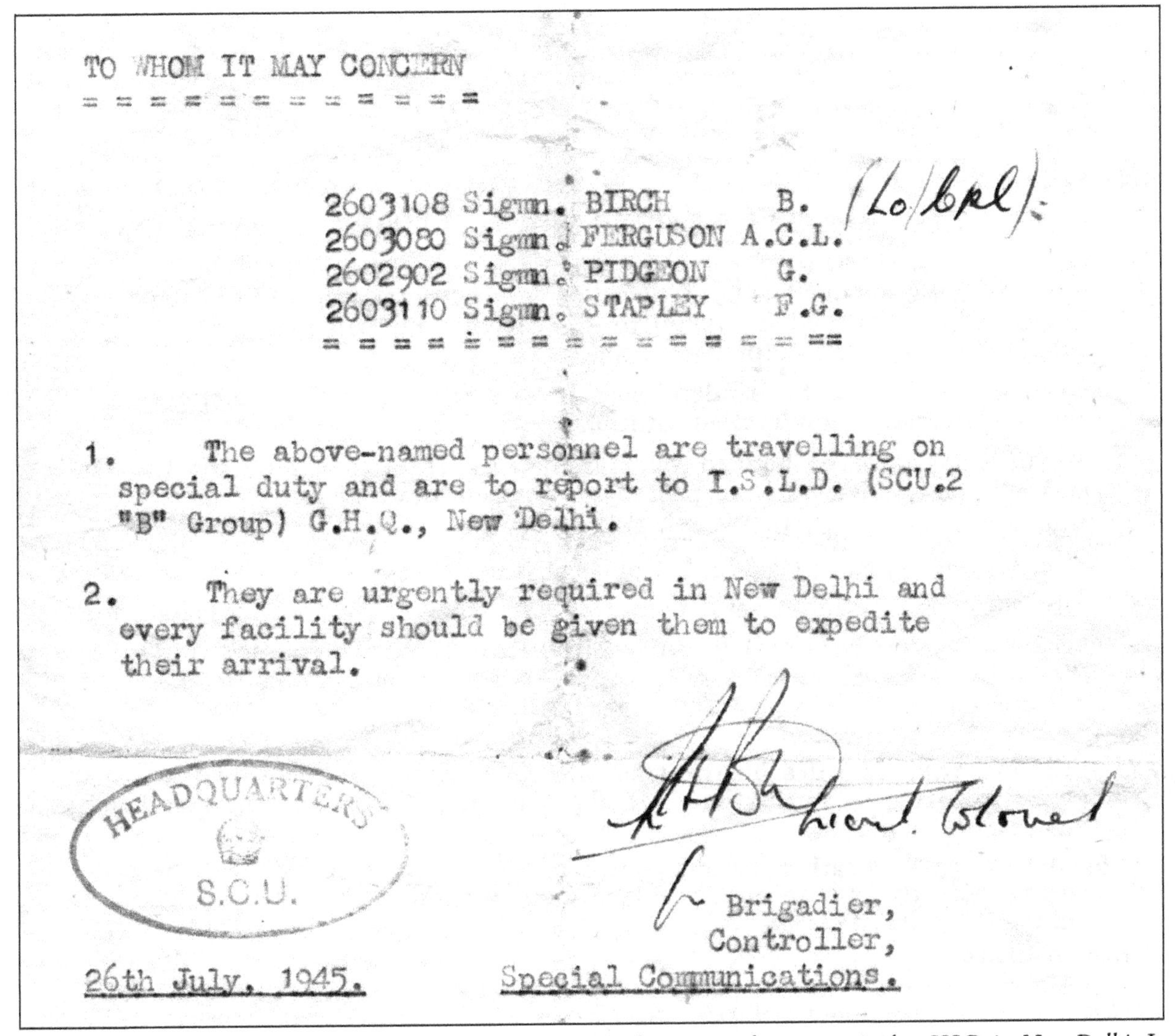

TO WHOM IT MAY CONCERN

2603108 Sigmn. BIRCH B. (Lo/Cpl).
2603080 Sigmn. FERGUSON A.C.L.
2602902 Sigmn. PIDGEON G.
2603110 Sigmn. STAPLEY F.G.

1. The above-named personnel are travelling on special duty and are to report to I.S.L.D. (SCU.2 "B" Group) G.H.Q., New Delhi.

2. They are urgently required in New Delhi and every facility should be given them to expedite their arrival.

HEADQUARTERS S.C.U.

Lieut. Colonel

Brigadier,
Controller,
Special Communications.

26th July, 1945.

An instruction 'To whom it may concern' to give us every facility to expedite our arrival at ISLD in New Delhi. It certainly worked in opening doors, wherever we went.

transport or for towing gliders. This one was Flight 7F.25 and its shiny silver fuselage had BOAC painted on its side. The date was 30 July, 1945 and this was one of the earliest flights by the revived civil airline.

The four of us went aboard and found ourselves in the company of admirals and generals as well as two or three important-looking civilians. How strange we four seemingly ordinary soldiers must have seemed to them! We took our seats and the plane took off on our flight to Delhi where, according to HQ at Whaddon, we were urgently required.

We lumbered slowly over the length of France at about 12,000 feet and around noon descended into Marseilles airport where we were taken to a restaurant that was being rebuilt. We sat for lunch at a table alongside about eight German generals who were also in transit. After a leisurely meal, we took off and flew during the afternoon towards Malta where we were to spend the night. The accommodation was some distance from the badly bomb-damaged airport; most of the buildings we saw on our journey were damaged to some extent – many were completely destroyed. After we had settled in, someone kindly took us for a drive round the island before the light faded.

Next day after breakfast, we reassembled at the airport then flew across to North Africa where we mainly

BRITISH OVERSEAS AIRWAYS CORPORATION

BOURNEMOUTH AIRPORTS.

NAME........ SGN. PIDGEON .... SERVICE......7F.25 DATE 30.7.45

PLEASE TAKE CAREFUL NOTE OF THE FOLLOWING INFORMATION

1. Your hats, coats, hand luggage, overnight baggage and small articles are your own responsibility, and should never leave your possession. Your registered baggage, however, will be taken to the Airport and will be waiting for you on the Customs bench for your departure.
2. Your Passports will be handed back to you when you are interviewed by the Security and Immigration Officials.
3. All Cameras must be handed over at the Airport for sealing by the Customs and will be handed back to passengers at their destination.
4. You will be accommodated at...Sandacres..........Hotel, Room No.....49....
5. You will be called at 0515...hours local time.
6. ...Breakfast.....will be served at the hotel ..... at....0600 hours local time

A Coach will leave the Hotel for the Airport at.......0630......hours local time Please be punctual as Motor Transport is limited and the Aircraft cannot be delayed.

GRATUITIES.

During your trip gratuities on the train and in all hotels are paid by the British Overseas Airways Corporation. In flight or in the Corporations rest houses they are not expected.

If for operational or weather reasons your departure is retarded you will be called correspondingly later so please do not worry, we will see that you are called in time for the Service.

S.L.P. 9023/6/45 P.T.O.

My British Overseas Airways Corporation service detail for Flight 7F.25 from Bournemouth Airport 30 July, 1945

followed the coast to Cairo. In the late afternoon sunshine, the pilot dipped the plane around the Pyramids so we had a splendid view of them. After landing we were taken to the Heliopolis Hotel which was a splendid edifice, so we four signalmen looked even more out of place. We each had super rooms, then gathered for a drink at one of the bars before going into the opulent dining-room, which was like something out of a Hollywood movie.

I recall one of the others asked for a Guinness and low and behold it appeared – cold and creamy. I was impressed by the service and amazed at the sheer variety of fine food offered to us. After dinner, we went out but the streets were a little daunting so after a drink or two, we made our way back towards the hotel. Just outside, we saw a civilian staggering in the doorway of a shop and clearly in distress. He had apparently been robbed but had also had more than his fair share of liquor. We took him to the hotel and arranged for a 'safe' taxi to take him to his home. Just before he left he told us he was the manager of Cairo Airport and on our departure we should ask for him. We did that and he waved us through some of the custom routines whilst our fellow passengers including the admirals and generals stood in line. He repeated his thanks for our help.

Next stop on this 'urgent' flight to Delhi was Baghdad, where we touched down for lunch. As I stepped off the gangway, clutching my warm battledress jacket, an elderly RAF sergeant said 'you won't want that jacket here sonny, its 125 degrees in the shade – if you can find any!' We were driven into Baghdad to the Regent Palace Hotel. After lunch there, we returned to the aircraft for the next part of our flight down to Sharjah in the

Fred Stapley and myself in Cairo en route to New Delhi. I had just changed into my khaki tropical kit that looks as if I had slept in it. Little wonder that the Admirals and Generals on our plane were puzzled by their fellow passengers!

Trucial States (now part of the United Arab Emirates) where we arrived around 11.00 pm local time. The desert airfield had a fortress-like rest house built by Imperial Airways in 1932 and guarded by Bedouins with ancient rifles – it was just like something out of *Beau Geste!* We were told the humidity was 90 per cent and the temperature 90 degrees to match. We had a meal of sorts but certainly did not feel like eating. We then took off and flew through a thunderstorm to Karachi in India where, to the best of my recollection, we landed at about 4.00 am.

Our 'urgent' flight had taken almost four days but we were followed by a number of SCU operators, flown in by RAF Transport Command. They took more than a week but it should be remembered that until this time, most military movements took place by troop ship. Four to five weeks would have been the norm via the Suez Canal and longer still, if one had to go round the Cape of Good Hope.

We transferred to Air India and finally arrived in Delhi where we were taken first to the ISLD office and then to the SCU11/12 unit out in the Delhi Cantonment to the east of the city. We were very tired and I collapsed into the (string) bed allotted to me, in what was just an army barrack-room without pretensions of any kind. The only difference I could see was that many of the chaps wore civilian clothes, and the military regime was not as severe as the stark army quarters indicated.

I was interviewed by Robin Addie and 'Tommy' Tucker who asked (it seemed) hundreds of questions about life at Whaddon and then finally talked about what they expected of me. Apparently, a team was being created to build a SCU wireless station at Manila to handle Ultra and Main Line traffic and I was to go to Calcutta immediately to join the team. The others were later to go to Okinawa, Rangoon (after it was recaptured) and Singapore. That was my understanding, following a talk we had between ourselves after the interviews.

We were taken to Delhi railway station and the four of us put in a sleeping compartment intended for six. The journey was comfortable and the train stopped at stations to enable you to stretch your legs and buy food. We ordered from one of the many uniformed waiters on the platform who brought our chosen meal out to us. As we had no time to finish before the train started, they then clung on the outside to collect our trays, and returned to their own station on the next train going back down the line!

At Calcutta's Howrah station, we had to telephone ISLD and a station wagon collected us for our journey over the great Hoogley Bridge and through this extraordinary city. Our home was to be at 36 Ballygunge Park Road, a large Victorian house in what had been an elegant quarter of the city in earlier times. There I met up with Douggie Frost and Alec Brazier, wireless engineers from my earlier days at Whaddon.

On our arrival in Calcutta, we were given the first news of the bomb dropped on Hiroshima, without comprehending the meaning of 'atomic' or the devastation it had caused. With such a hectic itinerary, it is now difficult to be sure, but we were probably travelling on the train when the bomb was dropped.

The second atom bomb was dropped on 9 August over Nagasaki, and Japan finally surrendered on 14 August. VJ day was celebrated thankfully, rather than wildly, since we were uncertain as to what the atom bomb was and sobered by the sheer number of casualties. However, there was great relief among us – and surely most others in the services – that we would not have to continue the war against the Japanese who had acquired such a bad reputation. After all, Calcutta was still almost a 'front line' town and certainly had large numbers of troops from Burma moving through it. Most of us went down to Chowringee, the main street of Calcutta, and there was a good party atmosphere, especially amongst the many troops down from Burma.

The sudden collapse of Japan changed the plans for Manila and some of the other stations. Robin Addie and Tommy Tucker arrived from Delhi and discussions obviously took place with the boss of SCU in India, Lieutenant Colonel Bill Sharpe. Captain Norman Walton was there after he had been in Africa and with him, also from 'A' detachment, was Major Mike Vivian. The Calcutta unit (still known as ISLD) was in the midst of building a powerful relay station near to Dum Dum Airport so as to improve traffic around the clock from Australia, the Philippines, China and the other proposed SLUs. This would handle Ultra and Main Line traffic. Whilst we all called the station 'Dum Dum' – as it was near Dum Dum airport – but it was actually at a place named Bajolah West, off the Barrackpore Road.

With the immediate plans shelved, we four were put on to working on the new transmitter station. The receivers and operators were already in place at Dhakuria Lake which was to the south-west of the city. We were later housed in huts there which meant we were less crowded than all those who were squeezed into the house in Ballygunge Park, a couple of miles away.

We rose at 5.30 am and worked at Dum Dum until about noon when it became too hot, then drove back to

The executives at SCU11/12 in Calcutta August 1945. From left to right Norman Walton, Mike Vivian, Robin Addie, Lt. Col. Bill Sharpe, Evelyn Watts and John 'Tommy' Tucker.

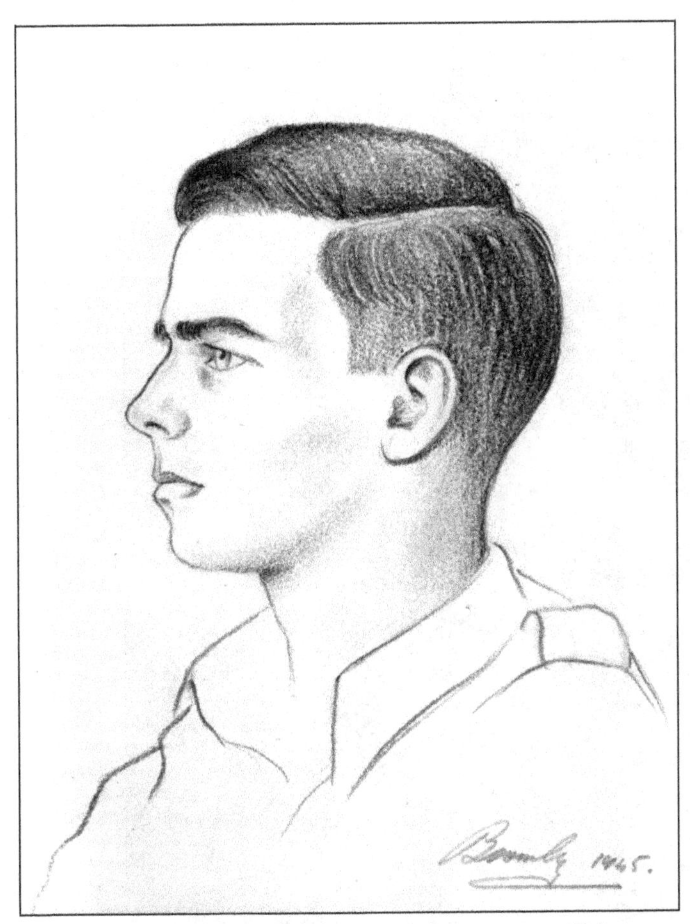

A portrait by Ken Bromley drawn whilst we were still living in Ballyunge Park Road, Calcutta, so it would have been in August – September 1945.

our billets for lunch, usually followed by a trip to the swimming pools at the Victoria Monument. On Thursdays, we always had a curry for lunch made from tins of Irish stew and we usually managed to find bottles of Murree beer to go with it. Then one either flopped down for a sleep or risked a swim in the lake some thirty yards from the door of the huts. There were all sorts of nasties in the lake from snakes to insects, causing ear and other infections.

With the war over, Section VIII was obviously in two minds about continuing the construction of the Dum Dum transmitter. I worked with several others, including Johnny Wills, Bill Stewart and Jimmy Price, on the aerial arrays. They were all Royal Signals linemen. I joined them for the experience but found my first climb up the 100-foot wireless mast quite daunting. There was very little to hang on to except the next rung up, about 15 inches away. Fine when you finally got to the top and you attached a belt – but then you had to climb down again!

I had a period working, I think, with Brian Birch and Fred Stapley, connecting the new landlines from the relays at Dhakuria into the transmitting room at Dum Dum. This included one never-to-be-forgotten all-night session when I soldered several hundred connections onto the new incoming board.

Over the years, the unit's store of wireless equipment had been distributed between the ISLD premises in Ballygunge Road, at Dum Dum, the workshops in Ballygunge Park Road, and Dhakuria. It was decided to bring them all back together at our base at Dhakuria Lakes. With not much to do otherwise, it was decided that I should be appointed i/c stores and so I followed in my father's footsteps. At the same time, it was decided the job required a jeep and trailer to move the goods about. That was a real bonus, making me one of the few men at Dhakuria with his own transport.

Just when the aerials were completed and the Calcutta station started full working to Main Line at Whaddon as planned, the British Government decided to quit India. We had lived through quite nasty riots but these often seemed to develop into fights between Muslims and Hindus rather than directly with the British, although we had to be careful.

In the centre of the city there was a great park called the Maidan; it contained the race course and the beautiful and vast Victoria Monument. During the height of the Japanese advance, the Maidan had been used as an emergency fighter airfield. Several of the major figures of the internal power struggle taking place in 1946 spoke at rallies there – Jinnah, Nehru and Gandhi, amongst others. We were told to keep away but the services swimming-baths were there, so I once heard Nehru speak passionately to a huge crowd.

SCU11/12 at Dhakuria in Calcutta in early 1946 as it closed down. Bill Sharpe and Mike Vivian are seated in the centre and there are many luminaries present in the picture. Bill Peat is standing far right and along the front seated on the ground – left to right – Ken Bromley our artist, Fred Stapley, Alec Brazier, Ken 'Zook' Howarth, Taffy Thomas, 'Fergie' Ferguson and so on!

One day, I received about eight large crates from Tom Kennerley who was in charge of our station at Kunming in China. When I unpacked them, I found twelve Whaddon MkVII sets in leather cases, some B2 wireless sets and a large collection of wireless spares of all kinds. One case contained 10 brand new US Army Smith and Wesson automatic pistols, complete with fine leather holsters and many boxes of ammunition. I asked Mike Vivian what should be done with it all and he told me to 'lose' all the wireless sets – without indicating how that could be achieved. However, I was instructed to take the MkVII leather cases, and the automatics to him in the morning. That night, a couple of friends rowed me out into the middle of the lake and the wireless sets were dropped quietly over the side into what we guessed was its deepest part.

Next morning, I dutifully took the automatics and the leather cases into the office but kept one MkVII case back for myself. It became my camera case and I used it as such until I presented it to David White at Bletchley Park in 1996. It is now there on display in Hut 1, complete with an early version of the MkVII agent's set, similar to those I had dumped in Dhakuria Lake.

One day, I was told to start to collect packing materials to send the unit's wireless equipment either back to Whaddon or down to Singapore. That was the first formal intimation I had of the closure of the stations in India, although the newspapers made it obvious we could not stay indefinitely. I made an inventory of the stores I held, the sets at our various sites, and was then told where they were to be sent. We could not find proper packing material of any kind, but then someone had a brilliant idea. The services in India and Burma wore KD (khaki drill) uniforms and its quartermasters were quite generous in replacing any worn out uniforms, socks, underwear or indeed any item of military apparel. That meant there was a virtual mountain of discarded uniforms over at Fort William and all I had to do was to take a requisition signed by Major Mike Vivian and help myself. My first trip was in our three-ton lorry which we filled to the brim and then put at the back of my stores under lock and key. The packing cases arrived and we started to fill them with wireless sets, valves and equipment of all kinds, well cushioned by discarded uniforms.

Unknown to me, the packing was being sabotaged by some of the men we employed to help. They unsealed some of the boxes, took out part of the clothing used as packing, and then exchanged it for new at their own quartermaster's stores. This happened mainly on the goods being returned to Whaddon which were despatched first, and I only discovered this was happening after the first shipment had left. Some of the sets eventually arrived at Whaddon loosely packed and damaged. We guarded the packed crates intended for Singapore much more closely but even then I found some had been tampered with when I unpacked them there later – we had not realised the lure of new clothing to the ordinary soldier.

In the preface, I indicated the book would only report the unit's affairs up to 1945 and give just a brief indication of my own story thereafter, leading to my demobilisation in 1947. I regret to find I have already gone well beyond VJ night in India in August, 1945. As I write this, I realise there is sufficient material for a book about my adventures in India and Singapore, but I feel I must bring my own tale to a fairly rapid close.

After the obligatory 'closing down' group photograph of SCU11/12 taken at Dhakuria in April 1946, the unit split up. The drivers and linemen were posted to ordinary army units around the city. The few Section VIII men, and some operators, went back to Delhi whilst Bill Sharpe, Norman Walton and Mike Vivian closed the doors on the Calcutta stations on which so much time and money had been spent. After a month or two in Delhi, we were posted to Singapore where accommodation had been found for us, and so back to Calcutta by train.

Six of us sailed from Calcutta in a small and very smelly troopship but as we went on board an officer stopped us and said we were to be the ship's police!. Although three of our party (Johnny Wills, Bill Stewart and Jimmy Price) were from the army – Freddie Stapley, Fergie Ferguson and myself were Section VIII – and totally shocked at being given army duties. However, we quickly forgot our worries when we found the exalted

Our team in Singapore towards the end of 1946. I am in the back row second left but those seated are a most interesting and important crew from earlier days. From the left: Frank Delzine a senior operator, then Wilf Lilburn, Bill Sharpe, 'Colly' Colborne and Bill Wort.

position of Ship's Police meant having in a cabin on the top deck, whilst others sweated away in the unsavoury bowels of the ship.

We stopped at Rangoon on the way to land some troops and collect others, including an SAS troop just out of the jungle. There were about six ships of various sizes anchored in the river at Rangoon, each simply swarming with Japanese prisoners-of-war. Whether these were temporary prisons or were to be used to convey them home, I have no idea.

Just outside Singapore we had to hoist the yellow fever flag as cases of smallpox were discovered in the ship and we were all vaccinated – yet again. Whilst in India one had every sort of inoculation you can think of – including for cholera – which was endemic in Calcutta during our stay. The Ship's Police came into their own at this time, wearing armbands with our title emblazoned on them, we had to ensure that the various groups took turns for inoculation. On reflection, I think this was my only real army function.

Eventually, we were allowed ashore and found we were billeted in a very nice house on an estate in the centre of the island called Adam Park, just off the Bukit Timah Road leading to the causeway and to Johore Bahru. There were two or three of us to a room but the rooms were large and airy and there was a balcony around the house. It was built on a slope terraced down to what had been a tennis court but the house and gardens had been the last line of defence earlier when the Japanese had invaded and much of the gardens were covered with the graves of members of an Australian machine gun platoon. Under the slope of the house, there were many messages of farewell to their families. Some were quite heart-breaking, since the writers knew the end was very near. I hope these messages have been preserved.

We managed to mark out a badminton court on the old tennis court area; it was used frequently, and I ran badminton tournaments from time to time. Singapore town was clean and attractive with several good cinemas, so we wanted for nothing. I played cricket for the unit's team which was known as Lord Killearn's Eleven, Killearn being the Acting High Commissioner. I continued to have the use of a jeep and as I was one of the few in the unit who could drive, I was roped in to take parties swimming at Johore Bahru, or down to Changi Point near the notorious Changi Jail. I hope I am conveying the idea of 'the good life' for certainly I thought it was so at the time, and still do today. The group picture taken at Adam Park at the end of 1946 which is centred around Bill Sharpe our boss is one I find quite fascinating. I am standing behind and slightly to one side of Wilf Lilburn with whom my journey in Section VIII had started, seven years earlier.

We had great parties, especially whenever a member of the unit went home. I must record one particular celebration which was given for Tom Kennerley and 'Taffy' Thomas, whom I had picked up at the airport the day before on their return from duty in Bangkok. During the evening, I had purchased some bronze cake knives from Taffy, and a sports jacket from Tom. He had obviously lived well in Bangkok as it was now difficult for him to button it up, but being almost brand new, I paid him 15 Straits Dollars for it. I greatly enjoyed asking him for my money back 'because it's worn out' when we met up again some fifty years later.

There was little to do at Adam Park and towards the end of 1946, I was getting ever more keen to get home. I was anxious to see my girl-friend Jane again and my father wanted me to join the family bathroom business inherited by my mother in 1945. Naturally, I wanted to be part of the business, instead of languishing in Singapore.

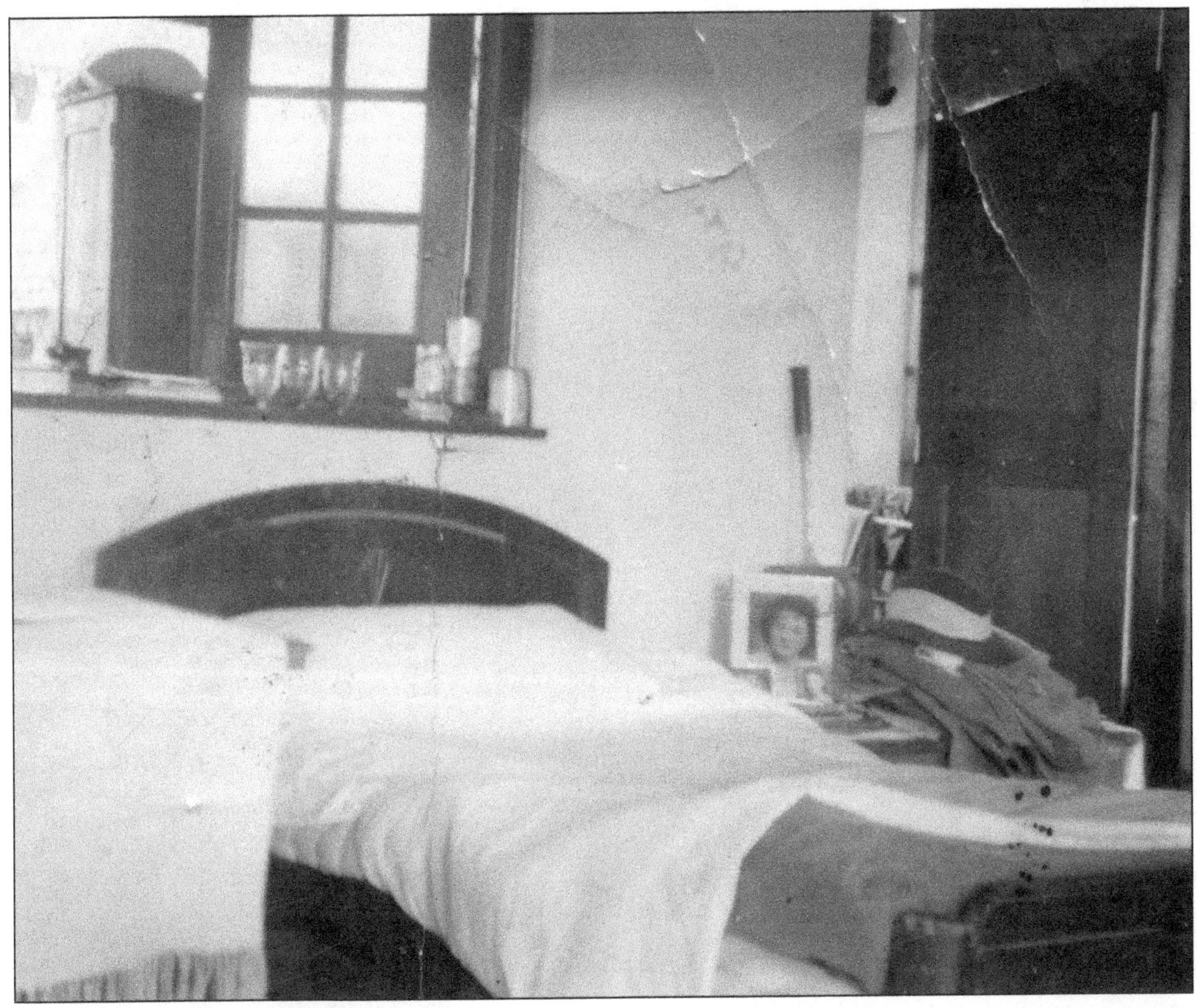

My bedroom at Adam Park Singapore with Jane's portrait on the bedside table

Father contacted friends still in the unit, especially Charlie Crocker, and I know from correspondence still in my files, that high level strings were pulled. In January 1947, I was told I would be leaving soon and so it turned out. I went home, via Bombay, with Fred Stapley and Fergie Ferugson on the RMS Andes, a largish troopship, arriving home at Southampton on March 7th 1947.

As a Special Enlistment, I was discharged under Kings Regulations. 1940 Para 390/XVIII/a 'His service being no longer required for the purpose for which he enlisted'. I was then given extensive leave based on my length of service, just as if I had been a real soldier, and the army sent me the pay to go with it. I regarded that as a real bonus!

The following fifty six years are the subject of my second and final book – 'Bathrooms, my family and Thomas Crapper.'

Appendix 1

Station Code Numbers

You will have read in 'The cast assembles' of Col. 'Micky' Jourdain. Before his departure, he gave Norman Walton his SIS Station Code Book and the main list is reproduced here. The last entry is in July 1940 so I suspect he left the unit soon after that date.

These SIS code numbers were a closely guarded secret at the time but they are mentioned in some detail, in Nigel West's book – MI6.

SUMMARY OF LEAVE—continued.

HOME STATION.

Period (inclusive)		Annual Leave	Sick Leave		Special Leave	Remarks
			With M.C.	W'out M.C.		
From	To	No of days	No. of days	No. of days	No. of days	

11000	SOFIA	41000	ATHENS
12000	BERLIN	42000	BERNE-GENEVA
13000	BRUSSELS.	43000	TALLINN
14000	BUCHAREST.	45000	PARIS
15000	BUDAPEST	48000	NEW YORK
17000	CAIRO	49000	MALTA
18000	ISTANBUL ANGORA	51000	GIBRALTAR
19000	COPENHAGEN	65000	SINGAPORE
21000	HELSINKI	72000	PANAMA
22116	AZORES	75000	MONTE VIDEO
23000	TOULOUSE	81000	ADEN
23500	MADRID.	82000	BAGHDAD
		83000	TEHERAN.
24000	LISBON	85000	FAROES
28000	SHANGHAI	88000	JERUSALEM
28002	HONG KONG	89000	CAIRO (Sir ... Petrie)
31000	RIGA	93000	OSLO
32000	ROME		
33000	The HAGUE.		
35000	BELGRADE.		
36000	STOCKHOLM. (CAIRO)		
38000	WARSAW.		

Inside the cover of his notebook, Colonel Jourdain had printed his family crest, and then each page showed the station name, its code and those operating from there

You will see from Jourdain's list for example, that 38000 is Warsaw, and 23000 is Toulouse. The first two figures indicate the country – thus *17*000 is Cairo. The Head of Station (the head of SIS at that location and usually the PCO – Passport Control Officer), is shown as 100 so would be designated *171*00. However, in SIS circles 17000 (Cairo), was described only as '17 Land' and nothing else.

His assistant would be *172*00 and so on. Agents recruited by 17200 would be numbered from zero, so his first would be 201 his second 202 and so on. These would be shown in the completed code as 17201 or 17202.

Further numbers indicates duties, so office staff were all designated *9*00, operators *5*0, and in chapter 28 you will read that Bill Miller's SIS number, whilst in Tangiers was 5695*2* – there being one other operator there when he arrived.

In Micky Jourdain's 'little book' each country has its page and in some cases the names and numbers of the Section VIII operator there. In Belgrade, in 'shorthand' we read – '950 H. B. Gerrish and '951 T James' but their full SIS numbers – based on Belgrade being 35000 ('35 Land') – would be 35950 and 35951.

Appendix 2

'Station X'

One of the remaining puzzles of our wartime units is how, in some circles, Bletchley Park came to be called 'Station X.'

The first time I heard the expression 'Station X' applied to Bletchley Park was in a recent book and a TV series of the same name. I had never heard it used in reference to BP during the time I worked at Whaddon – 1942 to 1945 – or whilst I lived in nearby Stony Stratford – 1940 to 1945. Father was at Whaddon during all that period and mother was a Red Cross Nurse in the clinic at BP in 1943 and 1944. At no time did they use the expression 'Station X' at home and indeed, the only reference to BP from my mother, was about her shift work there and the transport arrangements.

During a recent conducted tour of BP given by Ted Enever (one of the leading authorities on its history), he told my party that the expression 'Station X' was not used during the war to describe BP. He said it referred to the wireless station that was positioned in its tower. In his book 'Britain's Best Kept Secret' he mentions the tower at BP, and he goes on to say:-

'The radio room, small and cramped, was given the code name Station X. The radio's aerial was slung between the finials of the Mansion's Victorian roofline, before running to a tall cedar tree....'

Further down on that page, he goes on to say:

'The result of this policy thinking was that Station X, and the aerials spanning the trees and roof, were quickly dismantled and a new base established some seven miles to the South West at Whaddon Hall.'

The earliest reference I can find to 'Station X' is in R. V. Jones book 'Most Secret War' and here he says on page 59 – *'...I was told to go to Bletchley Park or 'Station X' as it was known,...'*

No reference to Station X is made by Winterbotham in either of his books.

In Ronald Lewin's excellent 'Ultra goes to War,' he has a chapter entitled 'Station X' but in the second paragraph he says:

'The other-wordly features of this unique establishment [Bletchley Park] was that whatever the names under which it came to be known – War Station, Room 47 of the Foreign Office, Station X, BP – nobody either inside or outside had a complete picture of what was happening there.'

This looks more like uncertainty by the writer looking back in 1978, rather than an authoritative view of what it was called at the time. Even so, it does not detract from the view that 'Station X' was the name just of the SIS wireless station, whilst it was housed at Bletchley Park.

In a paper by Pat Hawker, a leading authority and writer on the wartime wireless units, himself a member of RSS at Hanslope, he says inter alia.

'In the early months of WW2, the Barnes Station X moved first to Bletchley Park and then in 1940 split into two sections – the two way communications section moving to Beconsfield (later transferred to the BBC) and two-way communications links moving to Whaddon Hall, near Bletchley, which became the wartime HQ of Section VIII.'

One of the most widely accepted accounts of Bletchley Park is 'The Codebreakers – the inside story of Bletchley Park' written by Sir Harry Hinsley and Alan Stripp. In this, they listed the various names given to GC&CS at Bletchley Park and they included: War Station X, Room 47 Foreign Office, BP, The Park, GCHQ. They do not refer to just *'Station X'!*

Another leading authority on Bletchley Park is David White, Curator of its Wireless Museum in Hut 1. He unequivocally states that Station X was the SIS station built in the tower of Bletchley Park Mansion but entirely removed by early 1940.

None of my wartime colleagues have heard the expression used, other than as the name of one of our Section VIII wireless stations. The most senior of my wartime colleagues today is Edgar Harrison. He joined the unit in 1940, and went on after the war to become Chief Signals Officer of the whole of Diplomatic Wireless Service. He wrote to me recently and said:

'I never heard anyone refer to Bletchley Park as Station X. I can emphatically state that Gambier-Parry was referred to as XW. During my travels in the Middle East and elsewhere between September 1940 and the end of 1943 I had occasions to send messages to Gambier-Parry at Whaddon and to Jock Adamson at Cairo. They were respectively 'XW' and 'SOX'. I, as a member of Section VIII, had a dedicated number as did all the various countries and areas. For instance when I was sent to Russia from Cairo in 1941 the telegraphic report of my move would be "XW from SOX. 783 to 2200 land on 10th October." On setting up station in Russia I signalled "XW and SOX from 783. Station operational in 2200 land." The numbers 783 and 2200 are fictional.'

This confirms that 'X' was the pinnacle of the the whole SIS wireless network. For us that was Gambier-Parry's Section VIII HQ at Whaddon. In Chapter 28, Bill Miller based in the SIS station in Tangiers says, *'All transmissions with London, were addressed as 'XW' and went through Main Line at Whaddon Hall.'*

Undoubtedly the most authorative voice on the subject is that of John Darwin who, in his diary (extracts in Chapter 21), never refers to Bletchley Park as Station X, only as Bletchley, Bletchley Park, or War Station. His references to 'X' relate to its *wireless signal strength,* demonstrating that X was a *wireless* station.

Then how did this mix-up start? As I explained earlier, it was certainly not because it was the tenth (X) building purchased by SIS. At no time did Whaddon, or Hanslope, or any other SIS property that I know of, have a number allocated to it. If SIS properties were to be numbered, then where was Station nine IX or Station eleven XI?

Perhaps in using the term – 'War station X' – Hinsley and Stripp are nearest the mark, since I agree the property was indeed purchased by Admiral Sinclair as his 'War Station' to contain a number of his sections – including the new Section VIII and the existing Foreign Office/SIS Station X, being transferred up from Barnes.

The first time I heard the title had actually been used, came from Joyce Lilburn (Chapter 24), who told me that it was applied to material intended for Bletchley Park, whilst she was at the Naval Intelligence Department at the Admiralty. It is conceivable that their use of the name came from the combination of War Station and X mentioned by Hinsley and Stripp. Having put the name 'X' on the 'out tray' in 1939, might well have needed an Admiralty Order to remove it later, when it had become inaccurate!

In Joan Nicholls excellent book on the Y service, there is no mention of Station X, until you come to the last few pages. Here you will find reproductions of letters, showing an exchange of congratulatory letters (at the end of the war in Europe), between 'The Director of Station X' and the WOYG (War Office Y Group). Joan tells me the Y Service usually directed mail to Station X.

It seems that for a few, the name of the SIS wireless station X, though based at Bletchley Park for just a brief period, stuck to the place through the war.

This is the frontage of the old Barnes Police Station photographed in 2000 as part of the Barnes and Mortlake History Society Millennium project to record local buildings of interest. It faces directly onto the Thames and fronts the large complex of Police buildings that extended down Barnes High Street.

As you can see from the postcard picture in chapter 4 it had aerials on it, and we believe it likely that the Barnes wireless station X was first located in the very extensive buildings at the rear of the station, before moving into nearby Florence House. At the time of writing, the entire site is being redeveloped as apartments. However, as part of the planning arrangements, the frontage of the Police station itself will remain – a brick building near the Thames.

Appendix 3

VIs Wireless Interception and Organisation

The VI became adept in recognising the type of transmission which was likely to be of interest to Box 25. The call-sign might be the first point of recognition but as time went on it was often found that the call-signs were changed every day and perhaps, in any case, only a part of the procedure was picked up. The nature of the procedure was significant. Q-codes were commonly used by amateurs and service stations; these were three letters beginning with Q. QRM meant 'I am getting interference' and QSY 3745 indicated 'change frequency to 3745 kilocycles'.

The radio amateur was familiar with scores of these Q-codes and the German Abwehr used a particular procedure which the British interceptor learnt to recognise. For instance 'QSA NIL PSE CALL = K' meant 'Nothing heard, please call, I'm listening'. QSA meant 'Your signal strength is....' so NIL meant that he was inaudible; but radio amateurs did not use QSA in this manner. The following examples are from an original recorded on 19th December, 1941 where the section in brackets indicates what was actually transmitted:

1700 (CZE QSA0 PSE CALL=K) and later, (=SRI QSA0 QRX NEXT NW 73 GB VA) 5400 CW 3. 1700 is the time in GMT as noted by the V.I. CZE is the station call-sign. QSA0 means no signal strength to report hence no reply heard. SRI is sorry and QRX NEXT means that CZE will call again at the next prearranged time. The message concludes with, 'Now kind regards (73), goodbye(GB), am closing down (VA)'. 5400 is the frequency in kilocycles per second. Today it would be in kilohertz or perhaps given as 5.4 MHz. CW stands for continuous wave (Morse) and 3 is the signal strength on a scale of 1 to 5.

There are no German language abbreviations used and most of the transmission is international radio amateur-like procedure.

Another example on 5th July 1941 is YSN on 8450 Kc/s at 1815 sending the message [QTC CT 935/71 = AKRJD VURNT FHDAL VXTRS] etc. In this case 935 would be the message serial number and 71 the number of letters to follow. The 5-letter encyphered groups were sent at about 15 to 20 groups per minute. It is possible that many of the enemy operators were enlisted German radio amateurs. Certainly most of the procedure was amateur radio type except for the call-signs and messages in 5-letter code.

Regional Officers of RSS.

Home South, Leatherhead:	Capt. Sabine (later at Box 25) then Capt. Aubrey Johnson
Home North, Cambridge	(83 Regent St): Capt. Hall then Capt. Rolfe
South West, Exeter	(27 Dixs Field): D H Norton (?)
North West, Preston	(6 Jordan St): Capt. Walter Stanworth
Midlands, Leicester	Capt. Aubrey Johnson then Capt. A.E.Scarratt
Scotland, Stirling	(67 Port St): Capt. Wallace
Northern Ireland, Belfast	Capt. Joe Banham
Wales, Cardiff	Capt. Edmund Vale & D.Lowe (?)

This list is incomplete as there must have been a NE region. But assuming an average of only 200 VIs per region, estimated from their serial numbers, the total number of VIs exceeded 1500 and was put at 1700 by Kenneth Morton Evans.

Appendix 4

Bill Miller's Personal Code

An explanation by Bill Miller of his personal code used whilst in Tangier, see Chapter 28. The usual code method was by one time pads but the alternative was as follows and based on his selection of the Penguin book 'Poet's Pub' by Eric Linklater.

As Bill indicated in Chapter 28, one of the coding systems he was taught utilised a paperback Penguin book. At this time 1940, Penguin were the only paperbacks on the market. The coding system was a method of substitution.

To encode a message, one selected a random page from the book and recorded the page number. One then looked down the left hand side of the page, until you come to a line, where the first word or two, totalled around 12 or 15 letters and recorded the line number. Thus in the example, the page number was 145 and the line number was 31 which started 'Once again she'.

The page and line numbers, form the indicator group (always 5 figures), thus the example is 14531. In order to encode this figure, the user had his own key number, which was the only thing which had to be remembered. In Bill's case his personal number was 10817 (his house number and birthday). The indicator group and the key group were added together without carrying the digit forward.

Thus indicator group 14531
Personal key group 10817
added together ______
without carrying digit. 24348 = encoded indicator group

To decode the reverse procedure is applied thus:

Encoded indicator group received 24348
Take away key group 10817
without carrying digit, ______
leaves answer 14531 which is the actual, indicator group.

Example as indicated earlier
Page 145, line 31 = 'Once again she'
A grid is then formed, by using these words, written across the page, with a space in between each letter. Underneath these words, each letter of the words, is numbered it its alphabetical sequence. Thus 'A' would be 1, 'B' would be 2 and so on. If there were more than one similar letter in the words selected, then the letter on the left is numbered first and so on.

Thus words selected :-

O	N	C	E	A	G	A	I	N	S	H	E
11	9	3	4	1	6	2	8	10	12	7	5

Each letter numbered in its alphabetical order. The grid shown below is formed using the letters and numbers above. Text of message is then written, each letter occupying a square of the grid.

Text of example telegram. =
SS CORTEZ ARR BILBAO FM ARGENTINA WITH GERMAN EXPATS RETURNING GERMANY ENROL ARMED FORCES STOP UNIFORMED HITLER YOUTH DIRECTED COACHES FROM PORT TO ROAD TO OCC FRANCE.

Note. In this example the SS CORTEZ, is fiction. The remainder of the telegram is actual fact as Bill witnessed it. As he says, it was a bit of a shock seeing the Hitler Youth in March 1941.

O	N	C	E	A	G	A	I	N	S	H	E
11	9	3	4	1	6	2	8	10	12	7	5
S	S	C	O	R	T	E	Z	A	R	R	B
I	L	B	A	O	F	M	A	R	G	E	N
T	I	N	A	W	I	T	H	G	E	R	M
A	N	E	X	P	A	T	S	R	E	T	U
R	N	I	N	G	G	E	R	M	A	N	Y
E	N	R	O	L	A	R	M	E	D	F	O
R	C	E	S	S	T	O	P	U	N	I	F
O	R	M	E	D	H	I	T	L	E	R	Y
O	U	T	H	D	I	R	E	C	T	E	D
C	O	A	C	H	E	S	F	R	O	M	P
O	R	T	T	O	R	O	A	D	T	O	O
C	C	F	R	A	N	C	E	X	Y	Z	Z

Using the grid, to encode, form telegram of five figure groups reading downwards from Column 1, then Column 2, and so on. Thus first group would be R O W P G group 2 is L S D D H Group 3 is O A E M T.

Actual telegram would start with telegram number, total number of groups, with indicator group 24348 as first group. Thus:

Nr Grs 30. 24348
ROWPG LSDDH OAEMT TEROI RSOCC BNEIR EMTAT FOAAX
NOSEH CTRBN MUYOF YDPOZ TFIAG ATHIE RNRER TNFIR
EMOZZ AHSRM PTEFA ESLIN NNCRU ORCAR GRMEU LCRDX
SITAR EROOC OCRGE EADNE TOTYX.

(Note last X in final group has been added to make five figure group.)

Decode by use of decoded indicator group to ascertain page and line number, which would indicate 'ONCE AGAIN SHE' Count the numbers of *letters* in telegram which is 145 (29 x 5).

There are 12 letters in 'Once again she', thus grid will 12 across, and 12 down. Ignore the extra letter which was added to make five figure group. Thus grid formed by receiving station, would be the same as the grid of the sending station. Then write text of coded telegram within the grid, but downwards, starting from Column

1, then Column 2 and so on until grid is filled, then, plain language text of telegram can be read across, exactly the same as the original grid.

Code breakers would soon recognise that this was a substitution code, and would first of all try to ascertain what language the text was written. Thus as 'E' is the most common letter in English, we would cut out a number of 'E' s and abbreviate words in text – thus Germans could be encoded as Grmans.

There was a contingency plan, which was that if for any reason he mislaid, or was unable to gain access to his Penguin book, there was a reserve phrase, which could be used instead of a 'page and line' number phrase. He was asked to originate a phrase, which he could easily remember – it was never to be written down.

The phrase he chose was 'PRESTON DOWN AV', this being the address of a close friend who lived in Paignton. If he ever used this phrase to form a grid, the indicator group would be his personal key group – uncoded – 10817.

Therefore if Broadway received a telegram with this indicator, they would know that he was using the phrase Preston Down Av to form the grid. Needless to say, he never had to use it.

Thus to form a grid, using this phrase, it would be arranged, as follows:

P	R	E	S	T	O	N	D	O	W	N	A	V
8	9	3	10	11	6	4	2	7	13	5	1	12

Letters of Preston Down Av, being numbered in alphabetical order.

POET'S PUB

panies a trite reflexion, that facts are stranger than fiction and strong enough to ruin the prettiest plot.

It was clear that Mr. Wesson was a thief, and it was distressingly apparent that this actual thief had forestalled the more squeamish rifling to which she had incited Quentin. But the journalist in her rose and comforted her. Facts were facts and, what was more important, might become news. The secret of van Buren's discovery would be interesting to only a small part of the community, but the story of a desperate thief who had followed him across the Atlantic to rob him in a romantic English inn could be turned into news which would enthral a whole country. Nelly began to feel elated, and as her self-confidence came back she sturdily decided that to summon the police (especially the rustic police of Downish) and simply indicate a thief for them to arrest would be an anti-climax to this midnight adventure. She would wait till the morning and re-capture the documents herself. She might, though it was difficult to invent immediately the precise *modus operandi*, even arrest the thief herself. But certainly she would watch him. perhaps find out something about him to give body to her story, and in the end arrange a suitable *dénouement*. She must get to her room and think it out . . . but before she could do that she ought to restore the now unimportant portfolio under her arm to Saturday's table. There was no time for poetry with an international thief in the offing.

Once again she looked this way and the other, up and down the dimly lit corridor, and as she prepared to creep back with the sheaf of poetry she heard some distance away the elemental sound of water escaping to find its own level . . . and heard, too, more footsteps approaching. All her fears returned. She shrank back into the darkness and scarcely dared to peer out and see van Buren re-enter his room. She had never thought of van Buren. She had unconsciously assumed that he was in his bedroom sound

145

Appendix 5

Call Signs from Peking in the 1930s

Edgar Harrison was one of the Royal Signals troop in China that were used by the Foreign Office to man their wireless station during the 1930s. They had all been Army apprentices trained in W/T procedures. (See Chapter 31). The station had been built by the Royal Navy, and much of the work was connected with what was then, a most important Fleet in the British Navy. He has kindly created some examples of the call signs and procedures used.

Example 1.
You have a signal ready for transmission to the Governor Hong Kong. It will be handled by the Royal Navy.
GZ06 V HX8 X242
HX8 V GZ06 K
HX8 transmits message numbered say 10 and with a group count of 100
GZ06 receives this message and signals
HX8 V GZ06 R Nr 10 GRS 100 X241 VA to which HX8 replies
GZ06 V HX8 R X241 VA

Notes.
X242 = a message for you
X241 = nothing to communicate
VA = end of transmission
K = go ahead
R = received

Example 2.
The troopship Dilwara is outward bound to Chinwangtao with troop replacements for Tientsin and Peking. She gets in touch with HX8 using 'Q' code as follows:
HX8 de GDKB QRK? QTC K
GDKB de HX8 R QSA4 K
GDKB transmits telegrams. When received HX8 signals
GDKB de HX8 QSL nrs of telegrams received QRU VA
HX8 de GDKB R QRU VA

Notes.
QRK? = how are you receiving me
QSA = what is the strength of my signal (5 was me).
QTC = have telegrams for you
QSL = I have received telegrams
QRU = nothing to communicate
K, R and VA = as for 'X' code

Example 3.
Every evening at 2000 HX8 commences automatic transmitting using an endless perforated tape and 'Z'code as follows:
GLC de HX8 ZHC? ZKR GA?
communication being established GLC signals
HX8 de GLC ZOK ZHC? ZKR GA GA?
GLC de HX8 ZOK GA

Notes.
ZHC? = how are you receiving me
ZOK = am receiving you well
ZKR = i have the following traffic for you in group totals.
ZSC = scrambled code
FIGS = figure code
P/L = plain language
GA? = may I start transmitting?
GA = go ahead transmit

It almost goes without saying that each code consisted of a large number of 'X's, 'Q' s, and 'Z's.

In addition to the foregoing all ex-Boy Apprentice Tradesmen, having been trained at Liverpool Post Office, were conversant with the telegraphic system of the GPO.

Appendix 6

Agents' Sets

This list and technical descriptions were compiled by Pat Hawker, with additional comments from me. I have separated the photographs from the technical reports and given them a brief caption as I realise that the detailed notes will be of interest to only a few readers.

I understood from the late Brian Checkley, who worked at Hanslope in the 1980's, that full drawings and specifications of some of our wartime sets still exist in the MI6 Museum at Hanslope. However, they are unlikely to be released and it is therefore important to record Pat's descriptions here, or they will be lost to us.

Under this title you will see I have included the 'Coffin set' and the MkIX (Ascension) which were not handled by agents. However, they were used by members of our units, and in the case of Ascension directed to agents, as was the 'Coffin' on occasions. However, all the list down to the specials, were manufactured in our own workshops, the earliest at Barnes and Funny Neuk, but the vast majority later – at either Whaddon or Little Horwood.

Many transmitters, such as the MkIII (and the later Mk33), were in almost universal use by embassies in neutral countries, our mobile units such as SCU/SLUs, and some by agents. Transmitters-receivers like the MkV and the later MkVII 'suitcase sets' were designed for use by agents, but they were also held by our own units as a standby for the usual installation of HRO receiver and MkIII transmitter. You will have read that Edgar Harrison employed one of our agent suitcase sets when travelling with Mr. Churchill.

The following list includes only the more widely used Whaddon equipment since there were many 'specials' developed to meet particular requirements. There were also Mark numbers allocated to prototype designs that were later terminated. Again, most of the more popular models were made in various versions.

By 1939, there were transmitters installed at SIS outstations in Europe and by late 1939 some transmitters intended for use by 'stay-behind' agents had been distributed in Belgium, Holland and later France, Yugoslavia etc. These included transmitter-only units with 'broadcast-type' signal plans. An early AC/DC model, now in the 'Secret Warfare' section of the Imperial War Museum, is in the form of an aperiodic unit with no adjustable controls, meters or indicator lamps, in a square wooden box with a large mains dropper resistor mounted on the top panel, with two sets of sockets for crystals, mains input socket, terminals and sockets for antenna and earth. This was possibly the **MkIV** but this cannot be confirmed.

This particular transmitter was used by 'Service Clarence' headed by Walter Dewe and was provided before the invasion of Belgium in May 1940. See Chapter 14. The intention was that the return link would be concealed in the transmissions of a 'Black Broadcasting' station. There is no evidence that these stay-behind transmitters were ever successful in sending traffic. The mark numbers are unknown but they were possibly **MkI**, or **MkIV**.

MkI, MkII, early MkIII. These were all pre-Whaddon designs. We have no details for the for MkI. The MkII prototype was first built in a Bournemouth garage by Bob Hornby. Later work was carried out at the Barnes and Funny Neuk workshops. The MkII existed in two versions, transmitter only and transmitter plus receiver. A MkII was lost to the Germans during the operations leading to the Venlo incident in November 1939. Both versions of the MkII are held in the MI6 Museum but they are not accessible to the public.

Above: The famous MkIII transmitter used throughout the war in SCU/SLUs, in embassies, by agents and indeed in every aspect of Section VIII's work. This is the best known version with its wooden case with drop-down front.

MkIII. The is the most famous of the SCU sets. It was designed originally by Bob Hornby and then improved in joint research with Wilf Lilburn. It was used for SIS outstations, 'diplomatic' stations and mobile units including the mobile SCU/SLU ULTRA links with overseas Army Commands. There was an early suitcase version used as agent-set. Photocopies of Tinker Box version and photograph (Eindhoven) of the standard two-box version. No circuit diagrams exist. Some were supplied for SOE as well as SIS agents during 1940-42. It was later superseded by the Mk33.

The MkIII was put into production in the early workshops at Whaddon and was being made in lots of 50 to 100 when I joined the workshops in 1942. This was the wooden cased version and when the workshop at Little Horwood started up early in 1943 it was one of its main products.

The Mk33 followed the MkIII and was fundamentally the same but with three stages and with a metal front and casing.

MkIV. We believe this existed as agent-set (see above) but no details are known.

MkV. This was a relatively high-power transmitter-receiver as an agent and general purpose set in various forms, including suitcase and wooden box versions. This was the first set I worked on in the metal workshops at Whaddon and fair numbers were made. It was very bulky and heavy, and I am told its sheer size and weight were responsible for a number of agents being caught carrying it.

MkVII and MkVII/2 (Paraset). This was a simple transmitter-receiver in wooden box and later in a hinged cadmium metal case. Its great advantage was its small size compared with earlier trans/ceivers. It was usually carried in an attache case. The MkVII was started in production at

Whaddon but its obvious success, leading to a considerable increase in demand, coincided with the opening of the Little Horwood workshops where it became the main product – in conjunction with the MkIII.

MkIX. We believe this is the number allocated to the Ascension FM equipment designed at Whaddon for use in contacting agents on the ground from aircraft flying overhead. It was installed in aircraft flying from Tempsford and Hartford Bridge airfields – see Chapters 19 and 38.

MkX. A high power transmitter (813 PA) used for mobile units, main links etc. Also fitted (1943) at the marine base station for the clandestine vessels operating as Slocum's Navy from Cornwall and the Scilly Isles also the SCU9 mobile stations in 1944.

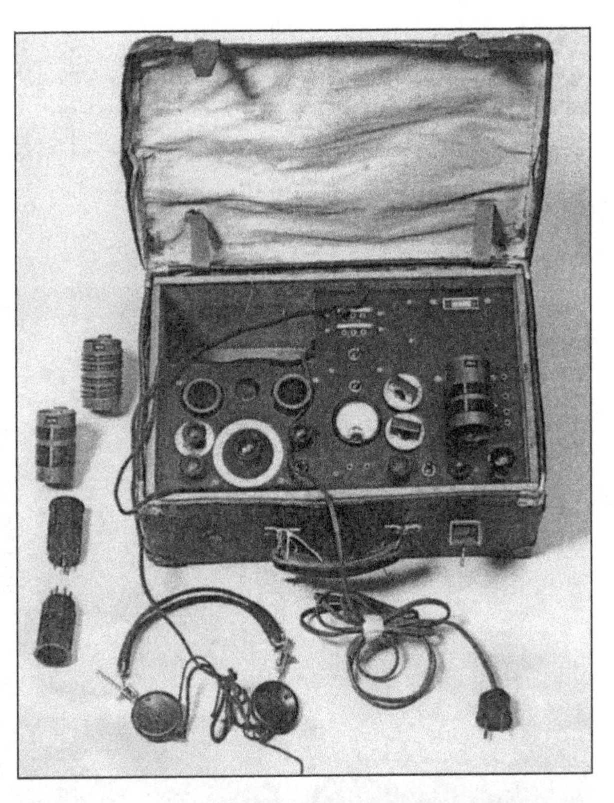

Right: The MkV was a true 'suitcase set' but rather too heavy. The plug-in coils are shown to the side. This picture is copied from a German Police handbook.

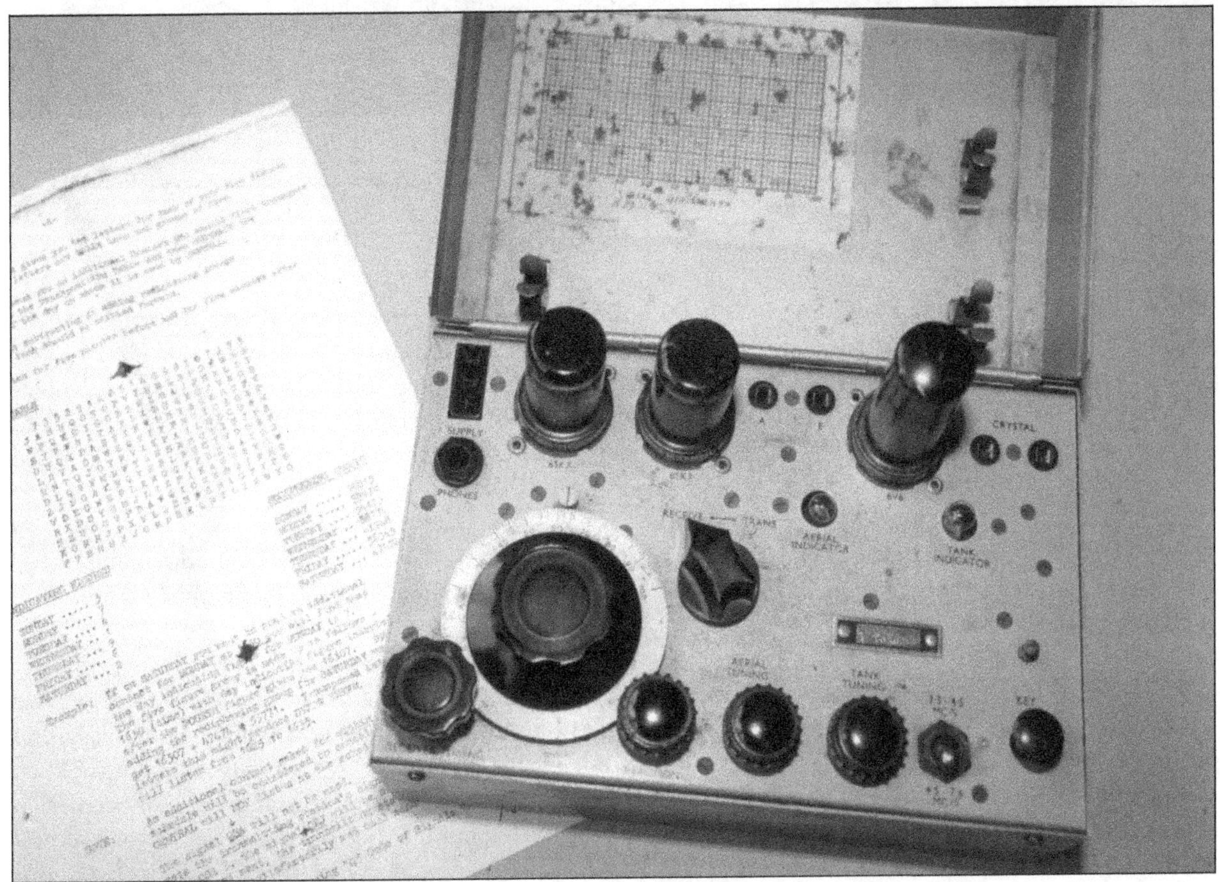

The MkVII/2 with signal plan was the 'Cash Box' version of the earlier MkVII (shown in chapter 14) and its name came from being cased in metal with a metal lid – instead of wood. It was also called the 'Paraset' and was made in large numbers at Little Horwood.

'Coffin set'. This was a combination of an HRO receiver complete with its coil box, and a Mk33 transmitter with power pack – all combined into a wooden case – hence the name 'Coffin Set'. The case was made at the Little Horwood workshops and the sets fitted into it there.

The 'Coffin' set, made up of Mk33 transmitter, HRO receiver and power pack.

Technical Descriptions

MkIII. This ubiquitous HF transmitter of about 25-30-watts output in wooden box. Made in several versions with early versions including single-stage crystal power oscillator (eg 1625), and also the two-stage 6V6',-807 version both with built-in power supply and with front panel displaying 160/80/40 metre (amateur) range switch. Early versions supplied to Czechs at Dulwich, and apparently to French ('Tinker Box'). Later standard version used two stages with 6V6/807 valves in two wooden boxes. Plug-in coils for transmitter with 807 used as power doubler on higher frequencies.

MkV. Believed to have included a transmitter-only version. 6V6 crystal oscillator and 832 power amp. valve. Used with 1-v-1 receiver with three 6SK7 valves. PSU rectifier 5Z3. Photocopy of wooden box version from a German police book.

MkVII and MkVII/2 (Paraset). 0-v-1 receiver (two 6SK7) and power oscillator (6V6). No meters but two pilot bulb indicators (oscillator and antenna current) and in-built 'silent' Morse key. Widely used as agent set from 1941-45. Full details available, including Lorain book. Separate PSU with alternative vibrator 6v PSU. Small calibration chart (in lid of metal version, separate for wooden version). Example in Imperial War Museum (wooden box version). Weight about 3.25 lb plus PSU about 6.51b. PSU for AC mains or a second version for 6V DC (vibrator unit).

MkIX. Ascension FM equipment (30 – 37MHz) master oscillator with PA providing some 25 or so watts o/p about 7 calibration points. In use for clandestine ships in late 1942. For agents, by about Summer 1943. PA valve 815 or 832 double-tetrode. Rectifier 5Z3. It is not known if models or circuit details still exist at Hanslope. A post-war model combining S-phone and MkIX ideas were developed by Steve Dorman.

MkX. Three stages with crystal oscillator but its circuit details are not available. About 150W output. Built-in high-voltage power supply.

MkXV. Two-stage transmitter (6L6 or 6F6 co; + 6L6 pa) PA could be used as a doubler for 9 to 18MHz and 1-v-1 receiver (three 6SK7) in wooden boxes used as agent equipment, often carried in a suitcase. Superseded the MkV. Also a pulser version of the PSU to enable transmitter to be used as a beacon. Details in Radio Bygones. About 15-20W output, 3 – 13MHz or 3.5 – 16MHz or 2.5 -18MHz. Three types of PSU in separate metal box: (1) Mains with 5T4 rectifier for the pulser version (no receiver but PSU in MkXV receiver-type box; (2) continental mains version with 6X5 rectifier; and (3) GBP Mark XV power unit for 6-V vehicle batteries, using Carter 417 XV Dynamotor. Introduced about 1943. Total weight about 45 lb or more.

MkXXI. Battery operated transmitter-receiver (agent) in separate units. Receiver 1-v-1 with three 1T4 valves. Transmitter co; – pa (1 S4 – 3A4). Transmitter output about 0.75W with 135V HT. Used 1.5V and 67.5V batteries with small crystal headphone. Introduced about 1944.

MkXXIV. (possibly earlier version as MkXIV): Compact aperiodic (no tuning) transmitter using two I I 7V heater pentode-rectifier valves (117L-GT) as rectifier, push-pull crystal oscillator transmitter for AC/DC mains operation. No mains transformer with diode sections used as mains rectifier. Associated small receiver with three miniature 117V heater valves.

Mk33. Transmitter, and a later version of MkIII but three-stages in a metal box. Plug-in PA tank coil with internal band switching for oscillator/driver-doubler stages. Coverage: 1.5 – 20MHz (1, 1.5 – 3 MHz; 2, 2.9 – 5.3 MHz; 3, 5.3 – 10.6MHz; 4, 10.6 – 20MHz). Valves: 1613 co; 1613 ba/fd; 807 pa. Separate PSU with 600-0-600V transformer and 83 rectifier. Input tappings for 100, 110, 120, 150, 200, 225 and 240V AC.

'Coffin set' (Mk?). Combination of a transmitter (version of MkIII with 807 PA) and HRO receiver in large wooden case. About 1944. Not intended as 'agent' equipment.

Mk119. HF transmitter-receiver built as separate units in hermetically sealed metal enclosures developed for use in Far East (1945). Transmitter coverage 1.5 – 20MHz in six switched bands. Superhet receiver with five bands; frequency coverage 0.5 – 20MHz in five bands: l, 0.5 – 1.05MHz; 2, 1.05 – 2.2MHz; 3, 2.2 -4.7MHz; 4, 4.7 – 10.0MHz; and 5,10.0 – 20.0MHz.

'Ascension' (generic name for Whaddon ground/air R/T equipment). Original unit low-power AM. But operational unit 35MHz with Frequency Modulation (10.7MHz IF). MkIX and possibly MkXI as airborne unit.

Special Sets

Many specials were made but some Mk design numbers did not go into production. Use was also made of sets manufactured elsewhere – like the OSS SSTR1, Anglo-Polish AP units and SOE S-phones, and the B-2 suitcase transmitter-receiver. It is possible that similar classes of equipment were given sequence Mark numbers e.g. MkIV may have in later versions been the MkXIV and then the MkXXIV. The MkV in later model the MkIV. The MkIII became the Mk33. Possibly there was a dry-battery operated MkI, then MkX1, then MkXX1, but this has not been confirmed.

'Lincoln'. May have been either Intelligence or SOE adaptor unit for plugging into receiver's audio output stage (compare German 'Grammo' unit). No details. Available in 1941-2.

'Manchester 6.1B'. A possible early 'stay-behind' agent-set for AC mains as reported used in Belgium in 1940. Could possibly be the set illustrated in the German Police Manual as the A.6 suitcase set where transmitter valve is given as 6FJE (probably a printing mistake). Receiver is stated to have 6V6 audio output stage. Almost certainly these valve types have been reversed, i.e. 6FJ7 receiver, 6V6 transmitter? . 6X5 rectifier. Size 210 by 190 by 105mm with general appearance not unlike the Polish AP sets.

At the time of writing, a number of the wireless sets used by Section VIII during the war are on view in Hut 1 at Bletchley Park, and in the reproduction of the 1939 SIS wireless station X – in the tower of the Mansion. Communication receivers purchased from USA and on show include AR87, AR88, HRO and the Hallicrafters.

Section VIII's own products include MkIII, MkVII (wooden cased model for use in a suitcase), MkVII/2 ('cash box' version), Mk33. Others on show are B1, B2, SOE's 'S' Phone', MCR1, and OP3 (Polish made set).

This picture is of the SCU9 station at Eindhoven in Holland, in late 1944. It is super in the context of SCU wireless equipment, as it shows a standard MkIII to the right behind the Morse key, an HRO with a MkVII/2 agents' set perched on top. The MkVII (in both forms) was usually carried by SCU mobile units as a backup.

Appendix 7

Letter to my Father

With the end of the war it was clear that the original plan to move Brian Birch, Freddie Stapley, 'Fergie' Ferguson and myself on to Manila, Okinawa, Rangoon and Singapore was abandoned. We were therefore set to work helping to finish the new wireless transmitting station at Dum Dum in Calcutta. The structure was already largely completed when we arrived in August 1945. The installation of wireless gear was finally finished around the end of that year, and the station fully operational in early 1946.

Brian and Freddie then started to work in our wireless repair shop at Dhakuria and I was put in charge of the wireless stores there – as described in Chapter 38.

Some time later, Major Robin Addie wrote to my father and the letter is reproduced here.

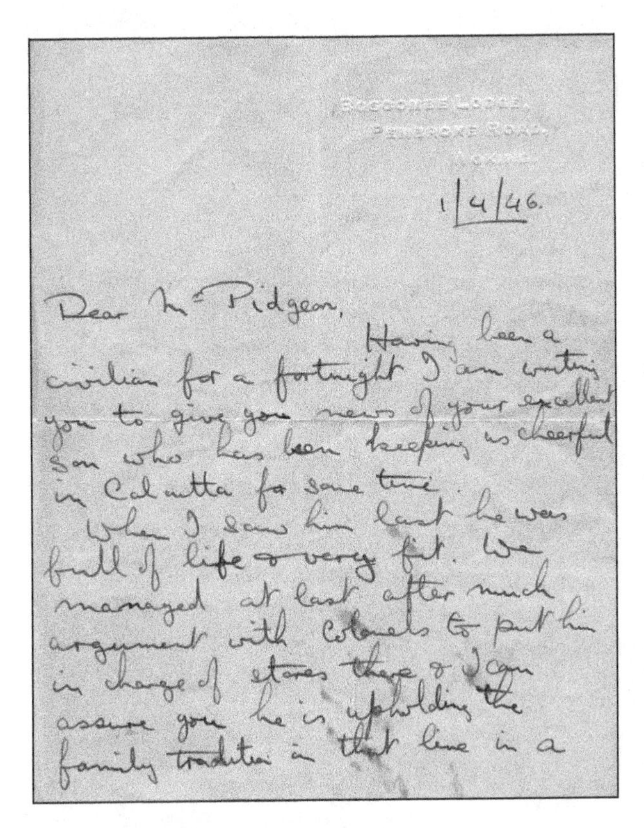

1/4/46.

Dear Mr Pidgeon,

Having been a civilian for a fortnight I am writing you to give you news of your excellent son who has been keeping us cheerful in Calcutta for some time.

When I saw him last he was full of life & very fit. We managed at last after much argument with Colonels to put him in charge of stores there & I can assure you he is upholding the family tradition in that line in a

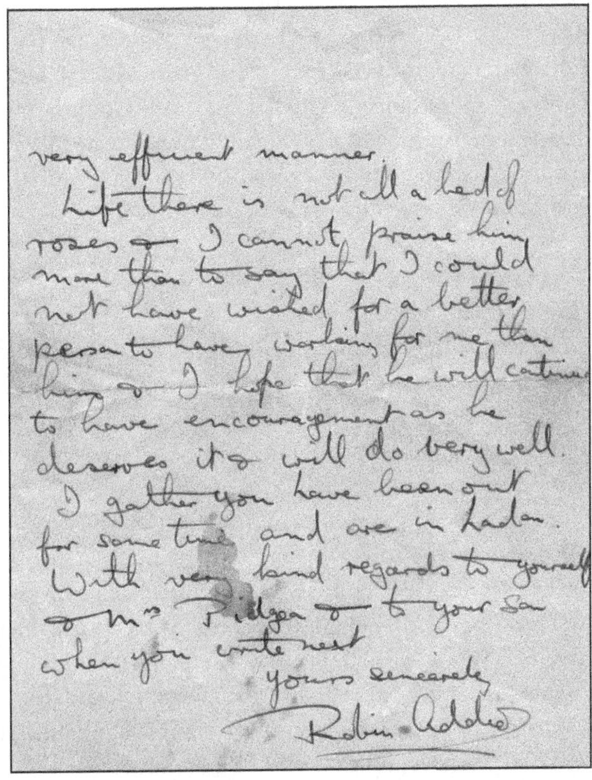

very efficient manner.

Life there is not all a bed of roses & I cannot praise him more than to say that I could not have wished for a better person to have working for me than him & I hope that he will continue to have encouragement as he deserves it & will do very well.

I gather you have been out for some time and are in London.

With very kind regards to yourself & Mrs Pidgeon & to your son when you write next

Yours sincerely

Robin Addie

Letter addressed to my father by Robin Addie for whom I worked in Calcutta in the period August 1945-early '46

Appendix 8

My Mother at Bletchley Park

In the last year or so before the war, volunteers came forward in their tens of thousands to become members of the ARP (Air Raid Precautions) organisation. Some, like my father, trained as Air Raid Wardens, others as auxiliary firemen in the AFS (Auxiliary Fire Service). Like others, from the Caterham School Scouts Troop, I was an ARP bicycle messenger. We wore our Scout's uniform with an ARP armband. My mother trained and became qualified as a Red Cross Nurse. She was based near Kenley but was at Croydon aerodrome in a surgical team, after the first German raid on the fighter station.

When we moved to Stony Stratford, she kept up her nursing under the local Red Cross commandant Mrs. Curwood, wife of the Baptist Minister. In 1942 – 1943 and again in 1944 – 1945 she worked part-time in the Clinic at Bletchley Park. I have no doubt her kindly manner, and ample bosom to cry on, were more good to young BP girls under stress and away from home, than much of the medicine prescribed.

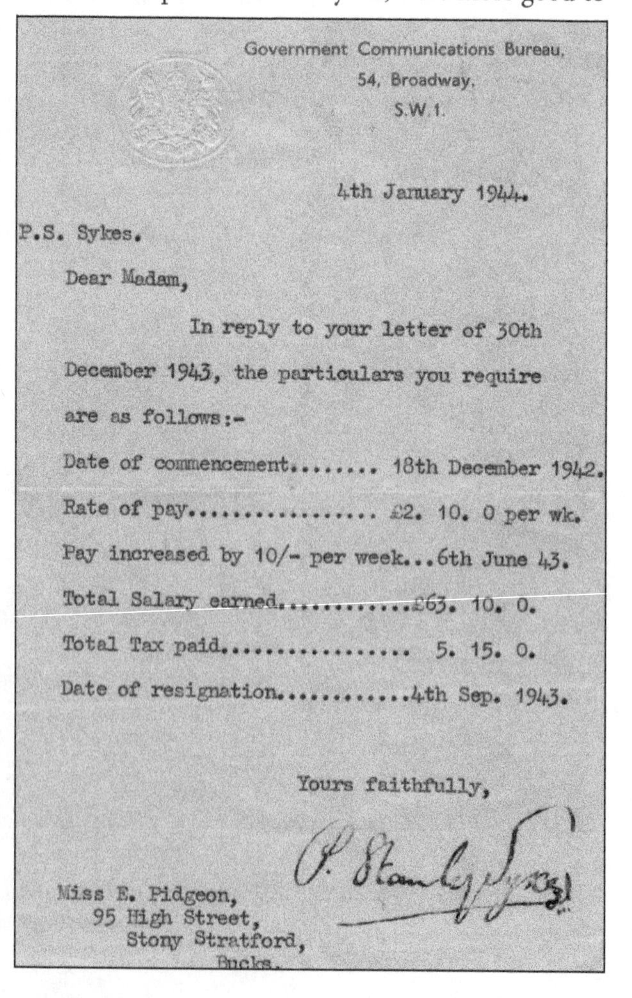
Government Communications Bureau,
54, Broadway,
S.W.1.

4th January 1944.

P.S. Sykes.

Dear Madam,

In reply to your letter of 30th December 1943, the particulars you require are as follows:-

Date of commencement........ 18th December 1942.
Rate of pay.................. £2. 10. 0 per wk.
Pay increased by 10/- per week...6th June 43.
Total Salary earned...........£63. 10. 0.
Total Tax paid................. 5. 15. 0.
Date of resignation...........4th Sep. 1943.

Yours faithfully,

P. Stanley Sykes

Miss E. Pidgeon,
95 High Street,
Stony Stratford,
Bucks.

The letter, reproduced here, is of great interest for a number of reasons. Firstly, it is from Commander Percival Stanley Sykes RN, Head of MI6 (Section VII) its finance division. Sykes was the Paymaster of the entire SIS – answerable only to 'C'.

Secondly, this personally signed letter is headed 'Government Communications Bureau' – yet another umbrella name for those working at Bletchley Park. The address on the letter – 54, Broadway, SW1 – was of course, the headquarters of SIS. In other correspondance the notepaper reads 'H.M. Government Communications Centre'.

The letter refers to her income tax for the first period of her work at BP, and is addressed to mother at our wartime home, 95 High Street, Stony Stratford. Her rate of pay was £2. 10. 0. per week (£2.50) increasing by 10/- per week (£0.50), in June after she had worked there six months. A workers wage was probably nearer £3.10. 0, suggesting that she was part time.

I believe, in the second period there, she negotiated a better wage but did the full day shift. As father and I went to catch the bus to Whaddon Hall around 8 am,

I recall she also left home to catch the Bletchley Park bus. This departed from near the Plough public house in Stony Stratford, with a host of other BP staff going on duty. There were dozens of buses and estate cars around, ferrying staff to the various places of work, from all the outlying villages.

However helpful his letter, I am sure that mother did not approve of being referred to as *Miss* E. Pidgeon – even by the SIS Paymaster – Commander Percival Stanley Sykes RN!

Epilogue

Richard Gambier-Parry

Richard Gambier-Parry – Head of MI6 (Section VIII) in the uniform of a Brigadier – as one usually saw him. Note, his World War I medals and his wings as a Royal Flying Corps pilot. At all times, he wore the Royal Welch Fusiliers 'Flash' from the back of his uniform collar, as I described earlier in Chapter 3 – 'Enter Gambier-Parry'. The photograph is, so far as I am aware, the only one of him in uniform, and is shown here by courtesy of Anthony Gambier-Parry – his great nephew.

In late 1945, he persuaded the Foreign Office to take over the SCU organisation in its entirety to handle its peacetime wireless communications. It was given the name Diplomatic Wireless Service (DWS), and he was its first Director. Whaddon's wireless operations was closed down in the winter of 1946/47 and the HQ moved to Hanslope Park.

In 1948, with the re-formation of the Territorial Army Brigadier Gambier-Parry established No 1 Special Communications Unit (TA). It had Lt. Col. Bill Sharpe as it's Commanding Officer, and was formed initially from those who had served in the SCU organisation. Brigadier Richard Gambier-Parry was appointed as its Honorary Colonel.
Richard Gambier-Parry became Director of Communications Foreign Office retiring in 1955. In the New Year's honours list of 1956 his great contributions, both in wartime and peace were acknowledged. He was appointed KCMG becoming Brigadier Sir Richard Gambier-Parry KCMG.

He and his wife Lisa, had a home in Malta but kept their cottage in the village of Milton Keynes which gave its name to the new town that developed in the area after the war. He was involved in local politics, was a good after dinner speaker, and popular in the local community which he served in many ways.

He died, after a long illness, at his home in Milton Keynes in June 1965, and a memorial service took place at All Saints Church on 26th June attended by a large congregation. In the address, Canon Curtis likened Gambier-Parry to Falstaff – 'ever a quick, and witty companion, with a gusto for life and living.'

Absolutely right; those of us who are left, and served in his organisation, will never forget this fine man.

The Government Communications Headquarters (GCHQ) arose from the Y Services, and the work done during the war at Bletchley Park. The Diplomatic Wireless Service (DWS), and Her Majesty's Government Communications Centre (HMGCC), are direct descendents of various aspects of Richard Gambier-Parry's organisation. We all owe something to his vision.

INDEX

Printed in June 2023
by Rotomail Italia S.p.A., Vignate (MI) - Italy